Optimal Control of Wind and Wave Energy Converters

Optimal Control of Wind and Wave Energy Converters

Guest Editors

Cristian Paul Chioncel
Gelu Ovidiu Tirian

MDPI

Basel • Beijing • Wuhan • Barcelona • Belgrade • Novi Sad • Cluj • Manchester

Guest Editors
Cristian Paul Chioncel
Department of Engineering Science
Babeș-Bolyai University
Cluj Napoca
Romania

Gelu Ovidiu Tirian
Engineering Faculty of Hunedoara
Politehnica University Timisoara
Timișoara
Romania

Editorial Office
MDPI AG
Grosspeteranlage 5
4052 Basel, Switzerland

This is a reprint of the Special Issue, published open access by the journal *Energies* (ISSN 1996-1073), freely accessible at: https://www.mdpi.com/journal/energies/special_issues/90R66FME8O.

For citation purposes, cite each article independently as indicated on the article page online and as indicated below:

Lastname, A.A.; Lastname, B.B. Article Title. *Journal Name* **Year**, *Volume Number*, Page Range.

ISBN 978-3-7258-6510-9 (Hbk)
ISBN 978-3-7258-6511-6 (PDF)
https://doi.org/10.3390/books978-3-7258-6511-6

Contents

About the Editors . . . . **vii**

Jacek Lukasz Wilk-Jakubowski, Lukasz Pawlik, Grzegorz Wilk-Jakubowski and Radoslaw Harabin
State-of-the-Art in the Use of Renewable Energy Sources on the Example of Wind, Wave Energy, Tidal Energy, and Energy Harvesting: A Review from 2015 to 2024
Reprinted from: *Energies* **2025**, *18*, 1356, https://doi.org/10.3390/en18061356 . . . . **1**

Junyang Xu and Pengcheng Nie
Optimal Control of Brushless Doubly Fed Wind Power Generator under Zero-Voltage Ride-Through
Reprinted from: *Energies* **2024**, *17*, 235, https://doi.org/10.3390/en17010235 . . . . **27**

Nikolaos Papadakis and Constantinos Condaxakis
An Experimental Performance Assessment of a Passively Controlled Wind Turbine Blade Concept: Part A—Isotropic Materials
Reprinted from: *Energies* **2024**, *17*, 3049, https://doi.org/10.3390/en17123049 . . . . **42**

Nikolaos Papadakis and Constantinos Condaxakis
An Experimental Performance Assessment of a Passively Controlled Wind Turbine Blade Concept: Part B—Material Oriented with Glass-Fiber-Reinforced Polymer
Reprinted from: *Energies* **2024**, *17*, 3286, https://doi.org/10.3390/en17133286 . . . . **72**

Cristian Paul Chioncel, Elisabeta Spunei and Gelu-Ovidiu Tirian
The Problem of Power Variations in Wind Turbines Operating under Variable Wind Speeds over Time and the Need for Wind Energy Storage Systems
Reprinted from: *Energies* **2024**, *17*, 5079, https://doi.org/10.3390/en17205079 . . . . **97**

Arkadiusz Małek, Agnieszka Dudziak, Andrzej Marciniak and Tomasz Słowik
Designing a Photovoltaic–Wind Energy Mix with Energy Storage for Low-Emission Hydrogen Production
Reprinted from: *Energies* **2025**, *18*, 846, https://doi.org/10.3390/en18040846 . . . . **118**

Alok Kumar and Atul Kelkar
Disturbance Rejection and Uncertainty Analysis in Wind Turbines Using Model Predictive Control
Reprinted from: *Energies* **2025**, *18*, 2504, https://doi.org/10.3390/en18102504 . . . . **140**

Pavlo Radiuk, Bohdan Rusyn, Oleksandr Melnychenko, Tomasz Perzynski, Anatoliy Sachenko, Serhii Svystun and Oleg Savenko
Criticality Assessment of Wind Turbine Defects via Multispectral UAV Fusion and Fuzzy Logic
Reprinted from: *Energies* **2025**, *18*, 4523, https://doi.org/10.3390/en18174523 . . . . **159**

About the Editors

Cristian Paul Chioncel

Cristian Paul Chioncel is an engineer and academic leader, currently Director of the Babes-Bolyai University Center in Resita. He has degrees in electronic and telecommunication, signal processing, and a PhD in Automation. Since 2002, he has progressed through the academic hierarchy to the position of Associate Professor, making significant contributions to both teaching and research. He has authored six specialized books and more than 180 scientific articles and has either coordinated or taken part in various national and international research initiatives. Mr. Chioncel is engaged in scientific events, professional organizations, and global partnerships, fostering innovation and education, cultivating organizational culture.

Gelu-Ovidiu Tirian

Gelu-Ovidiu Tirian is an engineer, economist, and academic leader, currently Dean of the Faculty of Engineering in Hunedoara, Politehnica University of Timișoara. He holds degrees in automation, industrial informatics, human resource management, and a PhD in Automation. Since 2002, he has advanced through the academic ranks to Associate Professor, contributing extensively to teaching and research. He is the author of 8 specialty books and over 160 scientific papers, and has coordinated or participated in numerous national and international research projects. Recipient of multiple excellence awards, he is actively involved in scientific events, professional associations, and international collaborations, promoting innovation, education, and institutional development.

energies

MDPI

Review

State-of-the-Art in the Use of Renewable Energy Sources on the Example of Wind, Wave Energy, Tidal Energy, and Energy Harvesting: A Review from 2015 to 2024

Jacek Lukasz Wilk-Jakubowski [1,2,*], Lukasz Pawlik [1], Grzegorz Wilk-Jakubowski [2,3] and Radoslaw Harabin [2,3]

1 Department of Information Systems, Kielce University of Technology, 7 Tysiąclecia Państwa Polskiego Ave., 25-314 Kielce, Poland; lpawlik@tu.kielce.pl
2 Institute of Crisis Management and Computer Modelling, 28-100 Busko-Zdrój, Poland; grzegorzwilkjakubowski@wp.pl (G.W.-J.); r.harabin@stsw.edu.pl (R.H.)
3 Institute of Internal Security, Old Polish University of Applied Sciences, 49 Ponurego Piwnika Str., 25-666 Kielce, Poland
* Correspondence: j.wilk@tu.kielce.pl

Abstract: Today, there is a marked increase in interest in the share of renewable energy sources in the energy mix, which benefits the environment. This also applies to the energy of sea (wave) current, as, without a doubt, the offshore area is becoming one of the leading areas of renewable energy, which translates into changes in energy production. This can be exemplified by the latest research in the context of theory, design, modeling, as well as application, control, and monitoring of wave turbines to enhance their performance. This article reviews the research in this context, systematizes information, identifies literature gaps, and presents future directions in this area. For this purpose, 3240 English-language publications from 2015 to 2024 were identified in the Scopus database. The data are analyzed according to the selected research domains. Some of them are review or conceptual in nature, while others are empirical in nature (experimental attempts and case studies). From this point of view, it becomes possible not only to systematize the state-of-the-art but also to identify future research prospects.

Keywords: energy safety; energy harvesting; wind energy; wave energy; ocean energy; tidal energy; offshore and floating wind turbines; wave energy converters; energy storage systems; energy efficiency; assessment and analyses

1. Introduction

In recent years, the need to combat climate change and ensure energy security on a global scale in the broadest sense has been noted. As a result, the efforts of many research centers are directed at increasing interest in the use of renewable energy sources, including the design and implementation of various types of ever newer and more efficient solutions. In this context, on a global scale, further changes in the structure of electricity produced can be expected as a result of the rapid increase in the amount of electricity generated from renewable sources relative to that generated by conventional sources, influenced, among other things, by ongoing government programs, including subsidies for investment in renewable energy technologies for the public sector [1–5]. In this regard, the wind and wave energy capture and conversion sector and the use of offshore and floating wind turbines, among others, play an important role, making it possible to transfer many investments from terrestrial areas to aquatic areas, which is the subject of various studies conducted in many academic and scientific centers around the world [6–10]. Certainly, a factor in favor of this is the stability of the energy supply, which becomes possible through the predictability of tidal, ocean, and wind energy [11]. Benefits in favor of the use of renewable sources are the inexhaustibility of the energy produced with them as well as their environmentally friendly nature (it is clean energy using wave and tidal/ocean energy and wind to produce electricity). Therefore, it is worth noting that offshore wave energy and tidal energy are noted for their positive environmental and human impact. Undoubted benefits arise, among other things, from the minimization of the CO_2 footprint as a result of not burning traditional fuels, such as coal, oil, and gas. Furthermore, the sector can serve as a complementary (backup) area to ensure energy safety in certain regions in the event of undesirable events, including natural disasters on land. In addition, the use of solutions installed on the seas or oceans has a positive impact on improving the aesthetics of terrestrial landscapes—free of wind turbines.

In practice, regardless of the energy source used (wind or wave), the challenge is to find a suitable location to ensure that the turbines function according to the characteristics of the wind or wave (maximum power point), perform the required expertise, and then perform the designs and implementation of technical solutions taking into account the applicable legal standards. Not without significance are the capacity of the grid and its continuous, increasingly intelligent monitoring, both of which translate into efficient distribution of power in the grid. Unsurprisingly, due to the location of the equipment and the benefits of using the marine or ocean area, it is becoming one of the key areas of renewable energy production, which, in turn, is reflected in the observed trends of change in the electricity generation market.

This review paper aims to provide a comprehensive review of research on the state-of-the-art in the use of renewable energy sources based on offshore wave energy and tidal energy and to identify possible research gaps and future research directions. In particular, this article focuses on the state-of-the-art in the use of renewable energy sources using the example of offshore wave energy and tidal energy based on publications from the last decade, i.e., from 2015 to 2024. Additionally, due to the review topic of this article, an analysis of optimal control and wave energy converters was performed as well as a review of the latest developments and scientific achievements in theory and practice, including the design, modeling, and practical use of offshore and floating wind turbines (in this regard, publications on control and monitoring of the operating condition, including, among others, energy storage systems, were analyzed in order to increase the efficiency of their operation). There were a total of 3240 results indexed in the Scopus database covering the last 10 years. It should be noted that there has been a significant increase in the number of publications during this period, which indicates the interest of researchers in this topic.

2. Materials and Methods

2.1. The Categories Used in This Study

To identify articles from the Scopus database on specific types of energy, an approach was used that will be described further, including the elements of system analysis. The selection of relevant categories (domains) was preceded by an extensive search by combining some specific search words into the title, abstract, and keywords of the publication according to the following scheme:

> TITLE-ABS-KEY (((offshore AND wave AND energy) OR (tidal AND energy))AND ((energy AND converter) OR (wave AND turbine))) AND PUBYEAR > 2014 AND PUBYEAR < 2025

As previously written, this resulted in 3240 records, whose publication data were imported from the Scopus database into a CSV (Comma-Separated Values) file and, additionally, into a text file. Consequently, the results can be viewed in terms of the information contained therein. This file has the following columns: "Authors", "Author full names", "Author(s) ID", "Title", "Year", "Source title", "Cited by", "DOI", "Link", "Abstract", "Author Keywords", "Index Keywords", "Document Type", "Publication Stage", "Open Access". To take into account the directions of interest of contemporary researchers, it was decided to select only those documents with a citation count of at least seven. The publications on energy efficiency and innovation, which translate into the development of technologies, were analyzed. In turn, those publications that were directly related to economics (economic analysis) were excluded from the analyses. As a result, 105 results were obtained and further reviewed. Each publication was reviewed for belonging to different categories. Documents that did not meet the specified criteria were rejected. Sometimes, it was difficult to assign a particular publication to a particular group of categories (it was necessary to read the entire article), or the publication belonged, by virtue of the subject matter undertaken, to several categories, which can be seen in the summary tables presented later in this manuscript. This is because, for example, information on the research methodology used did not always appear in the data analyzed from the source file. In turn, some of the documents could be assigned to multiple categories, as the topics addressed in them involved several categories. The data analysis process is illustrated in Figure 1.

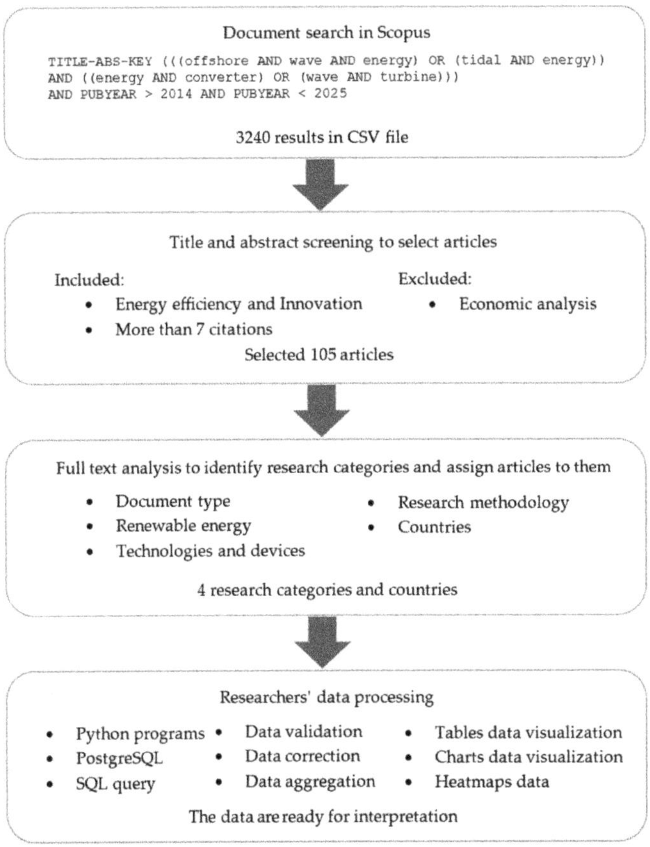

Figure 1. Data acquisition and preparation workflow.

All 105 results were analyzed for assignment to a specific document type (publication), subject category, research methodology, and country. This made it possible to show the directions and prospects of technology development with innovation and energy efficiency, taking into account the most cited publications. Partial data to identify the publication (its title, keywords, number of citations, and identifier) were recorded using the open-source relational database management system PostgreSQL (Postgre Structured Query Language). This paper includes tabular summaries to facilitate the document analysis process. An added value is the quantitative presentation of frequencies as well as cross-tables and thematic analysis of publications for each of the separated fields. The interrelationships of the systems are included in Figure 2.

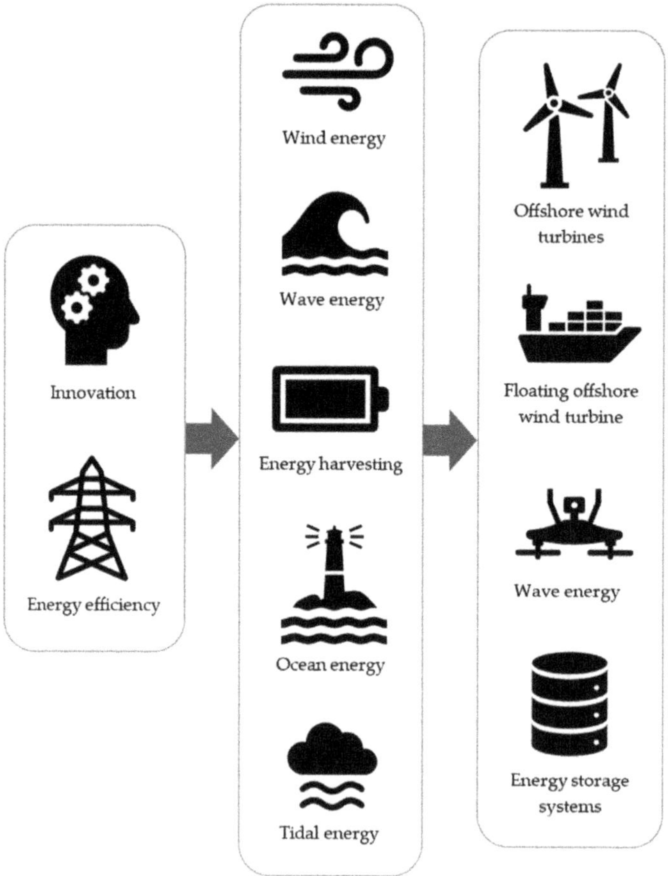

Figure 2. The interconnectedness of the systems.

In practice, the expectations of contemporary science are focused on energy efficiency and innovations, which translates into a growing interest in renewable energy sources, including wind, wave, ocean, tidal, and energy harvesting. This contributes to the development of technologies and equipment, including offshore wind turbines, floating offshore wind turbines, wave energy converters, and energy storage systems. The analysis carried out made it possible to assign publications to specific categories. The renewable energy category in this study does not contain strictly separate categories but should be viewed more as broader keywords. This includes the technologies and devices category. Within this group, the categories are not strictly separate. Offshore wind turbines can encompass both floating offshore wind turbines and fixed offshore wind turbines. Therefore, the authors included the broader category of offshore wind turbines and, within it, distinguished floating offshore wind turbines. This approach reflects the reality that many articles discuss both types within the same study. As can be seen in the table later in this article, the summed numbers in these categories exceed the total number of articles, indicating the overlap and integration of these subdomains in the literature. The results obtained were visualized using bar charts.

It should be noted that, for the purpose of this article, a classification was distinguished in which documents were assigned to one of three types of publications (document type). The first was conference paper, the second was journal article (paper), and the third and final was a subcategory other that included all other publications. In practice, this type included books or chapters in books, conference reviews, as well as reviews of scientific works, letters, editorials, erratas, published notes, short surveys, and documents containing various data from articles (data papers).

Subsequently, all articles found in the Scopus database were analyzed for division into two main broad domains (i.e., main categories) related to (1) renewable energy as well as (2) technologies and devices. Within the first domain, i.e., renewable energy, five subdomains were specified, understood as areas (subcategories): wind energy, wave energy, energy harvesting, ocean energy, and tidal energy. The work that belonged to the renewable energy domain concerned, in particular, (1) wind energy, including offshore wind turbines for its generation and design work; (2) wave energy, including conversion of wave energy into electricity, performance analyses, optimization and improvement of technical solutions, and cooperation of energy systems with other energy generation and distribution systems; (3) energy harvesting from waves and wind and improvement of efficiency of applied energy harvesting techniques; (4) ocean energy, including analyses of the application of various technologies for the use of ocean energy, among others, e.g., ocean thermal energy conversion; and (5) tidal energy, including extensive research on tidal turbines, their efficiency, as well as their environmental impact. In practice, the ocean energy category includes both wave energy and tidal energy. Often, articles categorized under ocean energy discuss aspects of both wave energy and tidal energy.

Within the second domain, i.e., technologies and devices, four areas of research interest were specified, understood as subdomains, in terms of which state-of-the-art was analyzed: offshore wind turbines, floating offshore wind turbines, wave energy converters, and energy storage systems. The documents that belonged to the technologies and devices domain dealt specifically with (1) offshore wind turbines, including designs, technical implementations, and their optimization; (2) floating offshore wind turbines, including refinement of applied solutions and integration of energy systems with other power generation and distribution systems; (3) wave energy converters, in particular analyses of the performance of different types of converters and research aimed at improving their efficiency; and (4) energy storage systems—this subcategory covered aspects of energy storage, including issues of compatibility of renewable energy systems.

In addition, in the analysis of documents from the Scopus database, attention was paid to the research methodology used. In this regard, depending on the research method applied, each publication was assigned to the appropriate subcategory: (1) experimental (laboratory, field); (2) literature analysis, including reviews of the literature and legal acts, semiotic analysis of books, chapters in books, papers, etc.; (3) case study (observation, field reports); and (4) conceptual (research aimed at developing new concepts or theories). In doing so, it should be noted that, depending on the subject matter covered and the research methods applied, each document was analyzed for belonging to more than one category and subcategory. In practice, this means that a document can be classified in more than one research domain. Similarly, if more than one research method was applied to a given document, several research methods were assigned to it, as exemplified by the summaries presented in the Section 3.

2.2. State-of-the-Art: A Review from 2015 to 2024

As previously mentioned, the presented review is primarily devoted to renewable energy as well as technologies and devices as the main domains (i.e., main categories). Within these domains, five and four subdomains were specified, respectively. In this section, all 105 articles from the Scopus database that meet the criteria described in the previous section are analyzed in terms of their covered topics. However, since many of these articles address multiple issues simultaneously (categories are not mutually exclusive), to avoid redundancy and unnecessary manuscript lengthening due to repeated references to the same articles analyzing different subcategories, each article is qualitatively analyzed in this section in relation to its alignment with the overall topic of this paper. Comprehensive summaries in the form of tables illustrating the quantitative summary of publications in 2015–2019 and 2020–2024 across all categories, the number of publications in 2015–2019 and 2020–2024 in different countries, and publications on renewable energy across other subcategories are presented in the Results section. Additionally, this section will include figures depicting the number of publications in 2015–2019 and 2020–2024 by document type; the number of publications in 2015–2019 and 2020–2024 within the first thematic category, i.e., renewable energy, divided into subcategories; the number of publications in 2015–2019 and 2020–2024 within the second thematic category, i.e., technologies and devices, divided into subcategories; the number of publications in 2015–2019 and 2020–2024 within the research methodology category, divided into subcategories; and the number of publications in 2015–2019 and 2020–2024 by country. This approach ensures that the articles undergo both qualitative and quantitative analysis. Additionally, a synthetic discussion and a critical analysis of the state-of-the-art research will complement the review. This aims to enhance the depth of the manuscript and provide a more comprehensive understanding of the current advancements and challenges in the field, which will be included in the next subsection.

Analyzing the state-of-the-art, an important and desirable issue related to the generation of renewable energy in the area of seas and oceans is to maximize its generation while reducing the movement of the platform on the water. Very good results are achieved by using different types of renewable energy systems: integrated floating wind, wave, and tidal energy systems [11,12]. Thanks to such synergy, it is possible to achieve a reduction in the tilt movements of the turbine, which translates into an increase in the amount of energy extracted. An undoubted advantage is the low failure rate of technical solutions. An example of the interaction of multiple renewable energy systems is the wind–wave hybrid system [12–15]. In practice, the reduction in unwanted oscillations in floating wind turbines is the subject of numerous studies [16–21]. The combination of wind energy and wave energy is presented in [22]. In turn, future trends and prospects for technology development are described, among others, in [19]. Many researchers have highlighted the problematic issue of reducing the efficiency of wind energy systems as a result of strong wind and wave power as well as the optimization of techniques to harness wind and wave power, for example, the optimization of wind turbines using aerodynamic thrust as well as mechanisms to control the position of the turbine to mitigate the wake effect, which can help improve the efficiency of the equipment, as described in [23–25]. In turn, the paper [26] presents an analytical model for analyzing the dynamic response of a combined wind turbine–damper system using the Simulink/Matlab environment.

In practice, wind speeds at sea are generally higher than on land and more uniform. A useful functionality of wind turbines is the ability to supply electricity to coastal areas as well as the lack of negative impact on the landscape. In addition to the advantages of their use, there are also disadvantages, including extreme loads during operation, which have a significant impact on the safety of system operation. In practice, all wind, wave, and seismic loads negatively affect the operation of offshore wind turbines, resulting in their energy efficiency as well as an increased risk of fatigue damage [27,28]. A linear model of a wind turbine controller to dampen rotor speed fluctuations caused by wave forcing is described in a paper on offshore wind and wave energy conversion [29]. The advantage of such a solution is to ensure a stable power supply to consumers in the coastal zone or on-island areas. Article [30] proposes a solution to the so-called average conditional exceedance rate method to minimize potential mechanical damage to the wind turbine. According to the concept presented in the article [31], the introduction of additional wave energy converters can increase energy production. As the authors write, much more energy can be obtained in this way than from a single floating wind turbine. The advantages of such a combination are less susceptibility to harmful weather conditions, lower costs, easier installation, and independent water depth inverter–turbine connections [32]. In general, wind and wave energy conversion is the subject of many works that describe generator optimization algorithms, such as [33–35]. Simultaneous generation of electricity from wind and wave energy can be carried out using a dedicated wave energy converter based on an oscillating column of water, which can affect sustainable energy production [36]. Some works, such as [37], describe an integrated energy conversion system through the use of a linear-rotating permanent magnet generator with a double stator. A promising device that, through the use of moored aerial systems, allows for one to harvest energy from wind blowing at heights inaccessible to traditional wind turbines is the design described in [38].

One solution to increase the use of wave energy as well as other renewable energy sources is to integrate an offshore wind turbine and a wave energy converter to produce potential energy [39]. The results presented in this work provide useful information in terms of optimizing integrated offshore wind–wave energy systems. The added value is the quantification of the benefits of optimized technical solutions. Since offshore wind turbines are systems with complex and highly nonlinear dynamics that are subjected to continuous loads, their control is an issue [17,40]. Today, smart control techniques are popular. To use renewable energy in the most optimal way, hybrid DC/AC microgrids (MG) are being modeled and designed for the power sector, which helps reduce the number of energy conversions [41]. The use of fuzzy logic is helping to improve the efficiency of energy harvesting [42–45]. Some of the work deals with the design of integrated wind–wave power generation platforms using numerical simulations. For example, the work [46] analyzed the effects of various factors on the hydrodynamic characteristics as well as the efficiency of wave energy conversion for an integrated energy platform. In the work [47], a torque compensation control strategy is included that takes into account the mitigation of large fluctuations in output power (due to the increase in response speed) as well as the use of a fuzzy inference system and optimization of the turbine power factor curve. On the other hand, an order reduction methodology for efficient estimation of the steady-state response of an offshore wind turbine is described in [48]. Control schemes to increase the efficiency of energy harvesting and the results of tests carried out in various scenarios can be found in [49,50]. Some works deal with the upgrading of available solutions for the construction of small wind turbines that can be applied locally with an energy storage system [51]. It turns out that offshore energy can make a significant contribution to the development of energy production and increase the reliability of mining complexes [52]. Structural works include procedures for determining the efficiency of pile driving for wind turbine foundations [53]. New trends and development prospects for renewable energy-powered desalination systems are included in [54,55].

In many publications, the authors highlight the need for alternative energy sources due to the worsening problem of global warming and the energy crisis [56,57]. The rapid consumption of conventional energy sources is also a factor in favor of this [58]. The energy generated from renewable resources (including wind and tidal power) is environmentally clean, which results in lower carbon emissions [59]. This is an issue of particular concern in Gulf countries, which are on the list of the world's largest carbon emitters [60]. Ultimately, the environmental impact can be further reduced by using technological improvements as well as increasing the efficiency and reliability of the clean energy solutions employed [61,62]. In practice, the subject of interest for contemporary researchers is the energy ecosystem in the context of specific applications [63]. As is well known, hydrogen is the cleanest propellant, whose combustion produces water. Hydrogen used in a fuel cell is characterized by high combustion efficiency, which translates into a lack of vibration and quiet operation of equipment. Researchers point out that hydrogen can be produced from solar energy, wind energy, coastal ocean energy, and bioenergy through chemical technology and water electrolysis using renewable energy sources [63,64].

In the case of wave energy, contemporary work includes oscillating columns (single chamber and multichamber), taking into account their advantages, service life, and performance (computer modeling is often used for this) [65], including the hydrodynamic characteristics of oscillating water column devices [66], multifunctional platforms [67], increasing energy production while reducing platform movements by combining various renewable energy systems [12], as well as energy harvesting system optimization techniques [68]. In recent years, modular solutions have become increasingly popular [60].

It should be noted that offshore renewable energy grids are currently the subject of much research [69,70]. Some works focus on articulated solutions [71], while others are conceptual and focus on optimization in terms of increasing the efficiency of energy harvesting [72–76], which is possible, among others, by negating the reactance of the system impedance. Engineering analyses often use numerical methods and employ Bayes' theorem (Bayesian optimization) [77]. Some designs are based on the use of triboelectric nanogenerators, which make it possible to acquire low-frequency and microamplitude wave energy, which is very difficult with traditional electromagnetic generators (the reason being the random and low-frequency nature of ocean waves) [78,79]. Many articles describe the latest developments in wave energy converter control [80–89]. In practice, the process of converting a renewable wave energy source into electrical energy plays an important role, which is reflected in the power capture performance of oscillating wave energy converters [90].

Undoubtedly, tidal energy is one of the potential sources of clean energy in both marine and ocean energy systems [91]. Contemporary developments related to ocean/tidal energy focus, among others, on the design and testing of ocean current turbines [92–95] as well as improving their operation [96]. Some of the work is simulation-based, such as [88,97]. The subject of research is also the analysis of future trends in the development of tidal current energy converters [98–100] or energy management systems [101]. The work related to tidal energy very often deals with hybrid solutions [102,103]. To automate and speed up the design process, a paper [104] proposes custom power converters based on design specifications, including solutions to control nonlinear loads (artificial intelligence can be useful [105]). On the other hand, work related to energy harvesting includes horizontal hydrodynamic power generation [106], design and development of horizontal marine current turbines [107], wave energy conversion [108,109], as well as optimization techniques [68,110] and integration of hybrid energy sources [63]. It should be noted that tidal and wave energy harvesting structures are currently the most advanced channels for ocean energy generation, with ocean energy devices in the early stages of development [111]. More work is expected in evaluations of applied technologies, the use of renewable energy, and the optimization of energy converters, including the application of artificial intelligence [112–116].

In conclusion, it is possible to see great potential and interest among researchers in the subject of renewable energy, with a particular focus on ocean energy and wave technologies, while at the same time, there is a need to continue research on energy storage and other less explored technologies used in the sector of renewable energy generated in the seas or oceans.

In the Section 3, quantitative summaries of the number of publications that meet the criteria are presented and described, allowing for us to see the prospective nature of the renewable energy sources analyzed, including harvesting as well as the directions of development of technologies and devices involving offshore wind turbines, floating offshore wind turbines, wave energy converters, and energy storage systems. In turn, the next subsection provides a technical discussion and critical analysis of the research on the state-of-the-art.

2.3. Technical Discussion and Critical Analysis

The analysis of the state-of-the-art in renewable energy technologies reveals several key trends and challenges. One significant trend is the increasing integration of multiple renewable energy systems, such as combined wind–wave hybrid systems, which aim to maximize energy generation while minimizing platform movements. This synergy not only enhances energy extraction but also reduces the failure rates of technical solutions. However, the optimization of these integrated systems remains a critical challenge, particularly in terms of mitigating the wake effect and improving the efficiency of energy converters [7,12,37,67].

Another notable trend is the development of advanced control strategies for offshore wind turbines and wave energy converters. Techniques such as aerodynamic thrust optimization and torque compensation control are being explored to enhance the stability and efficiency of these systems. Despite these advancements, the implementation of such control strategies in real-world scenarios is still in its nascent stages, requiring further research and validation [9,19,47,68,101].

The review also highlights the growing interest in modular solutions and the use of triboelectric nanogenerators for low-frequency wave energy harvesting. These innovations offer promising avenues for improving the efficiency and reliability of renewable energy systems. However, the scalability and long-term performance of these technologies need to be thoroughly investigated to ensure their viability for large-scale deployment [6,18,38,78,108].

In terms of research methodology, the predominance of conceptual and experimental approaches indicates that many solutions are still in the early stages of development. This underscores the need for more comprehensive case studies and literature analyses to evaluate the practical implementation and long-term impacts of these technologies. Additionally, the relatively low number of publications on energy storage systems suggests a critical gap in the research, highlighting the need for further exploration of efficient energy storage solutions to complement renewable energy generation [5,21,45,70,102].

Overall, while significant progress has been made in the field of renewable energy, several challenges and research gaps remain. Addressing these issues will require continued innovation, interdisciplinary collaboration, and substantial investment in research and development [1,4,40,71,111].

3. Results

Table 1 provides a summary covering the years of publication divided into two groups, 2015–2019 and 2020–2024, and the total number of publications for the given period. From a statistical point of view, when the *p*-values are significantly lower than 0.05, the null hypothesis can be rejected, which means that there are statistically significant differences between the distribution of the categories analyzed in the years 2015–2019 and 2020–2025, while otherwise, when the *p*-values are significantly higher than the assumed significance level, there is no basis to reject the null hypothesis, which means that no statistically significant differences are detected.

Table 1. Quantitative summary of publications in 2015–2019 and 2020–2024 in all categories.

Name	2015–2019	2020–2024	All Years	Share [%]	Chi-Square
Total	43	62	105	100.0	χ^2
Document type					
Conference paper	14	7	21	20.0	$\chi^2 = 8.03$ (df = 2, $p = 0.02$)
Journal article	28	50	78	74.29	
Other [a]	1	5	6	5.71	
Renewable energy [b]					
Wind energy	17	35	52	49.52	$\chi^2 = 3.43$ (df = 4, $p = 0.49$)
Wave energy	30	36	66	62.86	
Energy harvesting	33	37	70	66.67	
Ocean energy	42	48	90	85.71	
Tidal energy	17	19	36	34.29	
Technologies and devices [b]					
Offshore wind turbines	15	30	45	42.86	$\chi^2 = 4.95$ (df = 3, $p = 0.18$)
Floating offshore wind turbines	9	24	33	31.43	
Wave energy converters	28	30	58	55.24	
Energy storage systems	5	11	16	15.24	
Research methodology [c]					
Experiment	11	15	26	24.76	$\chi^2 = 0.49$ (df = 3, $p = 0.92$)
Literature analysis	4	5	9	8.57	
Case study	1	3	4	3.81	
Conceptual	27	39	66	62.86	

[a] Other document types are book, book chapter, review and conference review, letter, short survey, data paper, editorial, erratum, and note. [b] Each document analyzed can be classified into more than one research domain, such as renewable energy and technologies and devices, as well as subcategories (depending on its subject matter). [c] More than one research method can be applied in each analyzed document.

In Figure 3, based on the summary table, there are bar charts showing the number of documents of a certain type that had been published in two 5-year intervals (2015–2019 and 2020–2024), which is a total of ten years.

Based on Figure 3, it can be seen that the largest number of documents among the analyzed topics are scientific articles published in scientific journals as well as conference papers. Few publications in the field of renewable energy sources as well as technologies and devices accounted for the remaining publications, including books and chapters in books. It is worth noting that this trend is noticeable regardless of the analyzed time period of 5 years. In the period 2020–2024, there is an increased interest in the topic of offshore wave energy and tidal energy, which is exemplified primarily by the number of articles published in scientific journals. In 2020–2024, 50 papers were published versus 28 published papers in 2015–2019, which means almost a doubling of the number of publications over the last 5 years (an increase of 22 papers).

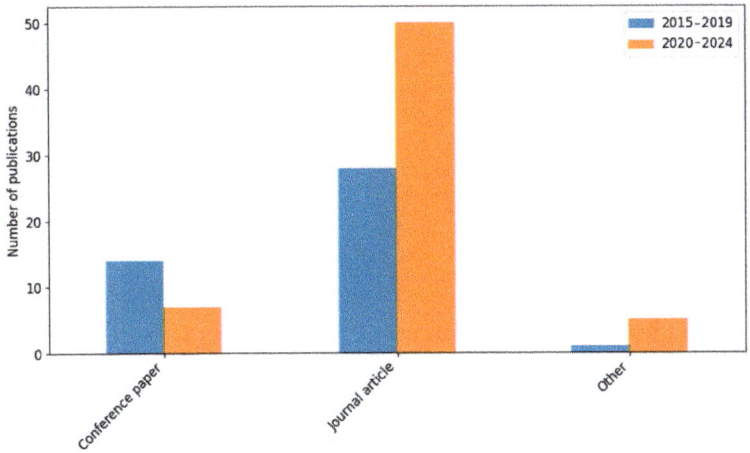

Figure 3. Number of publications in 2015–2019 and 2020–2024 in document type category.

The increased interest in the topic, especially considering the last five years, is also confirmed by the statistics in the form of bar charts in Figures 4 and 5, showing the number of publications in 2015–2019 and 2020–2024 that belong to the main thematic categories, i.e., renewable energy, as well as technologies and devices. In practice, it provides a quantitative analysis of frequency as well as a qualitative thematic analysis of publications in each of the identified fields by separate subcategories.

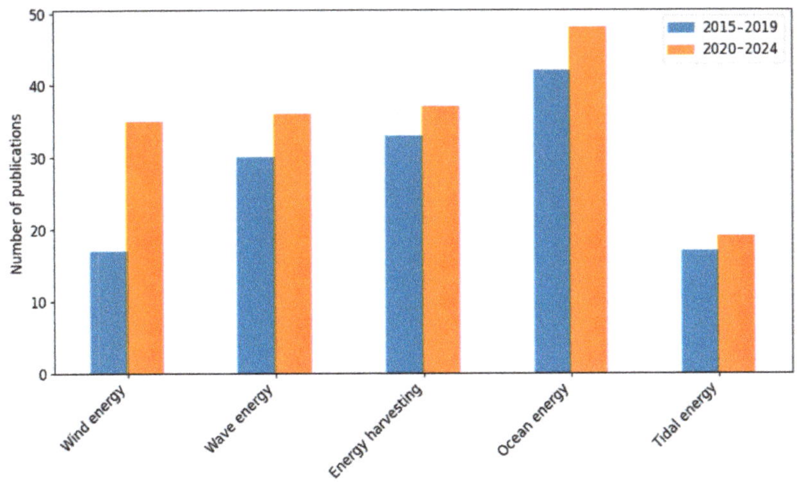

Figure 4. Number of publications in 2015–2019 and 2020–2024 that belong to the first thematic category, i.e., renewable energy, divided into subcategories.

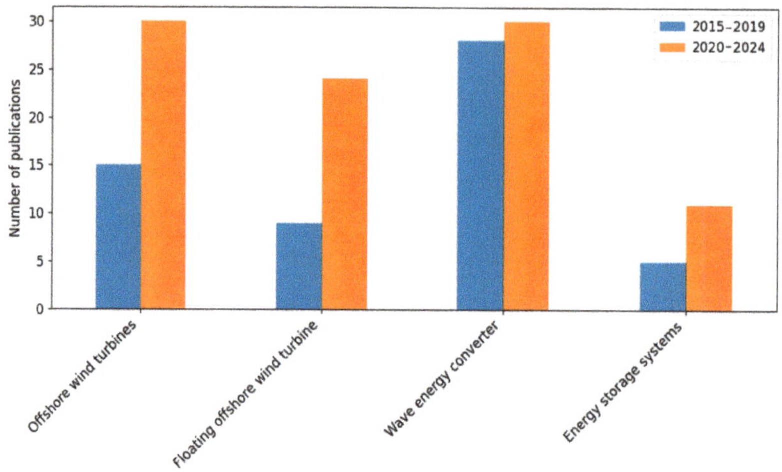

Figure 5. Number of publications in 2015–2019 and 2020–2024 that belong to the second thematic category, i.e., technologies and devices, divided into subcategories.

Although a clear increase in the number of publications in the renewable energy category can be seen over the last five years, the trend in the number of publications in each subcategory regardless of the time frame analyzed is similar. As Figure 4 shows, the topic most frequently covered within the renewable energy category was ocean energy (more than 48 publications in the last 5 years). Wind energy, wave energy, and energy harvesting were also topics of interest for researchers (more than 30 publications in each of these subcategories in the last 5 years). The topic that was the least frequently covered was tidal energy (there were less than 19 publications within this category in the analyzed time frame). However, regardless of the analyzed subcategory, one can observe an increase in the number of publications compared to 2015–2019 that ranges from 2 (tidal energy) to 18 (wind energy).

The results obtained for the technologies and devices category seem interesting against this background. As Figure 5 shows, offshore wind turbines and wave energy conversion were the most common topics within this category (a total of 60 publications in the last 5 years and 43 publications in 2015–2019). Floating offshore wind turbines have been the subject of 33 scientific papers in the last decade, while energy storage systems have been the subject of 16. Relating data from the last 5 years to the 2015–2019 period, a significant increase in the number of publications in each of the analyzed subcategories is noticeable. There was an increase in the number of publications on offshore wind turbines by 100% compared to the previous time period and floating offshore wind turbines by more than 260%, while the energy storage systems subcategory noted a 220% increase in the number of publications compared to 2015–2019. In practice, this makes it possible to observe a significant increase in researchers' interest in this topic. It should be noted that the most popular topic in this category was invariably wave energy conversion in 2015–2019, and the trend in the number of publications in each subcategory regardless of the analyzed time frame is similar. This may indicate the intensive development of technologies and devices for wind and wave energy conversion and the use of offshore and floating offshore wind turbines.

In Figure 6, bar charts show the number of publications in 2015–2019 and 2020–2024 that belong to the research methodology category, divided into four subcategories. Each time, depending on the research methodology used, the publications were assigned to at least one subcategory, experiment, literature analysis, case study, or conceptual, because more than one research method can be applied in each analyzed document (this applies when several research methods were used in a given document).

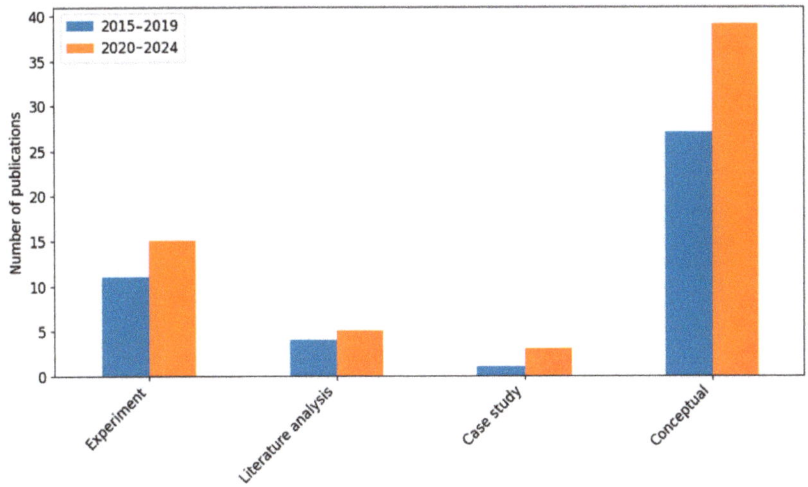

Figure 6. Number of publications in 2015–2019 and 2020–2024 that belong to the research methodology category, divided into subcategories.

The category containing conceptual works included all publications that conducted simulation studies (mainly in Matlab), publications involving mathematical modeling and using artificial intelligence to simulate physical phenomena. Based on the analysis of the results, it can be concluded that the most common methods used were conceptual and experimental as well as literature analysis. The smallest group of publications were those that used the case study method, which can be explained by the subject matter of the publication and the approach used to study reality.

The results, taking into account the countries of publication, look interesting against this background (see Table 2). It turns out that, in total, China had the highest number of publications between 2015 and 2024, namely 39 publications (more than twice as many as India, which ranked second, with 14 publications). Furthermore, a significant number of publications were recorded in Spain and the United Kingdom (13 publications each) as well as in the United States (10 publications). In the remaining countries, there was little interest in the analyzed topics, as evidenced by the number of papers published (fewer than 10 throughout the analyzed period). In practice, if an article has multiple authors affiliated with different countries, each of these countries is assigned to the article.

Table 2. Number of publications in 2015–2019 and 2020–2024 in countries.

Country	2015–2019	2020–2024	All Years	Share [%]	Chi-Square
All countries	43	62	105	100.0	χ^2
China	12	27	39	37.14	$\chi^2 = 12.99$ (df = 8, $p = 0.112$)
India	4	10	14	13.33	
Spain	6	7	13	12.38	
United Kingdom	9	4	13	12.38	
United States	4	6	10	9.52	
Italy	4	2	6	5.71	
Norway	0	5	5	4.76	
Portugal	3	2	5	4.76	
Other	8	13	21	20.00	

Figure 7, for illustrative purposes, includes bar charts showing publications in countries.

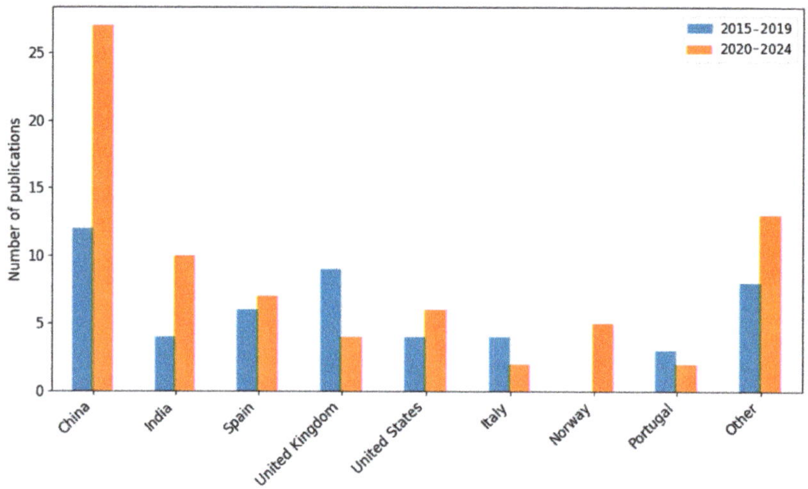

Figure 7. Number of publications in 2015–2019 and 2020–2024 divided into countries.

It should be noticed that the other category includes the following countries: France (4), Egypt (4), Australia (4), South Korea (3), The Netherlands (3), Brazil (3), Sweden (2), Ireland (2), Iran (2), Germany (2), Canada (2), Algeria (2), Singapore (1), Saudi Arabia (1), Russian Federation (1), Romania (1), Poland (1), Malaysia (1), Lebanon (1), Indonesia (1), Hong Kong (1), Greece (1), Denmark (1), and Bangladesh (1). For ease of reference, the number of publications in these countries is indicated in parentheses.

An analysis of the data in Table 2 and Figure 7 reveals some regularities. Between 2015 and 2024, a total of 105 publications were published in all countries, the majority of which were in 2020–2024 (60% of the total texts), which indicates a growing interest in the topic at that time, resulting in increased scientific activity. In practice, for 2015–2019, there were 43 publications, while for 2020–2024, the total number of publications increased to 62. As the analysis shows, there are no statistically significant differences in the number of publications between 2015 and 2019 and 2020 and 2024. The greatest interest in the topic in the last decade, regardless of the interval analyzed, was in China (a total of

39 publications, accounting for 37% of the total), which may be influenced by the country's geographic location and the advancement of authors from this country in this type of technology. As previously indicated, India ranked second, with 14 publications, accounting for just under 13% of the total. Spain and the United Kingdom shared third place (ex aequo), each with 13 publications. Notably, in both China and India, there has been a significant increase in the number of publications over the past five years. Specifically, China recorded a 225% increase, resulting in a final share of just over 37% of all publications, while India experienced a 250% increase, corresponding to a 13% share of all publications. Spain and the United Kingdom (each accounting for more than 12% of the total number of publications) also stand out positively in ongoing research. A broader analysis reveals that research is particularly concentrated in four countries—China, India, Spain, and the United Kingdom—as the work conducted there represents approximately 75% of all publications, confirming the strong interest in this research area within these countries.

Interestingly, against this background, the results summarized in Table 3 are related to publications by renewable energy in other thematic subcategories.

Table 3. Publications by renewable energy in other subcategories.

Name	Wind Energy	Wave Energy	Energy Harvesting	Ocean Energy	Tidal Energy	Total	Chi-Square
Total	52	66	70	90	36	105	χ^2
Technologies and devices							
Offshore wind turbines	45	31	31	39	5	45	$\chi^2 = 41.39$ (df = 12, $p = 4.22 * 10^{-5}$)
Floating offshore wind turbine	30	20	22	29	5	33	
Wave energy converter	25	58	57	57	12	58	
Energy storage systems	6	8	9	11	9	16	
Research methodology							
Experiment	9	15	21	23	9	26	$\chi^2 = 7.76$ (df = 12, $p = 0.8$)
Literature analysis	6	4	4	7	6	9	
Case study	3	4	3	4	1	4	
Conceptual	34	43	42	56	20	66	

As can be seen in Table 3, publications related to the main category of renewable energy were assigned to various subcategories (subdomains) related to wind energy, wave energy, energy harvesting, ocean energy, and tidal energy and then analyzed according to the areas of the second main category, that is, technologies and devices. This approach made possible a multifaceted analysis of publications from the Scopus database in terms of their affiliation with the subcategories of the two main areas. Furthermore, Table 3 provides a breakdown by the research methodology used, which is an added value.

Considering the first category, as can be seen from the summaries in Tables 1 and 3, the largest number of publications was devoted to ocean energy (90 publications), with slightly fewer devoted to energy harvesting (70 publications) and wave energy and wind energy (66 and 52 publications, respectively), suggesting the greatest interest among researchers in this issue. The smallest number of publications (36 publications) was recorded for tidal energy, which may indicate the need to develop this technology compared to other areas. For the second of the main categories, that is, technologies and devices, the highest number of publications was related to wave energy converters (58 publications). Offshore and floating wind turbines were slightly less popular among researchers (45 and 33 publications on this topic, respectively), allowing for us to see the high intensity of research conducted on these technologies. The least interest was recorded for energy storage systems (a total of 16 publications), indicating the relatively low popularity of solutions and analyses presented by researchers in this field. Consequently, more work is expected to be performed to develop renewable energy sources by providing efficient storage of energy from the seas and oceans. It should be noted that publications on energy storage systems are much less numerous in each of the subcategories of renewable energy, with the highest number recorded for the area of ocean energy (11 publications). The other subcategories of renewable energy sources had fewer than 10 publications (tidal energy—nine documents, wave energy—eight documents, and wind energy—six documents). In turn, nine publications were recorded for energy harvesting. On the basis of this, it can be concluded that energy storage systems are the subject of few scientific studies in relation to renewable energy, which suggests the need for continued future research in this area. The fact that subcategories are not mutually exclusive is evident, for example, in Figure 8, including publications by renewable energy in technologies and devices category. A heatmap is presented below.

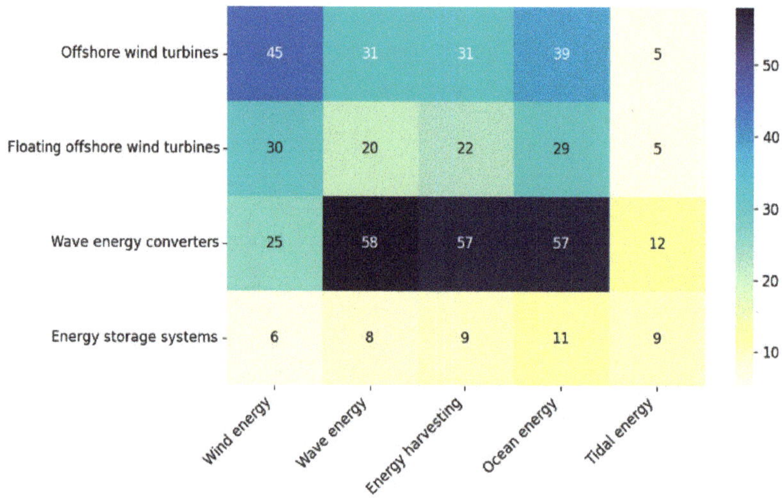

Figure 8. Publications by renewable energy in technologies and devices category.

Based on the data presented in Table 3 and Figure 8, it can be seen that a significant number of publications on wave energy converters were recorded for various subcategories of energy sources, particularly wave energy (58 publications) and ocean energy (57 publications). In addition, a large number of publications, i.e., 57, are works related to energy harvesting. This analysis leads to the conclusion that this technology is well developed and is being researched worldwide. In the case of offshore wind turbines, 45 publications were recorded in the wind energy subcategory, making it the main subject of research in this field. In contrast, slightly fewer publications were recorded for floating offshore wind turbines and wave energy converters—30 and 25 documents, respectively. It should be noted that a significant number of publications on offshore wind turbines were also recorded for the subcategories ocean energy (39 documents) and wave energy (31 documents). In addition, 31 papers are energy harvesting-related works. Relatively few publications were devoted to tidal energy (the least studied area), as exemplified by the smallest number of publications in this area (36 papers in total). Within this area, the least studied topics were offshore wind turbines and floating offshore wind turbines (five publications each), suggesting that these technologies are rarely the subject of new scientific papers. This may be due to the need for further development of the technology to increase its practical use. In practice, the predominance of wave energy converters in the ocean energy category, as shown in Figure 8, reflects the common practice of integrating multiple renewable energy sources within a single study.

Taking into account the research methodology (Table 3), the most frequently used research methods were conceptual (66 publications) and experimental (26 publications), which indicates the dominance of theoretical research and models in the field of renewable energy as well as their experimental nature. Since the research methodology based on literature analysis and case studies concerns a small number of documents (nine and four publications, respectively), it can be observed that relatively few publications represent the state-of-the-art, so there was a need to fill the literature gap in this area. A summary heatmap is presented in Figure 9.

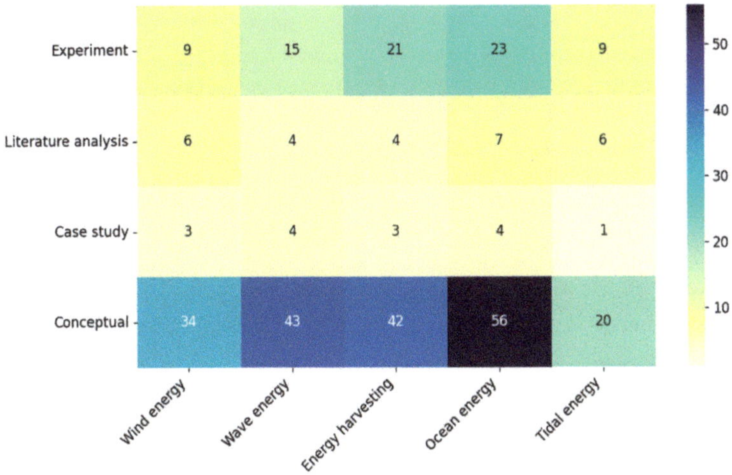

Figure 9. Publications by renewable energy in research methodology category.

Based on the heatmap in Figure 9, regarding the applied research methodology for the subcategories related to renewable energy analyzed, it can be seen that, of all the research methods for each of the subcategories of renewable energy, the conceptual approach dominates (for ocean energy—56 documents, for wave energy—43 documents, for energy harvesting—42 documents, while for wind energy—34 documents). The type of research methodology used makes it possible to note that the vast majority of documents are related to analysis and investigation of technological concepts. Also, in the case of experimental research, a similar relationship was observed (the largest group of publications were those on ocean energy—23 documents; slightly fewer publications were on energy harvesting—21 documents and wave energy—15 documents, while nine documents each were classified as wind energy and tidal energy). On the basis of this, it can be concluded that the experimental approach is often used. In turn, literature analysis and case studies have an insignificant share in scientific work, exemplified by the small number of published papers. In response to this, it was decided to conduct a comprehensive review of the state-of-the-art in the subject areas analyzed. As for case studies, the smallest number of publications was recorded in the areas of wind energy, energy harvesting, and tidal energy, suggesting that practical analyses of implemented projects are mostly in an early stage. It should be noted that only one publication was noted in the area of tidal energy. This means that tidal energy is the least studied area of all subcategories of renewable energy analyzed. Therefore, further scientific work, research projects, and investments in this area can be expected. In turn, ocean energy and wave energy are the most researched (state-of-the-art), which suggests their significant role in the renewable energy sector. Furthermore, energy harvesting and wind energy are the focus of contemporary researchers, with theoretical approaches dominating.

4. Unsolved Problems

Undoubtedly, interconnected economic, political, and market systems that emphasize energy efficiency and innovation are driving technological development, and this is also true for the renewable energy sector. This impact is immense. Consequently, in the last decade, there has been an increase in interest in work on the acquisition and use of renewable energy, including wind, wave, ocean, and tidal, and its harvesting. This, in turn, contributes to the development of technologies and equipment, including offshore wind turbines, floating offshore wind turbines, wave energy converters, and energy storage systems. This process is positively influenced by government subsidies and programs that provide incentives for the development of offshore and ocean wind farms, which is especially important for countries with access to seas or oceans. This is reflected in the significant activity of researchers from China, India, Spain, and the United Kingdom in the field of renewable energy sources. In practice, a considerable amount of work has been published in China and India, especially over the past five years. Technological developments are also influenced by advances in materials science (depending on the materials used, corrosion of components can be a concern) and electrical engineering, resulting in the construction of increasingly efficient and reliable energy converters. Technological developments are expected to continue to drive the change in the renewable energy market, positively impacting sustainability as well as the increasingly green future of the energy sector.

In the future, it will become possible to expand the research to include other major categories, such as environmental impact. Although this article, due to its finite volume and subject matter, omits economic, policy, and market analyses, among others, these issues are also relevant to the direction of technology development. It becomes possible to analyze works that include economic analysis, including the balance sheets of profits and losses resulting from the implementation of specific renewable energy systems, with particular attention to their environmental impact. An interesting area of research would be to focus scientific efforts on life cycle assessment so that the articles analyzed would address life cycle assessment of contemporary technologies for the renewable energy sector, noting that these are relatively new technologies, which translates into a limited number of available studies. A separate (and also not very numerous) group of publications is policy analyses, which include papers that deal with legislation and regulations applicable to the renewable energy sector, as well as market analyses, which deal with market research in the broadest sense, which is of particular interest in the context of forecasts for the development of technologies that use renewable energy sources. On this basis, analogous to the method presented, it will become possible to make compilations and heatmaps that will supplement the analysis of the state-of-the-art.

The renewable energy sector is poised for significant advancements and transformations in the coming years. Several key trends are expected to shape the future of this industry:

1. Integration of Multiple Renewable Energy Systems: The trend toward integrating various renewable energy systems, such as wind–wave hybrid systems, is expected to continue. This integration aims to maximize energy generation while minimizing platform movements, thereby enhancing overall efficiency and reliability [117].

2. Advanced Control Strategies: It is essential to develop advanced control strategies for offshore wind turbines and wave energy converters. Techniques such as aerodynamic thrust optimization and torque compensation control are being studied to improve stability and efficiency [118]. However, real-world implementation of these strategies will require further research and validation.
3. Modular and Scalable Solutions: Modular solutions and the use of triboelectric nanogenerators for low-frequency wave energy harvesting are gaining popularity. These innovations offer promising avenues for improving the efficiency and reliability of renewable energy systems [117]. The scalability and long-term performance of these technologies will be critical to large-scale deployment.
4. Energy Storage Systems: The relatively low number of publications on energy storage systems highlights a critical gap in the research. Efficient energy storage solutions are essential to complement renewable energy generation and ensure a stable energy supply [119]. Future research will likely focus on developing and optimizing these systems.
5. Policy and Market Dynamics: Government policies and market dynamics will continue to play a crucial role in the advancement of renewable energy technologies. Subsidies, incentives, and regulatory frameworks will drive innovation and investment in this sector [119]. Additionally, the growing demand for clean energy will push the industry towards more sustainable and efficient solutions.
6. Technological Innovations: Emerging technologies, such as artificial intelligence, blockchain, and advanced materials, will revolutionize the renewable energy sector. These innovations will enhance the efficiency, reliability, and integration of renewable energy systems [120].
7. Global Collaboration and Investment: International collaboration and investment will be crucial for the advancement of renewable energy technologies. Countries with significant renewable energy potential, such as China and the United States, are expected to lead the way in research, development, and deployment [121].

Overall, the future of renewable energy looks promising, with continuous advancements and innovations driving the industry towards a more sustainable and efficient energy landscape.

5. Conclusions

The implementation of smart solutions and remote monitoring significantly impacts efficient energy distribution. These systems optimize operations and maintain a constant energy supply, even under adverse conditions. Key issues include their integration with other energy sources to create hybrid power parks capable of operating during natural disasters. Satellite communications can then be used for data transmission from remote and difficult-to-reach locations [122–126], whose reception depends on a number of factors [127–135]. In practice, the spectrum of service integration and applied solutions based on conditions, including the occurrence of various emergencies, translates into a change in the way a given risk is identified, assessed, and controlled [136–146].

Public and private sector investments are driving many projects in seas and oceans. Over the last decade, numerous studies have focused on improving energy efficiency in the renewable energy sector. Research on new technologies and innovative solutions suggests many will be practically implemented in the future. Additionally, studies on the reliability of current power transmission lines are prevalent [147–150].

Ocean and wave energy research is prominent, highlighting their potential in the energy sector. Wave energy converters and offshore wind turbines are key research areas, while tidal energy and energy storage systems are less frequently studied, indicating a need for further development.

Conceptual and experimental methods dominate research, suggesting many solutions are in early testing stages. The limited number of literature analyses and case studies indicates a need for more in-depth evaluations of implemented solutions.

Renewable energy plays a transformative role in the global energy transition, promising a sustainable future. However, implementing innovative solutions requires costly research, public awareness, and extensive cooperation across political, economic, and social fields. While renewable energy sources in seas and oceans have a minimal carbon footprint, concerns about their impact on marine ecosystems persist. Efforts are being made to minimize this interference.

Key conclusions:

1. The location of systems in seas poses logistical and operational challenges. Implementing innovative technologies requires significant investment and cooperation.
2. Recent years have seen numerous studies on offshore wind turbines, wave energy, and energy harvesting, indicating their application potential. Tidal energy and energy storage systems, despite their potential, are less frequently researched.
3. Renewable energy in seas and oceans offers a minimal carbon footprint but raises concerns about ecosystem impact. Integrating different energy sources can create hybrid power parks that withstand extreme conditions, enhancing energy supply stability.

Author Contributions: Conceptualization, G.W.-J., R.H. and J.L.W.-J.; methodology, L.P. and J.L.W.-J.; software, L.P.; validation, L.P.; formal analysis, J.L.W.-J.; investigation, L.P.; resources, L.P.; data curation, L.P.; writing—original draft preparation, J.L.W.-J., G.W.-J. and L.P.; final writing—review and editing, J.L.W.-J.; visualization, L.P.; supervision, J.L.W.-J., G.W.-J. and R.H.; project administration, J.L.W.-J.; funding acquisition, J.L.W.-J. All authors have read and agreed to the published version of the manuscript.

Funding: This research received no external funding.

Institutional Review Board Statement: Not applicable.

Informed Consent Statement: Not applicable.

Data Availability Statement: Not applicable.

Conflicts of Interest: The authors declare no conflicts of interest.

References

1. Mazzucato, M.; Semieniuk, G. Financing Renewable Energy: Who Is Financing What and Why It Matters. *Technol. Forecast. Soc. Change* **2018**, *127*, 8–22. [CrossRef]
2. Johnstone, N.; Haščič, I.; Popp, D. Renewable Energy Policies and Technological Innovation: Evidence Based on Patent Counts. *Environ. Resour. Econ.* **2010**, *45*, 133–155. [CrossRef]
3. Carley, S. State renewable energy electricity policies: An empirical evaluation of effectiveness. *Energy Policy* **2009**, *37*, 3071–3081. [CrossRef]
4. Marques, A.C.; Fuinhas, J.A. Are public policies towards renewables successful? Evidence from European countries. *Renew. Energy* **2012**, *44*, 109–118. [CrossRef]
5. Przychodzen, W.; Przychodzen, J. Determinants of renewable energy production in transition economies: A panel data approach. *Energy* **2020**, *191*, 116583. [CrossRef]
6. Sugumar, B.K.; Anglani, N. A Novel Decision-Support Framework for Supporting Renewable Energy Technology Siting in the Early Design Stage of Microgrids: Considering Geographical Conditions and Focusing on Resilience and SDGs. *Energies* **2025**, *18*, 544. [CrossRef]
7. Cao, F.; Yu, M.; Liu, B.; Wei, Z.; Xue, L.; Han, M.; Shi, H. Progress of Combined Wind and Wave Energy Harvesting Devices and Related Coupling Simulation Techniques. *J. Mar. Sci. Eng.* **2023**, *11*, 212. [CrossRef]

8. Gao, Q.; Jiang, C.; Yang, Y.; Ritschel, U. Comparative Study of Numerical Methods for Predicting Wave-Induced Motions and Loads on a Semisubmersible. *J. Mar. Sci. Appl.* **2023**, *22*, 499–512. [CrossRef]
9. Zhang, F.; Zhu, W.; Zu, S.; Zhang, X.; Chen, J.; Wu, B.; Huang, J. Control Research on Active Pitch Control System for Horizontal-Axis Tidal-Current Turbine Generator. *Energies* **2025**, *18*, 764. [CrossRef]
10. López, M.; Rodríguez, N.; Iglesias, G. Combined Floating Offshore Wind and Solar PV. *J. Mar. Sci. Eng.* **2020**, *8*, 576. [CrossRef]
11. Jo, C.; Kim, D.; Hwang, S.; Goo, C. Shape Design of the Duct for Tidal Converters Using Both Numerical and Experimental Approaches (Pre-2015). *Energies* **2016**, *9*, 185. [CrossRef]
12. Li, L.; Gao, Y.; Yuan, Z.; Day, S.; Hu, Z. Dynamic Response and Power Production of a Floating Integrated Wind, Wave and Tidal Energy System. *Renew. Energy* **2018**, *116*, 412–422. [CrossRef]
13. Onea, F.; Rusu, L. Coastal Impact of a Hybrid Marine Farm Operating Close to the Sardinia Island. In Proceedings of the OCEANS 2015—Genova, Genova, Italy, 18–21 May 2015; pp. 1–7. [CrossRef]
14. Sousounis, M.C.; Shek, J.K.H.; Mueller, M.A. Modelling, Control and Frequency Domain Analysis of a Tidal Current Conversion System with Onshore Converters. *IET Renew. Power Gener.* **2016**, *10*, 158–165. [CrossRef]
15. Hallak, T.S.; Karmakar, D.; Soares, C.G. Hydrodynamic Performance of Semi-Submersible FOWT Combined with Point-Absorber WECs. In *Maritime Technology and Engineering*; CRC Press: Boca Raton, FL, USA, 2021; pp. 577–585. [CrossRef]
16. Galán-Lavado, A.; Santos, M. Analysis of the Effects of the Location of Passive Control Devices on the Platform of a Floating Wind Turbine. *Energies* **2021**, *14*, 2850. [CrossRef]
17. Serrano, C.; Sierra-Garcia, J.-E.; Santos, M. Hybrid Optimized Fuzzy Pitch Controller of a Floating Wind Turbine with Fatigue Analysis. *J. Mar. Sci. Eng.* **2022**, *10*, 1769. [CrossRef]
18. Aboutalebi, P.; M'zoughi, F.; Martija, I.; Garrido, I.; Garrido, A.J. Switching Control Strategy for Oscillating Water Columns Based on Response Amplitude Operators for Floating Offshore Wind Turbines Stabilization. *Appl. Sci.* **2021**, *11*, 5249. [CrossRef]
19. López-Queija, J.; Robles, E.; Jugo, J.; Alonso-Quesada, S. Review of Control Technologies for Floating Offshore Wind Turbines. *Renew. Sustain. Energy Rev.* **2022**, *167*, 112787. [CrossRef]
20. Jahangiri, V.; Sun, C. Integrated Bi-Directional Vibration Control and Energy Harvesting of Monopile Offshore Wind Turbines. *Ocean Eng.* **2019**, *178*, 260–269. [CrossRef]
21. Mitra, B.; Chowdhury, B.; Manjrekar, M. HVDC Transmission for Access to Off-shore Renewable Energy: A Review of Technology and Fault Detection Techniques. *IET Renew. Power Gener.* **2018**, *12*, 1563–1571. [CrossRef]
22. Michele, S.; Renzi, E.; Perez-Collazo, C.; Greaves, D.; Iglesias, G. Power Extraction in Regular and Random Waves from an OWC in Hybrid Wind-Wave Energy Systems. *Ocean Eng.* **2019**, *191*, 106519. [CrossRef]
23. Han, C.; Nagamune, R. Platform Position Control of Floating Wind Turbines Using Aerodynamic Force. *Renew. Energy* **2020**, *151*, 896–907. [CrossRef]
24. Tong, X.; Zhao, X.; Karcanias, A. Passive Vibration Control of an Offshore Floating Hydrostatic Wind Turbine Model. *Wind Energy* **2018**, *21*, 697–714. [CrossRef]
25. Verma, M.; Nartu, M.K.; Subbulakshmi, A. Optimal TMD Design for Floating Offshore Wind Turbines Considering Model Uncertainties and Physical Constraints. *Ocean Eng.* **2022**, *243*, 110236. [CrossRef]
26. Nazokkar, A.; Dezvareh, R. Vibration Control of Floating Offshore Wind Turbine Using Semi-Active Liquid Column Gas Damper. *Ocean Eng.* **2022**, *265*, 112574. [CrossRef]
27. Sun, C.; Jahangiri, V.; Sun, H. Adaptive Bidirectional Dynamic Response Control of Offshore Wind Turbines with Time-varying Structural Properties. *Struct. Control Health Monit.* **2021**, *28*, e2817. [CrossRef]
28. Machado, M.R.; Dutkiewicz, M.; Colherinhas, G.B. Metamaterial-Based Vibration Control for Offshore Wind Turbines Operating under Multiple Hazard Excitation Forces. *Renew. Energy* **2024**, *223*, 120056. [CrossRef]
29. Rasool, S.; Muttaqi, K.M.; Sutanto, D. A Novel Configuration of a Hybrid Offshore Wind-Wave Energy Conversion System and Its Controls for a Remote Area Power Supply. *IEEE Trans. Ind. Appl.* **2022**, *58*, 7805–7817. [CrossRef]
30. Xu, X.; Gaidai, O.; Naess, A.; Sahoo, P. Extreme Loads Analysis of a Site-Specific Semi-Submersible Type Wind Turbine. *Ships Offshore Struct.* **2020**, *15*, S46–S54. [CrossRef]
31. Li, Y.; Ong, M.C.; Wang, K.; Li, L.; Cheng, Z. Power Performance and Dynamic Responses of an Integrated System with a Semi-Submersible Wind Turbine and Four Torus-Shaped Wave Energy Converters. *Ocean Eng.* **2022**, *259*, 111810. [CrossRef]
32. Zhou, B.; Zhang, Z.; Li, G.; Yang, D.; Santos, M. Review of Key Technologies for Offshore Floating Wind Power Generation. *Energies* **2023**, *16*, 710. [CrossRef]
33. Daraz, A. Optimized Cascaded Controller for Frequency Stabilization of Marine Microgrid System. *Appl. Energy* **2023**, *350*, 121774. [CrossRef]
34. Zhao, C.; Thies, P.R.; Ye, Q.; Lars, J. System Integration and Coupled Effects of an OWT/WEC Device. *Ocean Eng.* **2021**, *220*, 108405. [CrossRef]
35. Chen, M.; Deng, J.; Yang, Y.; Zhou, H.; Tao, T.; Liu, S.; Sun, L.; Hua, L. Performance Analysis of a Floating Wind–Wave Power Generation Platform Based on the Frequency Domain Model. *J. Mar. Sci. Eng.* **2024**, *12*, 206. [CrossRef]

36. Lee, H.H.; Chen, G.-F.; Hsieh, H.-Y. Study on an Oscillating Water Column Wave Power Converter Installed in an Offshore Jacket Foundation for Wind-Turbine System Part I: Open Sea Wave Energy Converting Efficiency. *J. Mar. Sci. Eng.* **2021**, *9*, 133. [CrossRef]
37. Xu, L.; Zhang, C.; Zhu, X. Decoupling Control of a Dual-stator Linear and Rotary Permanent Magnet Generator for Offshore Joint Wind and Wave Energy Conversion System. *IET Electr. Power Appl.* **2020**, *14*, 561–569. [CrossRef]
38. Cherubini, A.; Vertechy, R.; Fontana, M. Simplified Model of Offshore Airborne Wind Energy Converters. *Renew. Energy* **2016**, *88*, 465–473. [CrossRef]
39. Shi, W.; Li, J.; Michailides, C.; Chen, M.; Wang, S.; Li, X. Dynamic Load Effects and Power Performance of an Integrated Wind–Wave Energy System Utilizing an Optimum Torus Wave Energy Converter. *J. Mar. Sci. Eng.* **2022**, *10*, 1985. [CrossRef]
40. Zambrana, P.; Fernández-Quijano, J.; Rubio, P.; Fernandez-Lozano, J.; Garcia, A. Development and Evaluation of Fuzzy Logic Controllers for Improving Performance of Wind Turbines on Semi-Submersible Platforms under Different Wind Scenarios. *Appl. Sci.* **2023**, *13*, 2422. [CrossRef]
41. Soundarya, G.; Sitharthan, R.; Sundarabalan, C.K.; Balasundar, C.; Karthikaikannan, D.; Sharma, J. Design and Modeling of Hybrid DC/AC Microgrid With Manifold Renewable Energy Sources. *IEEE Can. J. Electr. Comput. Eng.* **2021**, *44*, 130–135. [CrossRef]
42. Rubio, P.M.; Quijano, J.F.; López, P.Z.; Lozano, J.J.F.; Cerezo, A.G.; Casanova, J.O. Control Inteligente Para Mejorar El Rendimiento de Una Plataforma Semisumergible Híbrida Con Aerogenerador y Convertidores de Oleaje: Sistema de Control Borroso Para La Turbina. *Rev. Iberoam. De Automática E Informática Ind.* **2019**, *16*, 480. [CrossRef]
43. Belkhier, Y.; Achour, A.; Shaw, R.N. Fuzzy Passivity-Based Voltage Controller Strategy of Grid-Connected PMSG-Based Wind Renewable Energy System. In Proceedings of the 2020 IEEE 5th International Conference on Computing Communication and Automation (ICCCA), Greater Noida, India, 30–31 October 2020. [CrossRef]
44. Narasipuram, R.P. Optimal Design and Analysis of Hybrid Photovoltaic-Fuel Cell Power Generation System for an Advanced Converter Technologies. *Int. J. Math. Model. Numer. Optim.* **2018**, *8*, 245. [CrossRef]
45. Prashanth, V.; Rafikiran, S.; Basha, C.H.H.; Kumar, J.A.; Dhanamjayulu, C.; Kotb, H.; ELrashidi, A. Implementation of High Step-up Power Converter for Fuel Cell Application with Hybrid MPPT Controller. *Sci. Rep.* **2024**, *14*, 3342. [CrossRef] [PubMed]
46. Zhang, X.; Li, B.; Hu, Z.; Deng, J.; Xiao, P.; Chen, M. Research on Size Optimization of Wave Energy Converters Based on a Floating Wind-Wave Combined Power Generation Platform. *Energies* **2022**, *15*, 8681. [CrossRef]
47. Chen, H.; Sun, Y.; Cai, Y.; Charpentier, J.-F.; Han, J.; Aït-Ahmed, N.; Benbouzid, M. Improved Torque Compensation Control Based-Maximum Power Point Tracking Strategy for Large Scale Floating Offshore Wind Turbines. *Ocean Eng.* **2023**, *273*, 113974. [CrossRef]
48. Kluger, J.M.; Sapsis, T.P.; Slocum, A.H. A Reduced-Order, Statistical Linearization Approach for Estimating Nonlinear Floating Wind Turbine Response Statistics. In Proceedings of the ISOPE International Ocean and Polar Engineering Conference, ISOPE, Rhodes, Greece, 26 June–1 July 2016; p. ISOPE-I.
49. Al, M.; Fontanella, A.; van der Hoek, D.; Liu, Y.; Belloli, M.; van Wingerden, J.W. Feedforward Control for Wave Disturbance Rejection on Floating Offshore Wind Turbines. *J. Phys. Conf. Ser.* **2020**, *1618*, 022048. [CrossRef]
50. Aboutalebi, P.; M'zoughi, F.; Garrido, I.; Garrido, A.J. A Control Technique for Hybrid Floating Offshore Wind Turbines Using Oscillating Water Columns for Generated Power Fluctuation Reduction. *J. Comput. Des. Eng.* **2023**, *10*, 250–265. [CrossRef]
51. Abhinav, K.A.; Collu, M.; Ke, S.; Binzhen, Z. Frequency Domain Analysis of a Hybrid Aquaculture-Wind Turbine Offshore Floating System. In Proceedings of the ASME 2019 38th International Conference on Ocean, Offshore and Arctic Engineering, Glasgow, Scotland, UK, 9–14 June 2019. [CrossRef]
52. Ustinov, D.A.; Shafhatov, E.R. Assessment of Reliability Indicators of Combined Systems of Offshore Wind Turbines and Wave Energy Converters. *Energies* **2022**, *15*, 9630. [CrossRef]
53. Gómez, S.S.; Tsetas, A.; Metrikine, A.V. Energy Flux Analysis for Quantification of Vibratory Pile Driving Efficiency. *J. Sound Vib.* **2022**, *541*, 117299. [CrossRef]
54. García-Rodríguez, L.; Delgado-Torres, A.M. Renewable Energy-Driven Desalination: New Trends and Future Prospects of Small Capacity Systems. *Processes* **2022**, *10*, 745. [CrossRef]
55. Huang, Y.; Cheng, P.; Wan, D. Numerical Analysis of a Floating Offshore Wind Turbine by Coupled Aero-Hydrodynamic Simulation. *J. Mar. Sci. Appl.* **2019**, *18*, 82–92. [CrossRef]
56. He, K.; Ye, J. Dynamics of Offshore Wind Turbine-Seabed Foundation under Hydrodynamic and Aerodynamic Loads: A Coupled Numerical Way. *Renew. Energy* **2023**, *202*, 453–469. [CrossRef]
57. Kopperstad, K.M.; Kumar, R.; Shoele, K. Aerodynamic Characterization of Barge and Spar Type Floating Offshore Wind Turbines at Different Sea States. *Wind Energy* **2020**, *23*, 2087–2112. [CrossRef]
58. Panigrahi, B.K.; Thakura, P.R. Implementation of Cuk Converter with MPPT. In Proceedings of the 2017 Third International Conference on Advances in Electrical, Electronics, Information, Communication and Bio-Informatics (AEEICB), Chennai, India, 27–28 February 2017; pp. 105–110. [CrossRef]

59. Gorai, S.; Dasarathan, S.; Venkatarajan, S.; Sugavanam, V.; Kumar, G.R.P.; Sudhakaran, M. Investigation of Voltage Regulation in Grid–Connected PV System. *Indones. J. Electr. Eng. Comput. Sci.* **2020**, *19*, 1131. [CrossRef]
60. Eshra, N.M.; Amin, I. Hybrid Floating Power Station Driven by Renewable Energy for Saudi Arabia Coastal Areas. In Proceedings of the 2020 International Conference on Electrical, Communication, and Computer Engineering (ICECCE), Istanbul, Turkey, 12–13 June 2020; pp. 1–6. [CrossRef]
61. Chehaze, W.; Chamoun, D.; Bou-Mosleh, C.; Rahme, P. Wave Roller Device for Power Generation. *Procedia Eng.* **2016**, *145*, 144–150. [CrossRef]
62. Cheng, Y.; Song, F.; Fu, L.; Dai, S.; Yuan, Z.; Incecik, A. Experimental Investigation of a Dual-Pontoon WEC-Type Breakwater with a Hydraulic-Pneumatic Complementary Power Take-off System. *Energy* **2024**, *286*, 129427. [CrossRef]
63. Zhou, Y. Low-Carbon Transition in Smart City with Sustainable Airport Energy Ecosystems and Hydrogen-Based Renewable-Grid-Storage-Flexibility. *Energy Rev.* **2022**, *1*, 100001. [CrossRef]
64. Mayon, R.; Ning, D.; Sun, Y.; Ding, Z.; Wang, R.; Zhou, Y. Experimental Investigation on a Novel and Hyper-Efficient Oscillating Water Column Wave Energy Converter Coupled with a Parabolic Breakwater. *Coast. Eng.* **2023**, *185*, 104360. [CrossRef]
65. Elhanafi, A.; Macfarlane, G.; Ning, D. Hydrodynamic Performance of Single–Chamber and Dual–Chamber Offshore–Stationary Oscillating Water Column Devices Using CFD. *Appl. Energy* **2018**, *228*, 82–96. [CrossRef]
66. Elhanafi, A.; Fleming, A.; Macfarlane, G.; Leong, Z. Underwater Geometrical Impact on the Hydrodynamic Performance of an Offshore Oscillating Water Column–Wave Energy Converter. *Renew. Energy* **2017**, *105*, 209–231. [CrossRef]
67. Perez-Collazo, C.; Pemberton, R.; Greaves, D.; Iglesias, G. Monopile-Mounted Wave Energy Converter for a Hybrid Wind-Wave System. *Energy Convers. Manag.* **2019**, *199*, 111971. [CrossRef]
68. Rajaram, A.; Sathiyaraj, K. An Improved Optimization Technique for Energy Harvesting System with Grid Connected Power for Green House Management. *J. Electr. Eng. Technol.* **2022**, *17*, 2937–2949. [CrossRef]
69. Forestieri, J.N.; Farasat, M. Integrative Sizing/Real-Time Energy Management of a Hybrid Supercapacitor/Undersea Energy Storage System for Grid Integration of Wave Energy Conversion Systems. *IEEE J. Emerg. Sel. Top. Power Electron.* **2020**, *8*, 3798–3810. [CrossRef]
70. Wang, L.; Lin, M.; Tedeschi, E.; Engström, J.; Isberg, J. Improving Electric Power Generation of a Standalone Wave Energy Converter via Optimal Electric Load Control. *Energy* **2020**, *211*, 118945. [CrossRef]
71. Tay, Z.Y. Energy Extraction from an Articulated Plate Anti-Motion Device of a Very Large Floating Structure under Irregular Waves. *Renew. Energy* **2019**, *130*, 206–222. [CrossRef]
72. Narendran, K.; Murali, K.; Sundar, V. Investigations into Efficiency of Vortex Induced Vibration Hydro-Kinetic Energy Device. *Energy* **2016**, *109*, 224–235. [CrossRef]
73. Chen, S.; Jiang, B.; Li, X.; Huang, J.; Wu, X.; Xiong, Q.; Parker, R.G.; Zuo, L. Design, Dynamic Modeling and Wave Basin Verification of a Hybrid Wave–Current Energy Converter. *Appl. Energy* **2022**, *321*, 119320. [CrossRef]
74. Gaspar, J.; Calvário, M.; Soares, C. Pump and Gas Accumulator Based Phase Control of Wave Energy Converters. In *Renewable Energies Offshore*; CRC Press: Boca Raton, FL, USA, 2015; pp. 295–303. [CrossRef]
75. Rezanejad, K.; Soares, C.G. Hydrodynamic Investigation of a Novel Concept of OWC Type Wave Energy Converter Device. In Proceedings of the ASME 2019 38th International Conference on Ocean, Offshore and Arctic Engineering, Glasgow, Scotland, UK, 9–14 June 2019. [CrossRef]
76. Yang, J.; Zhang, D.; Chen, Y.; Liang, H.; Tan, M.; Li, W.; Ma, X. Design, Optimization and Numerical Modelling of a Novel Floating Pendulum Wave Energy Converter with Tide Adaptation. *China Ocean Eng.* **2017**, *31*, 578–588. [CrossRef]
77. Wu, J.; Yao, Y.; Li, W.; Zhou, L.; Göteman, M. Optimizing the Performance of Solo Duck Wave Energy Converter in Tide. *Energies* **2017**, *10*, 289. [CrossRef]
78. Yang, S.; Zhang, C.; Du, Z.; Tu, Y.; Dai, X.; Huang, Y.; Fan, J.; Hong, Z.; Jiang, T.; Wang, Z.L. Fluid Oscillation-Driven Bi-Directional Air Turbine Triboelectric Nanogenerator for Ocean Wave Energy Harvesting. *Adv. Energy Mater.* **2024**, *14*, 2304184. [CrossRef]
79. Shi, S.; Patton, R.J.; Abdelrahman, M.; Liu, Y. Learning a Predictionless Resonating Controller for Wave Energy Converters. In Proceedings of the ASME 2019 38th International Conference on Ocean, Offshore and Arctic Engineering, Glasgow, Scotland, UK, 9–14 June 2019. [CrossRef]
80. Tona, P.; Sabiron, G.; Nguyen, H.-N. An Energy-Maximising MPC Solution to the WEC Control Competition. In Proceedings of the ASME 2019 38th International Conference on Ocean, Offshore and Arctic Engineering, Glasgow, Scotland, UK, 9–14 June 2019. [CrossRef]
81. Ma, Y.; Li, T.; Zhang, X.; Zhang, L. Research on Coupling Prediction of Mooring Line Tension and Motion Response of Vertical Axis Floating Tidal Energy Converter. *J. Coast. Res.* **2015**, *73*, 496–504. [CrossRef]
82. Arena, F.; Malara, G.; Romolo, A. A U-OWC Wave Energy Converter in the Mediterranean Sea: Preliminary Results on the Monitoring System of the First Prototype. In *Renewable Energy Offshore*; Taylor & Francis Group: London, UK, 2015; pp. 417–421. [CrossRef]

83. Zhang, H.; Ding, X.; Zhou, B.; Zhang, L.; Yuan, Z. Hydrodynamic Performance Study of Wave Energy–Type Floating Breakwaters. *J. Mar. Sci. Appl.* **2019**, *18*, 64–71. [CrossRef]
84. Wan, C.; Yang, C.; Fang, Q.; You, Z.; Geng, J.; Wang, Y. Hydrodynamic Investigation of a Dual-Cylindrical OWC Wave Energy Converter Integrated into a Fixed Caisson Breakwater. *Energies* **2020**, *13*, 896. [CrossRef]
85. Silva, R.N.; Nunes, M.M.; Oliveira, F.L.; Oliveira, T.F.; Brasil, A.C.P.; Pinto, M.S.S. Dynamical Analysis of a Novel Hybrid Oceanic Tidal-Wave Energy Converter System. *Energy* **2023**, *263*, 125933. [CrossRef]
86. Cheng, Y.; Gong, J.; Zhang, J. Hydrodynamic Investigation on a Single-Point Moored Offshore Cage-Wave Energy Converter Hybrid System. *Ocean Eng.* **2024**, *299*, 116848. [CrossRef]
87. Bracco, G.; Casassa, M.; Giorcelli, E.; Giorgi, G.; Mattiazzo, G.; Passione, B.; Raffero, M.; Vissio, G.; Martini, M. Application of Sub-Optimal Control Techniques to a Gyroscopic Wave Energy Converter. *Renew. Energ. Offshore* **2015**, *1*, 265–269. [CrossRef]
88. Carrelhas, A.A.D.; Gato, L.M.C.; Henriques, J.C.C.; Falcão, A.F.O. Experimental Study of a Self-Rectifying Biradial Air Turbine with Fixed Guide-Vanes Arranged into Two Concentric Annular Rows. *Energy* **2020**, *198*, 117211. [CrossRef]
89. Sun, T.; Zhang, Z. Optimal Control and Performance Evaluation of an Inerter-Based Point Absorber Wave Energy Converter. *Ocean Eng.* **2022**, *259*, 111883. [CrossRef]
90. Xiao, X.; Xiao, L.; Peng, T. Comparative Study on Power Capture Performance of Oscillating-Body Wave Energy Converters with Three Novel Power Take-off Systems. *Renew. Energy* **2017**, *103*, 94–105. [CrossRef]
91. Zare, P.; Davoudkhani, I.F.; Zare, R.; Ghadimi, H.; Mohajeri, R.; Babaei, A. Maiden Application of Zebra Optimization Algorithm for Design PIDN-TIDF Controller for Frequency Control in Offshore Fixed Platforms Microgrid in the Presence of Tidal Energy. In Proceedings of the 2023 8th International Conference on Technology and Energy Management (ICTEM), Babol, Iran, 8–9 February 2023; pp. 1–7. [CrossRef]
92. Li, Y.; Liu, H.; Lin, Y.; Li, W.; Gu, Y. Design and Test of a 600-kW Horizontal-Axis Tidal Current Turbine. *Energy* **2019**, *182*, 177–186. [CrossRef]
93. Belkhier, Y.; Achour, A. An Intelligent Passivity-based Backstepping Approach for Optimal Control for Grid-connecting Permanent Magnet Synchronous Generator-based Tidal Conversion System. *Int. J. Energy Res.* **2021**, *45*, 5433–5448. [CrossRef]
94. Henriques, T.A.d.J.; Hedges, T.S.; Owen, I.; Poole, R.J. The Influence of Blade Pitch Angle on the Performance of a Model Horizontal Axis Tidal Stream Turbine Operating under Wave–Current Interaction. *Energy* **2016**, *102*, 166–175. [CrossRef]
95. Zheng, J.; Dai, P.; Zhang, J. Tidal Stream Energy in China. *Procedia Eng.* **2015**, *116*, 880–887. [CrossRef]
96. Wang, L.; Luo, X.; Feng, J.; Lu, J.; Zhu, G.; Wang, W. Method of Bionic Wavy Tip on Vortex and Cavitation Suppression of a Hydrofoil in Tidal Energy. *Ocean Eng.* **2023**, *278*, 114499. [CrossRef]
97. Junianto, S.; Mukhtasor, M.; Prastianto, R.W.; Jo, C.H. Effects of demi-hull separation ratios on motion responses of tidal current turbines-loaded catamaran. *Ocean Syst. Eng.* **2020**, *10*, 87–110. [CrossRef]
98. Song, K.; Kang, Y. A Numerical Performance Analysis of a Rim-Driven Turbine in Real Flow Conditions. *J. Mar. Sci. Eng.* **2022**, *10*, 1185. [CrossRef]
99. Khaligh, A.; Onar, O.C. *Energy Harvesting. Solar, Wind, and Ocean Energy Conversion Systems*, 1st ed.; CRC Press: Boca Raton, FL, USA, 2010. [CrossRef]
100. Si, Y.; Liu, X.; Wang, T.; Feng, B.; Qian, P.; Ma, Y.; Zhang, D. State-of-the-Art Review and Future Trends of Development of Tidal Current Energy Converters in China. *Renew. Sustain. Energy Rev.* **2022**, *167*, 112720. [CrossRef]
101. Barakat, M.; Tala-Ighil, B.; Chaoui, H.; Gualous, H.; Hissel, D. Energy Management of a Hybrid Tidal Turbine-Hydrogen Micro-Grid: Losses Minimization Strategy. *Fuel Cells* **2020**, *20*, 342–350. [CrossRef]
102. Srilakshmi, K.; Rao, S.P.C.; Deepika, G.; Thrinath, B.V.S.; Ramadevi, A.; Gadameddi, S.; Kumar, K.D.; Palanivelu, A. Performance Analysis of Artificial Intelligence Controller for PV and Battery Connected UPQC. *Int. J. Renew. Energy Res.* **2023**, *13*, 155–170. [CrossRef]
103. Srilakshmi, K.; Gaddameedhi, S.; Santhosh, D.T.; Yashaswini, N.; Valluri, N.; Reddy, J.G.P.; Kumar, N.M.G.; Naik, D.A. Design and Performance Analysis of Fuzzy Based Hybrid Controller for Grid Connected Solar-Battery Unified Power Quality Conditioner. *Int. J. Renew. Energy Res.* **2023**, *13*, 1–13. [CrossRef]
104. Fan, S.; Cao, N.; Zhang, S.; Li, J.; Guo, X.; Zhang, X. From Specification to Topology: Automatic Power Converter Design via Reinforcement Learning. In Proceedings of the 2021 IEEE/ACM International Conference on Computer Aided Design (ICCAD), Munich, Germany, 1–4 November 2021; pp. 1–9. [CrossRef]
105. Srilakshmi, K.; Sujatha, C.N.; Balachandran, P.K.; Mihet-Popa, L.; Kumar, N.U. Optimal Design of an Artificial Intelligence Controller for Solar-Battery Integrated UPQC in Three Phase Distribution Networks. *Sustainability* **2022**, *14*, 13992. [CrossRef]
106. Kim, E.S.; Bernitsas, M.M. Performance Prediction of Horizontal Hydrokinetic Energy Converter Using Multiple-Cylinder Synergy in Flow Induced Motion. *Appl. Energy* **2016**, *170*, 92–100. [CrossRef]
107. Elbatran, A.H.; Yaakob, O.; Ahmed, Y.; Abdallah, F. Augmented Diffuser for Horizontal Axis Marine Current Turbine. *Int. J. Power Electron. Drive Syst. (IJPEDS)* **2016**, *7*, 235. [CrossRef]

108. Azam, A.; Ahmed, A.; Yi, M.; Zhang, Z.; Tan, X.; Ali, A.; Li, N. A Self-Stabilizing Point Absorber Wave Energy Converter with a Top-Shaped Buoy and Non-Linear Power Take-off for Oceanographic Applications. *Ocean Eng.* **2023**, *288*, 116018. [CrossRef]
109. Liu, Z.; Houari, A.; Machmoum, M.; Benkhoris, M.-F.; Tang, T. An Active FTC Strategy Using Generalized Proportional Integral Observers Applied to Five-Phase PMSG Based Tidal Current Energy Conversion Systems. *Energies* **2020**, *13*, 6645. [CrossRef]
110. Ma, Y.; Hu, C.; Li, L. Hydrodynamics and Wake Flow Analysis of a Π-Type Vertical Axis Twin-Rotor Tidal Current Turbine in Surge Motion. *Ocean Eng.* **2021**, *224*, 108625. [CrossRef]
111. Uihlein, A. Life Cycle Assessment of Ocean Energy Technologies. *Int. J. Life Cycle Assess.* **2016**, *21*, 1425–1437. [CrossRef]
112. Serna, Á.; Yahyaoui, I.; Normey-Rico, J.E.; de Prada, C.; Tadeo, F. Predictive Control for Hydrogen Production by Electrolysis in an Offshore Platform Using Renewable Energies. *Int. J. Hydrogen Energy* **2017**, *42*, 12865–12876. [CrossRef]
113. Simonetti, I.; Crema, I.; Cappietti, L.; Safti, H.E.; Oumeraci, H. Site-Specific Optimization of an OWC Wave Energy Converter in a Mediterranean Area. In *Progress in Renewable Energies Offshore*; Taylor & Francis Group: London, UK, 2016. [CrossRef]
114. Rezanejad, K.; Gadelho, J.F.M.; López, I.; Carballo, R.; Soares, C.G. Improving the Hydrodynamic Performance of OWC Wave Energy Converter by Attaching a Step. In Proceedings of the ASME 2019 38th International Conference on Ocean, Offshore and Arctic Engineering, Glasgow, Scotland, UK, 9–14 June 2019. [CrossRef]
115. Mia, M.R.; Zhao, M.; Wu, H.; Palmer, H. Numerical Simulation of a Stationary Offshore Multi-Chamber OWC Wave Energy Converter. *Ocean Eng.* **2022**, *265*, 112546. [CrossRef]
116. Lu, K.-H.; Hong, C.-M.; Xu, Q. Recurrent Wavelet-Based Elman Neural Network with Modified Gravitational Search Algorithm Control for Integrated Offshore Wind and Wave Power Generation Systems. *Energy* **2019**, *170*, 40–52. [CrossRef]
117. World Economic Forum. 5 Smart #Renewable #Energy Innovations. Available online: https://www.weforum.org/stories/2023/09/renewable-energy-innovations-climate-emergency/ (accessed on 21 February 2025).
118. Bello, A.-R.O. The 8 Most Innovative Technologies in Renewable Energy. Available online: https://interestingengineering.com/science/most-innovative-technologies-in-renewable-energy (accessed on 21 February 2025).
119. Keefe, T.L.; Hardin, K.; Amon, C. 2025 Renewable Energy Industry Outlook. Available online: https://www2.deloitte.com/us/en/insights/industry/renewable-energy/renewable-energy-industry-outlook.html (accessed on 21 February 2025).
120. StartUs Insights. Top 10 Renewable Energy Trends in 2025. Available online: https://www.startus-insights.com/innovators-guide/top-10-renewable-energy-trends-2022 (accessed on 21 February 2025).
121. Derrick, M. Top 10: Energy Predictions for 2025. Available online: https://energydigital.com/top10/top-10-energy-predictions-for-2025 (accessed on 21 February 2025).
122. Kameda, S.; Okuguchi, T.; Eguchi, S.; Suematsu, N. Development of satellite-terrestrial multi-mode VSAT using software defined radio technology. In Proceedings of the Asia Pacific Microwave Conference (APMC 2014), Sendai, Japan, 4–7 November 2014.
123. Wilk-Jakubowski, J. Overview of broadband information systems architecture for crisis management. *Inform. Autom. Pomiary Gospod. Ochr. Sr.* **2020**, *10*, 20–23. [CrossRef]
124. Wilk-Jakubowski, J. Information systems engineering using VSAT networks. *Yugosl. J. Oper. Res.* **2021**, *31*, 409–428. [CrossRef]
125. Zieliński, R.J. *Satelitarne Sieci Teleinformatyczne*; Wydawnictwa Naukowo-Techniczne: Warszawa, Poland, 2009.
126. Suematsu, N.; Kameda, S.; Oguma, H.; Sasanuma, M.; Eguchi, S.; Kuroda, K. Multi-mode portable VSAT for disaster-resilient wireless networks. In Proceedings of the Asia Pacific Microwave Conference (APMC 2014), Sendai, Japan, 4–7 November 2014.
127. Wilk-Jakubowski, J. Broadband satellite data networks in the context of available protocols and digital platforms. *Inform. Autom. Pomiary Gospod. Ochr. Sr.* **2021**, *11*, 56–60. [CrossRef]
128. Zeng, L.; Zhang, C.; Qin, P.; Zhou, Y.; Cai, Y. One Method for Predicting Satellite Communication Terminal Service Demands Based on Artificial Intelligence Algorithms. *Appl. Sci.* **2024**, *14*, 6019. [CrossRef]
129. Alozie, E.; Abdulkarim, A.; Abdullahi, I.; Usman, A.D.; Faruk, N.; Olayinka, I.-F.Y.; Adewole, K.S.; Oloyede, A.A.; Chiroma, H.; Sowande, O.A.; et al. A Review on Rain Signal Attenuation Modeling, Analysis and Validation Techniques: Advances, Challenges and Future Direction. *Sustainability* **2022**, *14*, 11744. [CrossRef]
130. Wilk-Jakubowski, J. Measuring Rain Rates Exceeding the Polish Average by 0.01%. *Pol. J. Environ. Stud.* **2018**, *27*, 383–390. [CrossRef]
131. Mróz, M. *Satelitarne Zobrazowania Radarowe*; Polska Agencja Kosmiczna: Gdańsk, Poland, 2022.
132. Lewandowski, A. Radary w teledetekcji lasu. *Acta Agrophysica* **2005**, *6*, 31–40.
133. Wilk-Jakubowski, J. Predicting Satellite System Signal Degradation due to Rain in the Frequency Range of 1 to 25 GHz. *Pol. J. Environ. Stud.* **2018**, *27*, 391–396. [CrossRef]
134. Nkordeh, N.; Popoola, S.; Jefia, A.; Okeoghene, M.; Atayero, A.; Awomoyi, M.; Bobmanuel, I. The Impact of Satellite Communications on Environmental Hazard Control: Tool for the Realization of African Union Agenda 2063 Aspirations. *Int. J. Commun. Netw. Syst. Sci.* **2023**, *16*, 191–216. [CrossRef]
135. Wilk-Jakubowski, J. Total Signal Degradation of Polish 26-50 GHz Satellite Systems Due to Rain. *Pol. J. Environ. Stud.* **2018**, *27*, 397–402. [CrossRef]

136. Carreras-Coch, A.; Navarro, J.; Sans, C.; Zaballos, A. Communication Technologies in Emergency Situations. *Electronics* **2022**, *11*, 1155. [CrossRef]
137. Negi, P.; Pathani, A.; Bhatt, B.C.; Swami, S.; Singh, R.; Gehlot, A.; Thakur, A.K.; Gupta, L.R.; Priyadarshi, N.; Twala, B.; et al. Integration of Industry 4.0 Technologies in Fire and Safety Management. *Fire* **2024**, *7*, 335. [CrossRef]
138. Wilk-Jakubowski, G.; Harabin, R.; Skoczek, T.; Wilk-Jakubowski, J. Preparation of the Police in the Field of Counter-terrorism in Opinions of the Independent Counter-terrorist Sub-division of the Regional Police Headquarters in Cracow. *Slovak J. Political Sci.* **2022**, *22*, 174–208. [CrossRef]
139. Azarenko, O.; Honcharenko, Y.; Divizinyuk, M.; Mirnenko, V.; Strilets, V.; Wilk-Jakubowski, J.L. The influence of air environment properties on the solution of applied problems of capturing speech information in the open terrain. *Soc. Dev. Secur.* **2022**, *12*, 64–77. [CrossRef]
140. Šerić, L.; Stipanicev, D.; Krstinić, D. ML/AI in Intelligent Forest Fire Observer Network. In Proceedings of the 3rd EAI International Conference on Management of Manufacturing Systems, Dubrovnik, Croatia, 6–8 November 2018. [CrossRef]
141. Azarenko, O.; Honcharenko, Y.; Divizinyuk, M.; Mirnenko, V.; Strilets, V.; Wilk-Jakubowski, J.L. Influence of anthropogenic factors on the solution of applied problems of recording language information in the open area. *Soc. Dev. Secur.* **2022**, *12*, 135–143. [CrossRef]
142. Chen, Y.; Morton, D.C.; Randerson, J.T. Remote sensing for wildfire monitoring: Insights into burned area, emissions, and fire dynamics. *One Earth* **2024**, *7*, 1022–1028. [CrossRef]
143. Wilk-Jakubowski, G.; Harabin, R.; Ivanov, S. Robotics in crisis management: A review. *Technol. Soc.* **2022**, *68*, 101935. [CrossRef]
144. Jin, J.; Kim, S.; Moon, J. Development of a Firefighting Drone for Constructing Fire-breaks to Suppress Nascent Low-Intensity Fires. *Appl. Sci.* **2024**, *14*, 1652. [CrossRef]
145. Scott, R.; Nowell, B. Networks and Crisis Management. Oxford Research Encyclopedia of Politics. 2020. Available online: https://oxfordre.com/politics/view/10.1093/acrefore/9780190228637.001.0001/acrefore-9780190228637-e-1650 (accessed on 10 January 2025).
146. San-Miguel-Ayanz, J.; Ravail, N. Active Fire Detection for Fire Emergency Management: Potential and Limitations for the Operational Use of Remote Sensing. *Nat. Hazards* **2005**, *35*, 361–376. [CrossRef]
147. Chojnacki, A.L. *Analiza Niezawodności Eksploatacyjnej Elektroenergetycznych Sieci Dystrybucyjnych*; Wydawnictwo Politechniki Świętokrzyskiej: Kielce, Poland, 2013.
148. Chojnacki, A.L. Assessment of the Risk of Damage to 110 kV Overhead Lines Due to Wind. *Energies* **2021**, *14*, 556. [CrossRef]
149. Borecki, M.; Ciuba, M. Testing of Selected Surge Protection Devices in the Context of the Possibility of Ensuring the Reliability of Power Grids. *Energies* **2023**, *16*, 1445. [CrossRef]
150. Grąkowski, L.; Chojnacki, A.; Gębczyk, K.; Banasik, K. Statistical Analysis and Modeling of the Reliability of Overhead Low Voltage Lines. *Przegląd Elektrotechniczny* **2019**, *95*, 261–264. [CrossRef]

energies

MDPI

Article

Optimal Control of Brushless Doubly Fed Wind Power Generator under Zero-Voltage Ride-Through

Junyang Xu [1] and Pengcheng Nie [2,*]

[1] School of Naval Architecture, Ocean Energy Power Engineering, Wuhan University of Technology, Wuhan 430063, China; xujunyangd@whut.edu.cn
[2] State Key Laboratory of Advanced Electromagnetic Engineering and Technology, Huazhong University of Science and Technology, Wuhan 430074, China
* Correspondence: niepengcheng@hopewind.com

Abstract: In the grid-connected operation dynamics of brushless doubly fed generators (BDFGs), a dip in the grid voltage is equivalent to suddenly adding a reverse voltage source at the parallel node. By deriving the expressions of the transient current of power winding (PW), control winding (CW), and rotor winding (RW) of a BDFG in the complex frequency domain under a natural state, it was concluded that the overshoot and oscillation time are affected by the CW voltage, the drop degree and phase of the grid voltage, and the rotor speed. Therefore, an optimal control strategy is proposed. A state model with the CW current as the state variable was constructed using the Pontryagin minimum principle. The finite-time integral value of the square of the electromagnetic torque was set as the objective function to achieve the minimum value that could suppress the overshoot and oscillation of the electromagnetic torque, and the optimal CW voltage command value was directly solved to accelerate the convergence of the BDFG's physical quantities, thereby reducing the amplitude. Finally, the feasibility of the optimal control algorithm was verified using tests on an experimental platform.

Keywords: brushless doubly fed generators; zero-voltage ride-through; dynamic analysis; oscillations and overshoot; optimal control

1. Introduction

Brushless doubly fed electric generators without brushes and slip rings have superior characteristics, including low maintenance costs and high reliability [1–6]. Many scholars and scientific research institutions have studied this topic in the past few years [7–11]. Studies show that these generators outperform conventional brush-geared doubly fed induction generators (DFIGs) in the field of wind power generation [12–14].

For a BDFG in grid-connected operations, an instantaneous drop in the grid voltage will cause severe over-currents and oscillations in the BDFG windings.

Shao performed a dynamic analysis to investigate low-voltage ride-through (LVRT) events with symmetric drops set to 50% and zero [15]. A new reduced dynamic T-model was proposed in [16], and a detailed mathematical analysis was derived for the physical quantities of the PW and CW (flux, current, and voltage) subjected to voltage-dip conditions. However, the authors did not propose a feasible strategy for this condition.

Crowbars or series dynamic resistors were installed between the CW and converter to absorb the excess power in [17]. The disadvantage is that the system cost increases.

Regarding some control strategies without crowbars, the literature indicates that the active and reactive currents of the PW are a rated value and zero under a normal grid voltage. When an LVRT event occurs, the commanded values for the active and reactive currents are set to zero and a rated value, respectively, to output the reactive power to the grid [18,19]. However, large overshoots and oscillations occur in the system using these strategies.

In [20], the author pointed out that the control side of a brushless doubly fed generator has a large leakage inductance, which enhances its natural ability for low-voltage ride-through. Based on the impedance model and using the generalized Nyquist criterion, the influence of different short-circuit ratios, phase-locked loops, and current loop ratio parameters on operational stability was studied, and the Bode diagrams are provided in [21]. The author of [22] studied the role of virtual impedance and proposed a dynamic feedforward compensation to limit the forced component of the fault current in a DFIG and accelerate its natural component attenuation.

In [23], based on the expression of the transient voltage of the control winding, the flux change rates of the rotor winding, power winding, and control winding were added as feedforward compensation variables for the voltage command value. The advantage is that a transient fully decoupled control can be achieved, which can suppress the interference of these physical quantities with the control winding current when the grid voltage drops. The disadvantage is that the rotor flux of a BFDG cannot be directly obtained and it must be converted to make the control more complex. Therefore, the paper only provides a simulation, and it is difficult to conduct experimental verification. A virtual synchronous machine approach was proposed in [24], brushless doubly fed generators were analyzed and compared with the synchronous generators, and a current signal compensation method based on a virtual synchronous machine was proposed. The advantage is that this increases the system's moment of inertia, making the system respond slowly and reducing the impact of faults on the system. Simulation waveforms proved that the grid voltage drops to 30%. The disadvantages are poor stability, complex control, and simulation without experimental verification. A crowbarless LVRT control strategy based on flux linkage tracking for a BDFG under symmetrical voltage dips was proposed in [25]. A simple controller in a static reference frame rather than a rotating coordinate system was presented to implement flux linkage tracking. Based on the conventional control method, the amplitude and phase compensation of the CW flux were added to the control strategy. Meanwhile, a tracking coefficient, defined as the ratio of the CW flux to the PW flux, was proposed, and experiments were carried out using different coefficients. In the most severe case, with the maximum rotor speed and under zero-voltage dips, the experimental results showed that the CW current peak could be limited to twice the CW current rating, and the torque ripple was small during the fault. The advantage is that this belongs to the feedforward control of reverse compensation, and the fluctuation in the power winding magnetic flux is directly added to the control side, which can rapidly suppress the fluctuation in the control winding current. Moreover, the torque ripple during the fault is relatively small, but the article lacks an analysis of its effectiveness compared with other control strategies.

Given the deficiencies of the current research on the low-voltage ride-through of BDFGs, this paper further analyzes the dynamic operational characteristics of BDFGs when the grid voltage drops to 0 V. An optimal control strategy is proposed to improve the transient operational characteristics of BDFGs, which are based on the characteristics of the generator model during the fault period to solve the optimal CW voltage command value and minimize fluctuations in the electromagnetic torque to obtain a good transient response.

The main innovations of this paper are the transient characteristics of a BDFG under LVRT events and its optimal control. First, a full-order fault dynamics model for a BDFG with symmetrical voltage drops is established. It is found that the peak and oscillation duration of PW, CW, and RW currents are related to the CW voltage, the drop degree and phase of the grid voltage, and the generator speed. Then, an optimal control strategy is proposed to reduce system overshoot and accelerate convergence. This strategy can improve the system's anti-disturbance ability. Finally, the feasibility and effectiveness of the optimum control strategy are validated by comparing the experimental waveform data with the results of the conventional PI control strategy.

2. The Basic Operational Principle and Dynamic Mathematical Model of a BDFG

2.1. The Basic Structure and Operational Principle of a BDFG

A wound-rotor BDFG contains two stator windings: the PW and the CW. The two sets of windings have different pole counts. The rotor has only one set of winding, the RW, which is equivalent to two sets of windings with reverse-phase sequence connections. Figure 1 shows the schematic diagram of the windings in a BDFG [26,27].

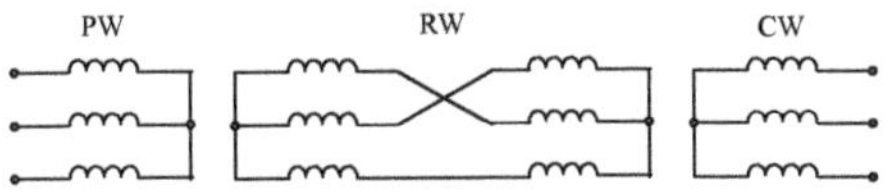

Figure 1. Equivalent schematic diagram of BDFG.

It is assumed that the voltages of the PW and CW are in the same phase sequence. As such, the following equations can be established:

$$\omega_r = \frac{\omega_p + \omega_c}{p_p + p_c} \tag{1}$$

$$\omega_{rp} = \frac{\omega_p p_c + \omega_c p_p}{p_p + p_c} \tag{2}$$

where p_p and ω_p are the pole pairs and electrical angular frequency of the PW, p_c and ω_c represent the corresponding parameters of the CW, ω_r is the mechanical angular velocity, and ω_{rp} is the electrical angular frequency of the RW current.

2.2. Dynamic Mathematical Model of a BDFG

The dynamic mathematical model of a wound-rotor BDFG is listed in a PW synchronous speed coordinate system [28]. The motor convention is used for both the PW and CW sides.

$$\boldsymbol{u}_p = R_p \boldsymbol{i}_p + p\boldsymbol{\psi}_p + j\omega_p \boldsymbol{\psi}_p \tag{3}$$

$$\boldsymbol{u}_c = R_c \boldsymbol{i}_c + p\boldsymbol{\psi}_c - j\omega_c \boldsymbol{\psi}_c \tag{4}$$

$$\boldsymbol{u}_r = R_r \boldsymbol{i}_r + p\boldsymbol{\psi}_r + j\omega_{rp} \boldsymbol{\psi}_r = 0 \tag{5}$$

$$\boldsymbol{\psi}_p = L_p \boldsymbol{i}_p + M_{pr} \boldsymbol{i}_r \tag{6}$$

$$\boldsymbol{\psi}_c = L_c \boldsymbol{i}_c + M_{cr} \boldsymbol{i}_r \tag{7}$$

$$\boldsymbol{\psi}_r = L_r \boldsymbol{i}_r + M_{pr} \boldsymbol{i}_p + M_{cr} \boldsymbol{i}_c \tag{8}$$

$$T_e = p_p M_{pr} \mathrm{Im}(\boldsymbol{i}_p \boldsymbol{i}_r{}^*) + p_c M_{cr} \mathrm{Im}(\boldsymbol{i}_c \boldsymbol{i}_r{}^*) \tag{9}$$

$$T_m - T_e = J\frac{d\omega_r}{dt} \tag{10}$$

where $\boldsymbol{u}_p$, $\boldsymbol{u}_c$, and $\boldsymbol{u}_r$ are the PW, CW, and RW voltages, $\boldsymbol{\psi}_p$, $\boldsymbol{\psi}_c$, and $\boldsymbol{\psi}_r$ are the flux linkage values of the PW, CW, and RW, and $\boldsymbol{i}_p$, $\boldsymbol{i}_c$, and $\boldsymbol{i}_r$ are the PW, CW, and RW currents. The parameters R_p, R_c, and R_r represent the resistances of the PW, CW, and RW, p represents the differential operators, and the superscript * represents the conjugate of the complex vectors. The parameters L_p, L_c, and L_r are the self-inductance values for the PW, CW, and RW, M_{pr} is the mutual inductance between the PW and RW, and M_{cr} is the mutual inductance between the CW and RW. T_e and T_m are the electromagnetic and mechanical torques, and J is the moment of inertia.

3. Transient Analysis of an LVRT Event

Using a grid voltage orientation strategy, the PW voltage vector is oriented with the d-axis. Then, according to (3)–(8), the vector diagram of a BDFG in a PW synchronous speed rotating coordinate system can be depicted as follows [29].

In Figure 2, the expression of the vector in the dq-axis coordinate system is $\boldsymbol{F} = f_d + jf_q$, where $\boldsymbol{F}$ can represent any vector, and φ is the grid voltage phase.

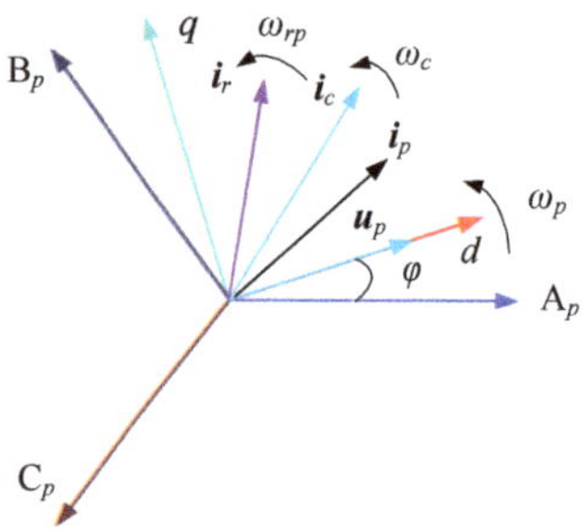

Figure 2. Vector diagram of the rotating coordinate system.

For a BDFG in grid-connected operation, an instantaneous symmetrical drop in the grid voltage at the grid connection point of the wind farm is equivalent to the sudden addition of a reverse voltage source at the connection point, which inevitably leads to severe oscillations and overshoot in the PW, CW, and RW currents and electromagnetic torque. The worst case is that the grid voltage suddenly drops to 0 V, which is equivalent to a PW short circuit fault occurrence.

If the winding currents after the grid voltage fault must be solved, the steady state current before the fault and the transient current after the fault need to be calculated according to the superposition principle.

When the BDFG is in grid connection and stable operation, the differential terms in (3)–(8) are 0 and the steady-state equations are obtained. The steady-state equivalent circuit is shown in Figure 3.

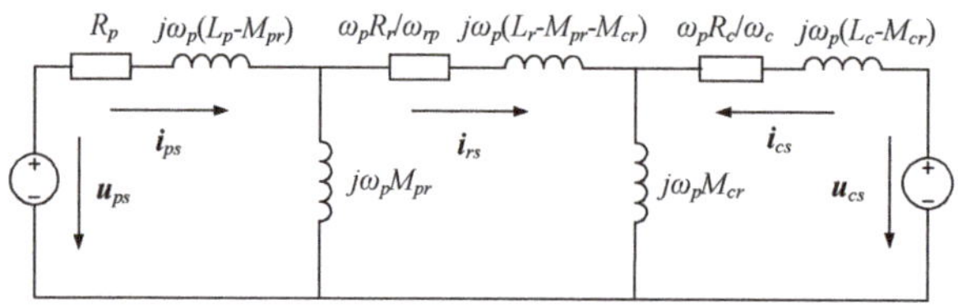

Figure 3. Steady-state equivalent circuit diagram of BDFG.

The analytical expressions of the PW, CW, and RW currents of the BDFG in steady-state operation are obtained as follows:

$$\begin{cases} \boldsymbol{i}_{ps} = \frac{\boldsymbol{u}_{ps}}{Z_{ps1}} + \frac{\boldsymbol{u}_{cs}}{Z_{ps2}} \\ \boldsymbol{i}_{cs} = \frac{\boldsymbol{u}_{ps}}{Z_{cs1}} + \frac{\boldsymbol{u}_{cs}}{Z_{cs2}} \\ \boldsymbol{i}_{rs} = \frac{\boldsymbol{u}_{ps}}{Z_{rs1}} + \frac{\boldsymbol{u}_{cs}}{Z_{rs2}} \end{cases} \tag{11}$$

In (11), $\boldsymbol{u}_{ps}$, $\boldsymbol{u}_{cs}$, $\boldsymbol{i}_{ps}$, $\boldsymbol{i}_{cs}$, and $\boldsymbol{i}_{rs}$ represent the steady-state voltages and currents of the PW, CW, and RW. The expressions of the steady-state impedances Z_{ps1}, Z_{ps2}, Z_{cs1}, Z_{cs2}, Z_{rs1}, and Z_{rs2} are shown in Appendix A. To study the natural characteristics of generators under grid voltage drops, the following assumptions must be made: (1) there is no sudden change in the rotor speed before and after grid voltage drops; (2) the CW side voltage does not change before and after voltage drops.

The dynamic equivalent circuit of the BDFG in the case of an instantaneous drop in the grid voltage is shown in Figure 4.

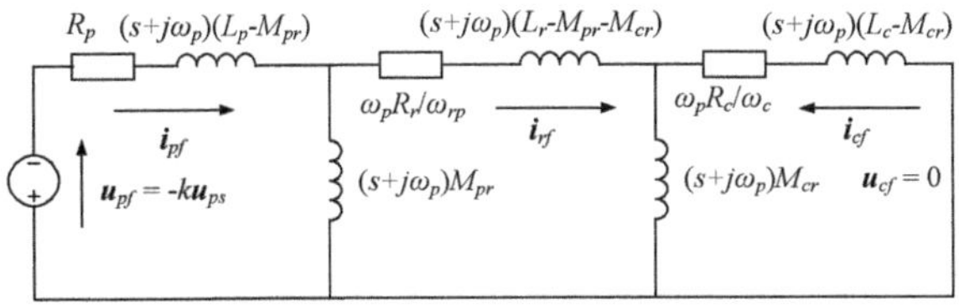

Figure 4. Grid fault equivalent circuit diagram of BDFG.

In the case of an instantaneous symmetrical drop in the grid voltage, in (3)–(8), $u_p = u_{pf}$ is a sudden reverse grid voltage, and $u_{pf} = -ku_{ps}$, where k is the voltage drop degree coefficient of the PW. When $k = 1$, a reverse voltage with the same amplitude as the original grid voltage is suddenly added at the parallel node, that is, the grid voltage instantaneously drops to 0 V. Since the CW voltage is assumed to be constant, according to Thevenin's theorem, when analyzing the impact of grid voltage fault, the CW voltage is 0 V, such that $u_c = u_{cf} = 0$. Then, i_{pf}, i_{cf}, and i_{rf} represent the PW, CW, and RW currents generated by the grid fault. Via Laplace transformation and its inverse transformation, the solutions of the grid fault currents i_{pf}, i_{cf}, and i_{rf} are as follows:

$$\begin{cases} i_{pf} = \frac{u_{pf}}{Z_{pf0}} + \frac{u_{pf}}{Z_{pf1}} e^{-\frac{t}{T_p}} e^{-j\omega_p t} + \frac{u_{pf}}{Z_{pf2}} e^{-\frac{t}{T_c}} e^{j\omega_c t} + \frac{u_{pf}}{Z_{pf3}} e^{-\frac{t}{T_r}} e^{-j\omega_{rp} t} \\ i_{cf} = \frac{u_{pf}}{Z_{cf0}} + \frac{u_{pf}}{Z_{cf1}} e^{-\frac{t}{T_p}} e^{-j\omega_p t} + \frac{u_{pf}}{Z_{cf2}} e^{-\frac{t}{T_c}} e^{j\omega_c t} + \frac{u_{pf}}{Z_{cf3}} e^{-\frac{t}{T_r}} e^{-j\omega_{rp} t} \\ i_{rf} = \frac{u_{pf}}{Z_{rf0}} + \frac{u_{pf}}{Z_{rf1}} e^{-\frac{t}{T_p}} e^{-j\omega_p t} + \frac{u_{pf}}{Z_{rf2}} e^{-\frac{t}{T_c}} e^{j\omega_c t} + \frac{u_{pf}}{Z_{rf3}} e^{-\frac{t}{T_r}} e^{-j\omega_{rp} t} \end{cases} \quad (12)$$

In (12), Z_{pf0} is the impedance of the direct current (DC) steady-state component of the PW fault current. Z_{pf1}, Z_{pf2}, and Z_{pf3} are the impedances of the attenuated fundamental frequency, CW frequency, and RW frequency components of the PW fault current. The attenuated component of the fault current indicates that the flux linkage of the PW, CW, and RW does not suddenly change due to the principle of flux linkage conservation, so an oscillating current with an attenuation trend is generated in the PW circuit via these flux linkages. As Z_{cf0}, Z_{cf1}, Z_{cf2}, and Z_{cf3} are the relevant impedances of the CW fault current, and Z_{rf0}, Z_{rf1}, Z_{rf2}, and Z_{rf3} are the relevant impedances of the RW fault current, the meanings of these parameters can be analogized. The detailed expressions of these impedances are shown in Appendix A. The transient impedances are related to the BDFG's parameters.

Furthermore, T_p, T_c, and T_r are the attenuation time constants of the PW, CW, and RW The expressions are described in (13).

$$\begin{cases} T_p = \frac{L_pL_cL_r - L_pM_{cr}^2 - L_cM_{pr}^2}{(L_cL_r - M_{cr}^2)} \frac{1}{R_p} = \frac{L_{p-short}}{R_p} \\ T_c = \frac{L_pL_cL_r - L_pM_{cr}^2 - L_cM_{pr}^2}{(L_pL_r - M_{pr}^2)} \frac{1}{R_c} = \frac{L_{c-short}}{R_c} \\ T_r = \frac{L_pL_cL_r - L_pM_{cr}^2 - L_cM_{pr}^2}{L_pL_c} \frac{1}{R_r} = \frac{L_{r-short}}{R_r} \end{cases} \quad (13)$$

In (13), $L_{p\text{-}short}$, $L_{c\text{-}short}$, and $L_{r\text{-}short}$ are the short inductances of the PW, CW, and RW. These short-circuit inductors consist of leakage inductance circuits and mutual inductance. The circuit diagrams are shown in Figure 5.

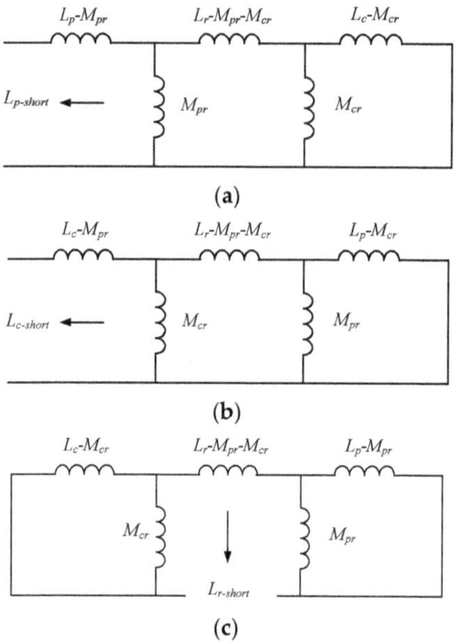

Figure 5. Short inductances of PW, CW, and RW. (**a**) PW short inductance, (**b**) CW short inductance, and (**c**) RW short inductance.

The transient impedances are related to the BDFG's parameters. The attenuation coefficients of the PW, CW, and RW are determined for a manufactured BDFG.

According to the superposition principle, the expressions of the PW, CW, and RW currents after the grid voltage fault occurs are the linear superpositions of the steady-state component and the dynamic component.

$$\begin{cases} i_p = \frac{u_{cs}}{Z_{ps2}} + \frac{u_{pf}}{Z_{pf1}} e^{-\frac{t}{T_p}} e^{-j\omega_p t} + \frac{u_{pf}}{Z_{pf2}} e^{-\frac{t}{T_c}} e^{j\omega_c t} + \frac{u_{pf}}{Z_{pf3}} e^{-\frac{t}{T_r}} e^{-j\omega_{rp} t} \\ i_c = \frac{u_{cs}}{Z_{cs2}} + \frac{u_{pf}}{Z_{cf1}} e^{-\frac{t}{T_p}} e^{-j\omega_p t} + \frac{u_{pf}}{Z_{cf2}} e^{-\frac{t}{T_c}} e^{j\omega_c t} + \frac{u_{pf}}{Z_{cf3}} e^{-\frac{t}{T_r}} e^{-j\omega_{rp} t} \\ i_r = \frac{u_{cs}}{Z_{rs2}} + \frac{u_{pf}}{Z_{rf1}} e^{-\frac{t}{T_p}} e^{-j\omega_p t} + \frac{u_{pf}}{Z_{rf2}} e^{-\frac{t}{T_c}} e^{j\omega_c t} + \frac{u_{pf}}{Z_{rf3}} e^{-\frac{t}{T_r}} e^{-j\omega_{rp} t} \end{cases} \tag{14}$$

Thus far, the analytical expressions of the PW, CW, and RW currents of the BDFG have been solved in the PW synchronous vector space under an instantaneous symmetrical drop in the grid voltage. The analytical expressions in the three-phase static coordinate system can be obtained via a rotation transformation of the complex form. The simulations will be conducted in the following sections.

Based on (11), (12), and (14), the overshoot of the PW, CW, and RW currents is related to the CW voltage u_c, the degree of grid voltage drops k, the phase of the grid voltage vector u_p (that is, φ in Figure 2), and the rotor speed ω_r. The convergence speed is related to the parameters of the BDFG. The grid voltage drops inevitably, causing a sudden increase and oscillations in the PW, CW, and RW currents. In the actual brushless doubly fed wind power generation system, the drop degree and phase of the grid voltage cannot be controlled, and the rotor speed will not suddenly change due to the inertia of the generator. Only the CW voltage of the BDFG can be changed via certain control methods to reduce overshoot and accelerate convergence, and this function can be realized via optimal control.

4. The Optimal Control Strategy

According to modern control theory, combined with the characteristics of the BDFG model during grid voltage drop faults, its input is the voltage and phase of the CW, the initial conditions are determined, the fault duration is determined, and the terminal is free. If the objective function is set as the finite-time integral value of the square of the electromagnetic torque T_e at this time, making it achieve the minimum value, it could suppress the oscillation of the electromagnetic torque. The model can be established via the Pontryagin minimum principle in optimal control, and the optimal CW voltage command value can be directly solved.

The special structure of the BDFG's rotor is a self-closing short circuit with no wiring leads. The RW only serves to transfer a magnetic field and energy and hardly provides active power. Moreover, the RW current is generally not considered in the control loop. Therefore, ignoring the RW's resistance and utilizing the 0 V rotor voltage condition, (3)–(10) are simplified into a reduced-order model of the BDFG described in (15)–(17). Motor conventions are used on both the power side and the control side.

$$\boldsymbol{u}_p = R_p\boldsymbol{i}_p + (p + j\omega_p)(L_{sp}\boldsymbol{i}_p + M_{pc}\boldsymbol{i}_c) \tag{15}$$

$$\boldsymbol{u}_c = R_c\boldsymbol{i}_c + (p - j\omega_c)(L_{sc}\boldsymbol{i}_c + M_{pc}\boldsymbol{i}_p) \tag{16}$$

$$T_e = (p_p + p_c)M_{pc}\mathrm{Im}(\boldsymbol{i}_p\boldsymbol{i}_c{}^*) \tag{17}$$

The relationship between the parameters in (15)–(17) and the parameters in (3)–(10) is

$$\begin{cases} L_{sp} = L_p - \frac{M_{pr}{}^2}{L_r} \\ L_{sc} = L_c - \frac{M_{cr}{}^2}{L_r} \\ M_{pc} = \frac{M_{pr}M_{cr}}{L_r} \end{cases} \tag{18}$$

According to mathematical models (15)–(17), under the condition of the PW voltage orientation, taking the *dq*-axis components of the CW current as the state variable and the *dq*-axis components of the CW and PW voltages as the input, the state equation is obtained, as shown in (19):

$$p\begin{bmatrix} i_{cd} \\ i_{cq} \end{bmatrix} = \lambda_m \begin{bmatrix} -\frac{R_pM_{pc}{}^2}{L_{sp}} - R_cL_{sp} & -(\omega_p - \omega_{re})M_{pc}{}^2 + \omega_cL_{sp}L_{sc} \\ (\omega_p - \omega_{re})M_{pc}{}^2 - \omega_cL_{sp}L_{sc} & -\frac{R_pM_{pc}{}^2}{L_{sp}} - R_cL_{sp} \end{bmatrix} \begin{bmatrix} i_{cd} \\ i_{cq} \end{bmatrix} + \lambda_m L_{sp} \begin{bmatrix} u_{cd} \\ u_{cq} \end{bmatrix} - \lambda_m M_{pc} \begin{bmatrix} u_{pd} \\ 0 \end{bmatrix} + \lambda_m \begin{bmatrix} -\omega_{re}M_{pc}\psi_{pq} \\ \frac{R_pM_{pc}\psi_{pq}}{L_{sp}} \end{bmatrix} \tag{19}$$

where ω_{re} is the equivalent electrical angular velocity of the BDFG:

$$\omega_{re} = \omega_r(p_p + p_c) \tag{20}$$

and the expression of λ_m is

$$\lambda_m = \frac{1}{L_{sp}L_{sc} - M_{pc}{}^2} \tag{21}$$

Under steady-state conditions, the relationship between the *d*-axis component of the PW voltage u_{pd} and the *q*-axis component of the PW flux linkage ψ_{pq} is

$$u_{pd} = -\omega_p\psi_{pq} \tag{22}$$

Since the parameters of the generator system are determined after the instantaneous symmetrical drop in the grid voltage, the generator system described in (19) is considered scleronomous. Because u_{pd} is a constant, u_{cd} and u_{cq} are free variables. Therefore, when discussing optimal control, the input term of u_{pd} can be ignored. The state space Equation (19) is reconstructed as follows:

$$\frac{dx}{dt} = \boldsymbol{Ax} + \boldsymbol{Bu} = f[x, \boldsymbol{u}, t] \tag{23}$$

$$\begin{cases} x = \begin{bmatrix} i_{cd} & i_{cq} \end{bmatrix}^T \\ u = \begin{bmatrix} u_{cd} & u_{cq} \end{bmatrix}^T \end{cases} \tag{24}$$

$$A=\lambda_m \begin{bmatrix} -\frac{R_p M_{pc}{}^2}{L_{sp}} - R_c L_{sp} & -(\omega_p - \omega_{re}) M_{pc}{}^2 + \omega_c L_{sp} L_{sc} \\ (\omega_p - \omega_{re}) M_{pc}{}^2 - \omega_c L_{sp} L_{sc} & -\frac{R_p M_{pc}{}^2}{L_{sp}} - R_c L_{sp} \end{bmatrix} \tag{25}$$

The mathematical model shown in (23) is the research object of optimal control. After the model is established, the initial condition, constraints condition, and performance function are determined. Then, the optimal control input must be solved.

According to the optimal control theory, the objective function is the finite-time integral value of the square of the electromagnetic torque T_e. Then, the minimum value principle is used.

The starting condition is

$$x(t_0) = x_0 = \begin{bmatrix} i_{cd0} & i_{cq0} \end{bmatrix}^T \tag{26}$$

The control constraints condition is

$$-U \le u \le U,\ U = \begin{bmatrix} u_{cmax} & u_{cmax} \end{bmatrix}^T \tag{27}$$

where u_{cmax} is the maximum allowable CW voltage amplitude. Both u_{cd} and u_{cq} cannot exceed the maximum value of the CW voltage amplitude. This is because the CW voltage is limited by the capacity of power electronic devices, so there is a maximum value for u_{cd} and u_{cq}.

The terminal time t_f is determined, and the state $x(t_f)$ is free:

$$\lambda(t_f) = 0 \tag{28}$$

The performance function is

$$J = \int_{t_0}^{t_f} L[x(t), u(t), t] dt \tag{29}$$

The expression of L is

$$L[x(t), u(t), t] = T_e{}^2 \tag{30}$$

According to the approximate expression of T_e described in (17), the square of the electromagnetic torque is

$$T_e^2 = \left[\frac{3 u_{pd} M_{pc}}{2 \omega_p L_{sp}} (p_p + p_c) \right]^2 \begin{bmatrix} i_{cd} & i_{cq} \end{bmatrix} \begin{bmatrix} 1 & 0 \\ 0 & 0 \end{bmatrix} \begin{bmatrix} i_{cd} \\ i_{cq} \end{bmatrix} = x^T Q x \tag{31}$$

The Hamiltonian function is taken as

$$H = L[x(t), u(t), t] + \lambda^T f[x, u, t] \tag{32}$$

The necessary conditions for achieving optimal control are the optimal control u^*, the optimal trajectory x^*, and the optimal covariance vector λ^* meeting the state Equation (33) and the costate Equation (34):

$$\dot{x} = \frac{\partial H}{\partial \lambda} = Ax + Bu \tag{33}$$

$$\dot{\lambda} = -\frac{\partial H}{\partial x} = -2Qx - A\lambda \tag{34}$$

The system of equations could be theoretically solved with the simultaneous Equations (26), (28), (33), and (34). According to the characteristics of the matrices L and Q, only the d-axis component i_{cd} of the CW current is included, and it is independent of the q-axis component i_{cq}. Then, the initial value of u_{cq} can be used, and only the optimal solution of u_{cd} needs to be solved:

$$\begin{cases} u_{cd-opt} = -(\frac{R_p M_{pc}{}^2}{L_{sp}{}^2} + R_c)\frac{i_{cd0}}{1-e^{-(\lambda_1-\lambda_2)t_f}}[e^{\lambda_2 t} - e^{-(\lambda_1-\lambda_2)t_f} e^{\lambda_1 t}] - [\frac{1}{L_{sp}}(\omega_p - \omega_{re})M_{pc}{}^2 + \omega_c L_{sc}]i_{cq0} \\ u_{cq-opt} = [\frac{1}{L_{sp}}(\omega_p - \omega_{re})M_{pc}{}^2 + \omega_c L_{sc}]i_{cd0} - (\frac{R_p M_{pc}{}^2}{L_{sp}{}^2} + R_c)i_{cq0} \end{cases} \quad (35)$$

In (35), $u_{cd\text{-}opt}$ and $u_{cq\text{-}opt}$ are the optimal CW voltage reference values, and λ_1 and λ_2 are the characteristic roots of the system matrix A, expressed as

$$\lambda_{1,2} = -\frac{R_p M_{pc}{}^2}{L_{sp}} - R_c L_{sp} \pm j\sqrt{(\frac{R_p M_{pc}{}^2}{L_{sp}} + R_c L_{sp})^2 - 4\omega_c{}^2(M_{pc}{}^2 - L_{sp}L_{sc})^2} \quad (36)$$

Considering that the exponential function is difficult to express in the algorithm, the instruction values of $u_{cd\text{-}opt}$ and $u_{cq\text{-}opt}$ in the program are given equivalent sectional fixed values, which can restrain the oscillations of the electromagnetic torque oscillation.

The switching logic between conventional PI control and optimal control is shown in Figure 6, where the superscript * represents the instruction value.

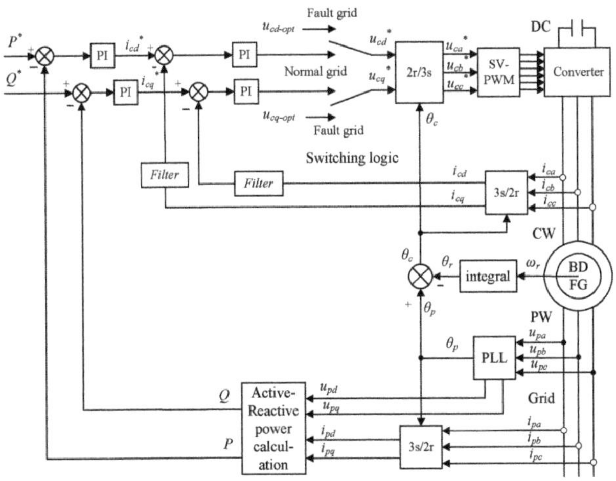

Figure 6. Optimal control structure of the BDFG.

The optimal control strategy is only enabled during the instantaneous voltage drop period in the power grid, and the optimal command values $u_{cd\text{-}opt}$ and $u_{cq\text{-}opt}$ of the optimal CW are directly given at this time. The working states before the drop and after the power grid fault are both closed-loop dual PI control strategies.

5. Experimental Results

Figure 7 shows a brushless doubly fed wind power generation system platform. The prime mover is a variable-frequency induction motor (VFIM), which simulates the wind turbine. The speed of the motor ω_r, that is, the speed of the brushless doubly fed generator, is controlled with a general-purpose inverter. The system is equipped with an AC source, which is a grid simulator that can generate grid voltage drops.

Figure 7. Experimental platform of a brushless doubly fed power generation system.

The parameters of the BDFG in Figure 7 are shown in Table 1.

Table 1. Prototype machine parameters for experiments.

Parameter	Value	Parameter	Value	Parameter	Value	Parameter	Value
Machine rating	4 kW	PW rated current	6 A	L_p	0.6659 H	M_{cr}	0.1841 H
PW/CW pole pairs	1/3	CW rated current	10 A	L_c	0.1898 H	R_p	2.5 Ω
PW rated voltage	380 V (50 Hz)	Operating speed	600~1200 r/m	L_r	0.8442 H	R_c	2.3 Ω
CW rated voltage	230 V (30 Hz)	Rated torque	80 N·m	M_{pr}	0.6547 H	R_r	2.9 Ω

The algorithm period of the system is 160 us, the sampling period is 160 us, and the IGBT switching period is 6.25 kHz. The root mean square of the grid voltage phase–phase is 380 V, and the frequency is 50 Hz. The rotor speed is regulated at 900 rpm. The active power is 4 kW, and the reactive power is 0 kvar.

First, the system waveform under traditional PI control is observed and analyzed. The overshoot and oscillations in the physical quantities of the generator are obvious, and the duration reaches 1.3 s. As seen in Figure 8b, the change in the PW current is very obvious, where its peak value increases from 8.5 A to 32 A, an increase of approximately 3.7 times. Figure 8c shows that the CW current amplitude is 15 A before the grid voltage drops and suddenly rises to 32 A, increasing to 2.1 times the rated value, which is quite bad for a converter. At the same time, Figure 8d shows that i_{cq} is synchronous with i_{pd}, but i_{cd} changes inversely with i_{pq}, such that $i_{pd} = i_{cd}$ and $i_{pq} = -i_{cq}$. There are severe oscillations in the four dq components. Figure 8e shows that the maximum rotor speed ω_r reaches 1000 rpm, and the minimum speed is 820 rpm. Figure 8f shows a large fluctuation in the torque T_e, which changes from −50 N·m to 50 N·m. Figure 8g shows a fluctuation in the PW's active and reactive power.

The system waveform under the optimal control strategy is shown in Figure 9. As seen in Figure 9b, the peak value of the PW current increases to 25.5 A from 8.5 A and stabilizes after three cycles at 15 A. Figure 9c shows that the oscillations in the CW current cease after a period, and the stable value is the same as the rated value. Its maximum amplitude is only 1.7 times its rated value of 25 A. That is, the current of the converter does not exceed two times the rated value, which is within the bearing range of the IGBT. Figure 9d shows that i_{pd} increases to 16 A from 8.5 A, and i_{cq} increases to 12 A from 7.5 A. However, they quickly stabilize. Then, i_{pq} drops to −9 A from 0 A but can be quickly stabilized. Furthermore, i_{cd} changes very little and stays near 12.5 A. Finally, Figure 9e demonstrates that the maximum rotor speed ω_r reaches 960 rpm, and the minimum speed is 870 rpm. Figure 9f shows a minor fluctuation in the torque T_e. Figure 9g shows that the fluctuation in the PW's active and reactive power is small.

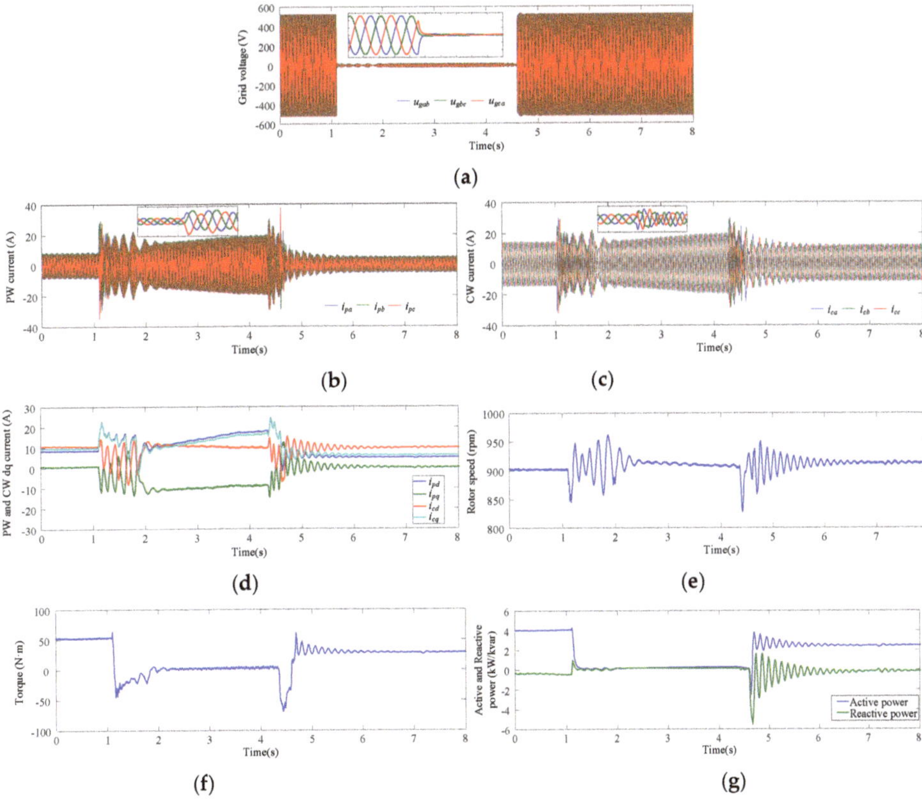

Figure 8. (**a**) Grid voltage, (**b**) PW current, (**c**) CW three-phase current, (**d**) PW and CW current dq components, (**e**) rotor speed, (**f**) electromagnetic torque, and (**g**) active and reactive power under PI control.

Comparing Figures 9 and 10, it can be observed that the optimal control and flux linkage tracking control are comparable. The dynamic performance indicators of the PI control, flux linkage tracking, and optimal control are shown in Table 2. The numerical values of the PI and optimal control methods are on the left and right, respectively.

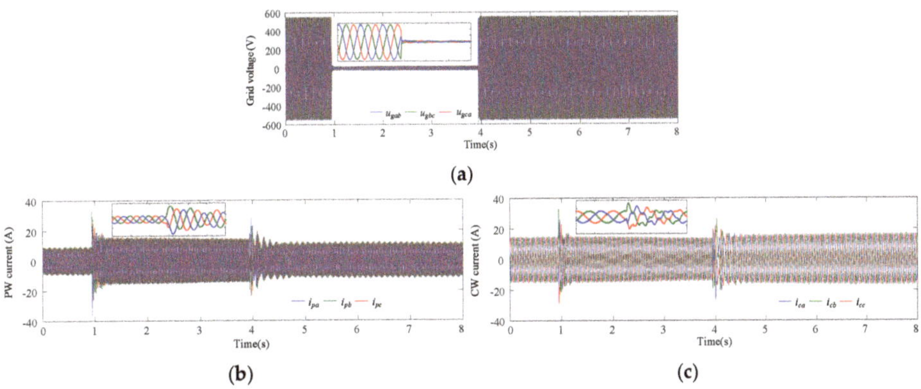

Figure 9. *Cont.*

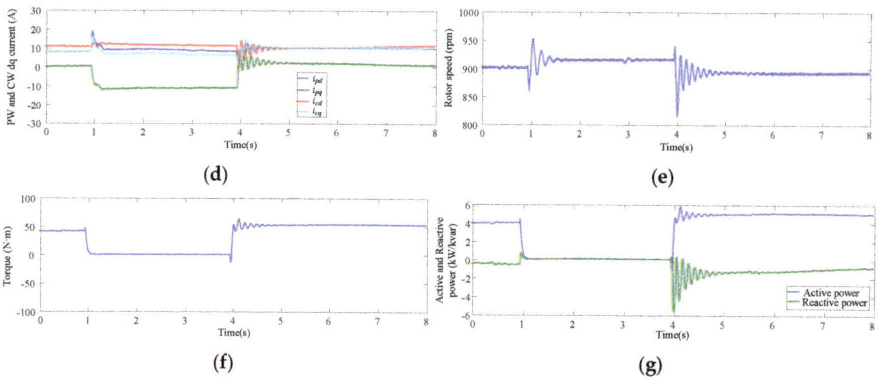

Figure 9. (**a**) Grid voltage, (**b**) PW current, (**c**) CW three-phase current, (**d**) PW and CW current dq components, (**e**) rotor speed, (**f**) electromagnetic torque, and (**g**) active and reactive power under optimum control.

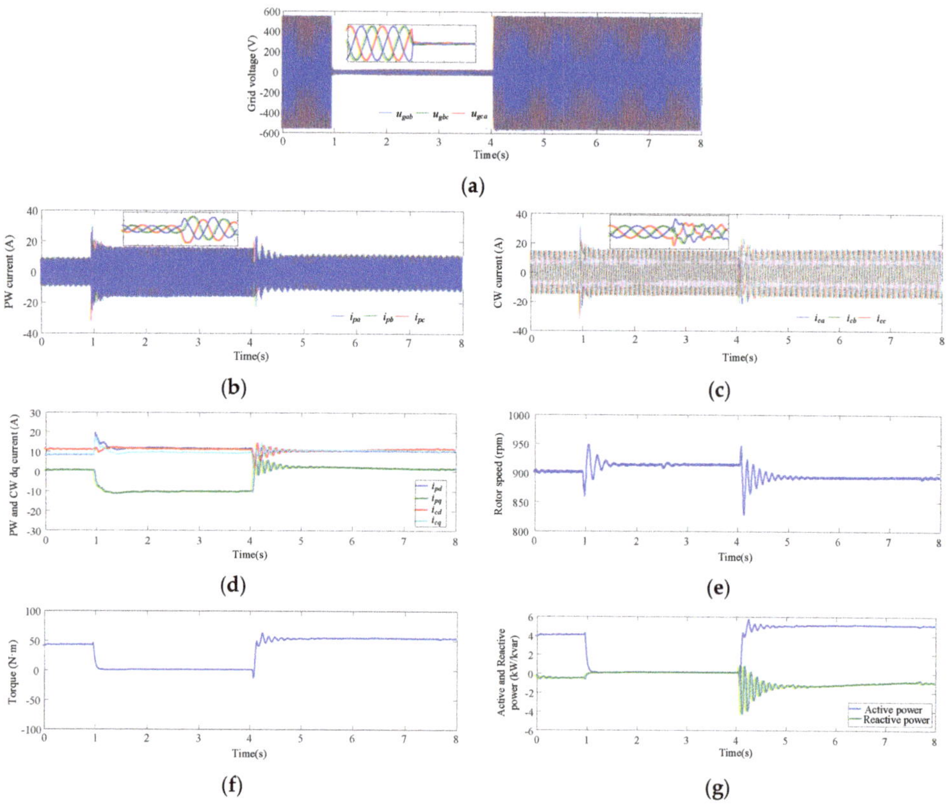

Figure 10. (**a**) Grid voltage, (**b**) PW current, (**c**) CW current, (**d**) PW and CW current dq components, (**e**) rotor speed, (**f**) electromagnetic torque, and (**g**) active and reactive power under flux linkage tracking.

Table 2. Quantity comparison between PI control, flux linkage tracking, and optimal control.

Variable	Extremum	Extremum Time	Regulating Time	Overshoot
i_{pd}	−15 A/20 A/15.5 A	0.2 s/0.02 s/0.02 s	1.3 s/0.9 s/0.8 s	275%/100%/100%
i_{pq}	−24 A/−12 A/−12 A	0.05 s/0.03 s/0.02 s	1.3 s/0.9 s/0.8 s	189%/112%/112%
i_{cd}	25 A/13 A/16 A	0.1 s/0.03 s/0.02 s	1.3 s/0.9 s/0.8 s	150%/33%/33%
i_{cq}	−15 A/19 A/18 A	0.05 s/0.03 s/0.02 s	1.3 s/0.9 s/0.8 s	275%/100%/100%
T_e	70 N·m/70 N·m/70 N·m	0.05 s/0.03 s/0.02 s	1.3 s/0.9 s/0.8 s	600%/248%/250%
n_r	1000 rpm/958 rm/960 rm	0.1 s/0.03 s/0.02 s	1.3 s/0.9 s/0.8 s	8.7%/4.8%/4.4%

6. Conclusions

In this paper, the transient characteristics of a BDFG under grid faults were revealed by solving the analytical expression of the BDFG's stator and rotor winding currents during a transient voltage drop in the power grid. It was found that the system's overshoot was related to the CW voltage. Furthermore, an optimal control method was proposed to suppress the oscillation in the generator's electromagnetic torque during faults. This strategy can reduce the peak values of the PW, CW, and RW currents and shorten the convergence time.

The experimental results show that the overshoot of the physical quantities of the generator under flux linkage tracking control and optimal control are significantly smaller than those under the traditional PI control strategy when dealing with voltage drops. Moreover, the oscillation duration is shortened, and the anti-disturbance ability of the whole control system is enhanced. The performance indicators such as the transient response, the current and torque overshoot, and the convergence time of advanced algorithms such as flux linkage tracking and optimal control are significantly better than those of traditional PI control. The optimal control and flux linkage tracking control are comparable. Therefore, the effectiveness and practicability of the optimal strategy have been verified.

Author Contributions: J.X. designed the principles of the overall work, proposed the control strategies, and prepared the initial draft of this paper. P.N. implemented software programming, proposed some technical comments, and edited the final draft of this paper. All authors have read and agreed to the published version of the manuscript.

Funding: This research received no external funding.

Data Availability Statement: Data is contained within the article.

Conflicts of Interest: The authors declare no conflicts of interest.

Appendix A

$$Z_m{}^3 = j\omega_p\omega_c\omega_{rp}(L_pL_cL_r - L_pM_{cr}{}^2 - L_cM_{pr}{}^2) + j\omega_pL_pR_cR_r - j\omega_cL_cR_pR_r + j\omega_{rp}L_rR_pR_c + R_pR_cR_r$$
$$+\omega_p\omega_cL_pL_cR_r - \omega_p\omega_{rp}(L_pL_r - M_{pr}{}^2)R_c + \omega_c\omega_{rp}(L_cL_r - M_{cr}{}^2)R_p$$

$$Z_{ps1} = \frac{Z_m{}^3}{R_rR_c - j\omega_cL_cR_r + j\omega_{rp}L_rR_c + (L_pL_r - M_{cr}{}^2)\omega_c\omega_{rp}}$$

$$Z_{ps2} = \frac{Z_m{}^3}{M_{pr}M_{cr}\omega_p\omega_{rp}}, Z_{cs1} = \frac{Z_m{}^3}{M_{pr}M_{cr}\omega_{rp}\omega_c}$$

$$Z_{cs2} = \frac{Z_m{}^3}{R_pR_r + j\omega_pL_pR_r + j\omega_{rp}L_rR_p - (L_pL_r - M_{pr}{}^2)\omega_p\omega_{rp}}$$

$$Z_{rs1} = \frac{Z_m{}^3}{j\omega_{rp}M_{pr}(R_c - j\omega_cL_c)}, Z_{rs2} = \frac{Z_m{}^3}{j\omega_{rp}M_{cr}(R_p + j\omega_pL_p)}$$

$$\begin{cases} L_m{}^3 = L_p L_c L_r - L_p M_{cr}{}^2 - L_c M_{pr}{}^2 \\ \lambda_1 = -\frac{1}{T_p} - j\omega_p, \lambda_2 = -\frac{1}{T_c} + j\omega_c, \lambda_3 = -\frac{1}{T_r} - j\omega_{rp} \end{cases}$$

$$\begin{cases} Z_{pf0} = \dfrac{-L_m{}^3\lambda_1\lambda_2\lambda_3}{R_r R_c - j\omega_c L_c R_r + j\omega_{rp} L_r R_c + (L_p L_r - M_{cr}{}^2)\omega_c\omega_{rp}} \\ Z_{pf1} = \dfrac{L_m{}^3(\lambda_3 - \lambda_1)(\lambda_1 - \lambda_2)}{-(L_c L_r - M_{cr}{}^2)(\lambda_1 - \lambda_2 - \lambda_3) - \frac{\lambda_2\lambda_3 L_m{}^3}{Z_{pf0}}} \\ Z_{pf2} = \dfrac{L_m{}^3(\lambda_1 - \lambda_2)(\lambda_2 - \lambda_3)}{-(L_c L_r - M_{cr}{}^2)(\lambda_2 - \lambda_1 - \lambda_3) - \frac{\lambda_1\lambda_3 L_m{}^3}{Z_{pf0}}} \\ Z_{pf3} = \dfrac{L_m{}^3(\lambda_2 - \lambda_3)(\lambda_3 - \lambda_1)}{(L_c L_r - M_{cr}{}^2)(\lambda_3 - \lambda_1 - \lambda_2) - \frac{\lambda_1\lambda_2 L_m{}^3}{Z_{pf0}}} \end{cases}$$

$$\begin{cases} Z_{cf0} = \dfrac{-L_m{}^3\lambda_1\lambda_2\lambda_3}{M_{pr} M_{cr}\omega_{rp}\omega_c} \\ Z_{cf1} = \dfrac{L_m{}^3(\lambda_3 - \lambda_1)(\lambda_1 - \lambda_2)}{-\lambda_1 M_{pr} M_{cr} - M_{pr} M_{cr}(j\omega_{rp} - j\omega_c) - \frac{\lambda_2\lambda_3 L_m{}^3}{Z_{cf0}}} \\ Z_{cf2} = \dfrac{L_m{}^3(\lambda_1 - \lambda_2)(\lambda_2 - \lambda_3)}{-\lambda_2 M_{pr} M_{cr} - M_{pr} M_{cr}(j\omega_{rp} - j\omega_c) - \frac{\lambda_1\lambda_3 L_m{}^3}{Z_{cf0}}} \\ Z_{cf3} = \dfrac{L_m{}^3(\lambda_2 - \lambda_3)(\lambda_3 - \lambda_1)}{-\lambda_3 M_{pr} M_{cr} - M_{pr} M_{cr}(j\omega_{rp} - j\omega_c) - \frac{\lambda_1\lambda_2 L_m{}^3}{Z_{cf0}}} \end{cases}$$

$$\begin{cases} Z_{rf0} = \dfrac{-L_m{}^3\lambda_1\lambda_2\lambda_3}{j\omega_{rp} M_{pr}(R_c - j\omega_c L_c)} \\ Z_{rf1} = \dfrac{L_m{}^3(\lambda_3 - \lambda_1)(\lambda_1 - \lambda_2)}{-\lambda_1 L_c M_{pr} + M_{pr}(R_c + j\omega_{rp} L_c - j\omega_c L_c) - \frac{\lambda_2\lambda_3 L_m{}^3}{Z_{rf0}}} \\ Z_{rf2} = \dfrac{L_m{}^3(\lambda_1 - \lambda_2)(\lambda_2 - \lambda_3)}{-\lambda_2 L_c M_{pr} + M_{pr}(R_c + j\omega_{rp} L_c - j\omega_c L_c) - \frac{\lambda_1\lambda_3 L_m{}^3}{Z_{rf0}}} \\ Z_{rf3} = \dfrac{L_m{}^3(\lambda_2 - \lambda_3)(\lambda_3 - \lambda_1)}{-\lambda_3 L_c M_{pr} + M_{pr}(R_c + j\omega_{rp} L_c - j\omega_c L_c) - \frac{\lambda_1\lambda_2 L_m{}^3}{Z_{rf0}}} \end{cases}$$

References

1. Xiong, F.; Wang, X. Design of a low-harmonic-content wound rotor for the brushless doubly fed generator. *IEEE Trans. Energy Convers.* **2014**, *29*, 158–168. [CrossRef]
2. Kong, M.; Wang, X.; Li, Z. Asynchronous operation characteristics and soft-starting method for the brushless doubly-fed motor. *IET Electr. Power Appl.* **2017**, *11*, 1276–1277. [CrossRef]
3. Zhu, L.; Zhang, F.; Jin, S.; Ademi, S.; Su, X.; Cao, W. Optimized Power Error Comparison Strategy for Direct Power Control of the Open-Winding Brushless Doubly Fed Wind Power Generator. *IEEE Trans. Sustain. Energy* **2019**, *10*, 2005–2006. [CrossRef]
4. Abdi, S.; Abdi, E.; McMahon, R. A New Iron Loss Model for Brushless Doubly-Fed Machines With Hysteresis and Field Rotational Losses. *IEEE Trans. Energy Convers.* **2021**, *36*, 3221–3229. [CrossRef]
5. Kan, C.; Zheng, J.; Wang, X.; Wang, Q. Design and Starting Performance Study of a Soft-Start Brushless Doubly Fed Machine. *IEEE Trans. Energy Convers.* **2022**, *37*, 97–106. [CrossRef]
6. Chen, X.; Wang, X.; Kong, M.; Li, Z. Design of a Medium-Voltage High-Power Brushless Doubly Fed Motor With a Low-Voltage Fractional Convertor for the Circulation Pump Adjustable Speed Drive. *IEEE Trans. Ind. Electron.* **2022**, *69*, 7720–7731. [CrossRef]
7. Sadeghi, R.; Madani, S.M.; Ataei, M. A New Smooth Synchronization of Brushless Doubly-Fed Induction Generator by Applying a Proposed Machine Model. *IEEE Trans. Sustain. Energy* **2018**, *9*, 371–379. [CrossRef]
8. Wang, X.; Lin, H. DC-link current estimation for load-side converter of brushless doubly-fed generator in the current feed-forward control. *IET Power Electron.* **2016**, *9*, 1703–1704. [CrossRef]
9. Yan, X.; Cheng, M. A Robustness-Improved Control Method Based on ST-SMC for Cascaded Brushless Doubly Fed Induction Generato. *IEEE Trans. Ind. Electron.* **2021**, *68*, 7061–7070. [CrossRef]
10. Xu, L.; Cheng, M.; Wei, X.; Yan, X.; Zeng, Y. Dual Synchronous Rotating Frame Current Control of Brushless Doubly Fed Induction Generator Under Unbalanced Network. *IEEE Trans. Power Electron.* **2021**, *36*, 6712–6723. [CrossRef]
11. Ge, J.; Xu, W.; Liu, Y.; Xiong, F. Novel Equivalent Circuit Model Applicable to All Operation Modes for Brushless Doubly Fed Induction Machines. *IEEE Trans. Ind. Electron.* **2022**, *69*, 12540–12549. [CrossRef]
12. Ruiz-Cruz, R.; Sanchez, E.N.; Loukianov, A.G.; Ruz-Hernandez, J.A. Real-Time Neural Inverse Optimal Control for a Wind Generator. *IEEE Trans. Sustain. Energy* **2019**, *10*, 1172–1182. [CrossRef]
13. Zeng, Y.; Cheng, M.; Wei, X.; Zhang, G. Grid-Connected and Standalone Control for Dual-Stator Brushless Doubly Fed Induction Generator. *IEEE Trans. Ind. Electron.* **2021**, *68*, 9196–9205. [CrossRef]

14. Gowaid, I.A.; Abdel-Khalik, A.S.; Massoud, A.M.; Ahmed, S. Ride-Through Capability of Grid-Connected Brushless Cascade DFIG Wind Turbines in Faulty Grid Conditions—A Comparative Study. *IEEE Trans. Sustain. Energy* **2013**, *4*, 1002–1013. [CrossRef]
15. Shao, S.; Abdi, E.; McMahon, R. Dynamic Analysis of the Brushless Doubly-Fed Induction Generator during Symmetrical Three-Phase Voltage Dips. In Proceedings of the 2009 International Conference on Power Electronics and Drive Systems (PEDS), Taipei, China, 2–5 November 2009.
16. Sadeghi, R.; Madani, S.M.; Lipo, T.A.; Kashkooliet, M.R.A.; Ataei, M.; Ademi, S. Voltage-Dip Analysis of Brushless Doubly Fed Induction Generator Using Reduced T-Model. *IEEE Trans. Ind. Electron.* **2019**, *66*, 7510–7512. [CrossRef]
17. Tohidi, S.; Hashem, O.; Zolghadri, M.R.; Shao, S.; Tavner, P. Analysis and Enhancement of Low-Voltage Ride-Through Capability of Brushless Doubly Fed Induction Generator. *IEEE Trans. Ind. Electron.* **2013**, *60*, 1149–1150. [CrossRef]
18. Abdi, E.; McMahon, R.; Malliband, P. Performance analysis and testing of a 250 kW medium-speed brushless doubly-fed induction generator. *IET Renew. Power Gener.* **2013**, *7*, 636–637. [CrossRef]
19. Long, T.; Shao, S.; Malliband, P.; Abdi, E.; McMahon, R. Crowbarless Fault Ride-Through of the Brushless Doubly Fed Induction Generator in a Wind Turbine Under Symmetrical Voltage Dips. *IEEE Trans. Ind. Electron.* **2013**, *60*, 2836–2837. [CrossRef]
20. Betz, R.E. Comparison of Rotor Side Converter Protection for DFIGs and Brushless Doubly Fed Reluctance Machines under Fault Conditions. In Proceedings of the 2021 23rd European Conference on Power Electronics and Applications (EPE'21 ECCE Europe), Ghent, Belgium, 6–10 September 2021; pp. 1–10.
21. Wang, Z.; Xu, H.; Li, Z. Small Signal Impedance Modeling and Stability Analysis of BDFIG Tied into Weak Power Grid. In Proceedings of the 2021 IEEE 4th International Electrical and Energy Conference (CIEEC), Wuhan, China, 28–30 May 2021; pp. 1–6.
22. Wang, S.; Shang, L. Fault Ride Through Strategy of Virtual-Synchronous-Controlled DFIG Based Wind Turbines Under Symmetrical Grid Faults. *IEEE Trans. Energy Convers.* **2020**, *35*, 1360–1371. [CrossRef]
23. Huang, J.; Li, S. Analytical expression for LVRT of BDFIG with enhanced current control to CW and reactive power support from GSC. *Elsevier Electr. Power Energy Syst.* **2018**, *98*, 243–255. [CrossRef]
24. Lu, M.; Chen, Y.; Zhang, D.; Su, J.; Kang, Y. Virtual Synchronous Control Based on Control Winding Orientation for Brushless Doubly Fed Induction Generator (BDFIG) Wind Turbines Under Symmetrical Grid Faults. *Energies* **2019**, *12*, 319. [CrossRef]
25. Zhang, A.; Chen, Z.; Gao, R.; Wang, J.; Ma, Z.; Wang, S.; Wang, Y. Crowbarless Symmetrical Low-Voltage Ride Through Based on Flux Linkage Tracking For Brushless Doubly Fed Induction Generators. *IEEE Trans. Ind. Electron.* **2019**, *67*, 7606–7616. [CrossRef]
26. Sun, L.; Chen, Y.; Peng, L.; Kang, Y. Numerical-based frequency domain controller design for stand-alone brushless doubly fed induction generator power system. *IET Power Electron.* **2017**, *10*, 588–598. [CrossRef]
27. Shipurkar, U.; Strous, T.D.; Polinder, H.; Ferreira, J.A.; Veltman, A. Achieving Sensorless Control for the Brushless Doubly Fed Induction Machine. *IEEE Trans. Energy Convers.* **2017**, *32*, 1612–1613. [CrossRef]
28. Liu, Y.; Xu, W.; Zhu, J.; Blaabjerg, F. Sensorless Control of Standalone Brushless Doubly Fed Induction Generator Feeding Unbalanced Loads in a Ship Shaft Power Generation System. *IEEE Trans. Ind. Electron.* **2019**, *66*, 741–742. [CrossRef]
29. Ouyang, J.; Tang, T.; Diao, Y.; Li, M.; Yao, J. Control method of doubly fed wind turbine for wind speed variation based on dynamic constraints of reactive power. *IET Renew. Power Gener.* **2018**, *12*, 973–974. [CrossRef]

energies

MDPI

Article

An Experimental Performance Assessment of a Passively Controlled Wind Turbine Blade Concept: Part A—Isotropic Materials

Nikolaos Papadakis * and Constantinos Condaxakis

Power Plant Synthesis Laboratory, Department of Mechanical Engineering, Hellenic Mediterranean University, 71410 Heraklion, Greece; condax@hmu.gr
* Correspondence: npap@hmu.gr; Tel.: +30-2810-379722

Abstract: This paper is the first part of a two-part series, which presents preliminary findings on a novel flexible curved wind turbine blade designed for passive control, comparing its aerodynamic performance and behavior against a conventional straight blade. Characterized by its ability to twist around its longitudinal axis under bending loads, the flexible curved blade is engineered to self-regulate in response to varying wind speeds, optimizing power output and enhancing operational safety. This design utilizes inherent elasticity and specific geometric configurations to develop torsional loads, resulting in continuous adjustment of the blade's pitch angle via twist–bend deformation. The study focuses on a comparative analysis conducted in a wind tunnel, testing both a small-scale model of the conventional blade and the flexible curved blade of equivalent diameter. Results indicate that the flexible curved blade concept successfully moderates its rotational speed and power output at higher wind speeds and demonstrates the capability to start generating power at lower wind speeds and stabilize power effectively, aligning with sustainability goals by potentially reducing reliance on active control systems. Despite promising outcomes, passive control mechanisms did not activate at the designed wind speeds, revealing a misalignment between expected and actual performance and underscoring the need for further refinements in blade design and control settings. Additionally, the power coefficient (Cp) versus tip speed ratio (TSR) comparison showed that flexible curved blades operate within a lower TSR range and exhibit controlled capping of power under high wind conditions, marked by a distinctive 'hook-like' feature in C_p behavior. This study confirms the feasibility of designing and manufacturing passively controlled wind turbine blades tailored to specific performance criteria and underscores the potential of such technology. Future work, to be detailed in a subsequent paper, will explore further optimizations and the use of Glass Fiber-Reinforced Polymer (GFPR) composite materials to enhance blade flexibility and performance.

Keywords: passive control; aero elastic tailoring; adaptive blades; curved blades

1. Introduction

1.1. Context and Importance of Renewable Energy

Over the past five years, global energy dynamics have undergone significant turbulence, emphasizing the vital role of renewable energy sources. The energy market has exhibited remarkable volatility, with major events reshaping both economic and environmental landscapes. For instance, during the COVID-19 pandemic, oil prices dramatically plummeted into negative territory, exposing the fragility of conventional energy markets to global disruptions [1]. Similarly, the recent conflict in Ukraine has led to a sharp increase in gas prices, intensifying strains on the global energy supply chain and underscoring the geopolitical risks tied to fossil fuel dependence.

In contrast to these fluctuations, renewable energy technologies such as wind and photovoltaic systems have demonstrated resilience and cost-effectiveness, attributable to technological advances and economies of scale. The levelized cost of electricity (LCOE)

from these sources has consistently decreased over the past decade, increasingly aligning their competitiveness with that of traditional energy sources [2]. Wind energy, in particular, has achieved significant maturity, characterized by improved turbine efficiencies, larger capacity installations, and an increase in offshore projects. The rapid expansion of wind energy projects worldwide signifies a collective movement towards a sustainable and reliable energy supply.

The escalating impacts of climate change further necessitate the shift towards renewable energy, necessitating a reduction in carbon emissions and a pivot to cleaner energy alternatives. The progress in wind and photovoltaic technology not only presents an economic opportunity but is also critical for achieving energy security, environmental sustainability, and global collaboration in addressing climate volatility.

Significantly, global electricity generation is increasingly reliant on renewable sources. The International Energy Agency (IEA) highlighted that, as of 2013, the European Union was projected to raise its renewable electricity share from 6.9% in 2011 to 23.1% by 2035 [3]. However, these figures were quickly surpassed, with updates in 2022 showing that by 2015, the share had already reached 22.8%, and by 2027, it is anticipated to increase to 38.1% [4]. These estimates are corroborated by Eurostat data indicating that by 2020, renewable energy's contribution to electricity generation in the EU had already escalated to 39.0% [5]. In the United States, as of October 2023, approximately 10.3% and 3.4% of electricity production were attributed to wind and solar energy, respectively [6]. The growing reliance on renewable sources, particularly wind energy, underscores a pivotal shift in energy strategies globally, setting the stage for a detailed discussion on the advancements and context of wind energy in the following section.

1.2. Small Wind Energy Converter Systems

The broad adoption of wind energy necessitates a significant reduction in energy production costs. This reduction is largely contingent on optimizing the performance of wind energy converter systems (WECS). A crucial element in making wind energy more attractive and competitive is minimizing both initial investment and ongoing maintenance costs.

Figure 1 presents the levelized cost of energy (LCOE) for offshore wind turbines, including substantial contributions from both the turbine itself and its maintenance. These costs are roughly equivalent, largely due to the logistical challenges and accelerated wear from the marine environment [7]. This parity is largely due to the logistical challenges of servicing offshore wind turbines, the monitoring systems costs and the accelerated wear caused by the salty marine environment.

Ponta [8] asserts that the evolution of wind power relies on overcoming the limitations of current blade technology. He proposes the adoption of smart, self-controlling blades that utilize mode coupling to bypass active control systems. This innovation leads to lighter, simpler turbines that are responsive to changing conditions and could significantly lower both construction and operational costs.

The philosophy of passive control in wind energy systems is predicated on the turbine's ability to respond to excitation loads and the blade's capacity to function as a feedback mechanism, adapting to operational conditions to meet design criteria. The primary design objectives include maximizing power yield, minimizing operational loads, and optimizing control within the WECS. Implementing a passive pitch control philosophy can drastically reduce blade aerodynamic loads and overall structural stress. Turbines equipped with passive pitch-controlled blades could potentially eliminate the need for complex, moving parts found in actively controlled variable pitch blades. Instead, these passive pitch-controlled blades could adjust their geometry in response to aerodynamic loads, combining the benefits of variable pitch blades—such as responsiveness, maximization of energy yield, and eventually reduced energy cost—with the simplicity of manufacturing akin to fixed pitch blades. An additional advantage of passive pitch control blades is their quicker response time, presenting a viable solution for enhancing the feasibility of small WECS in particular.

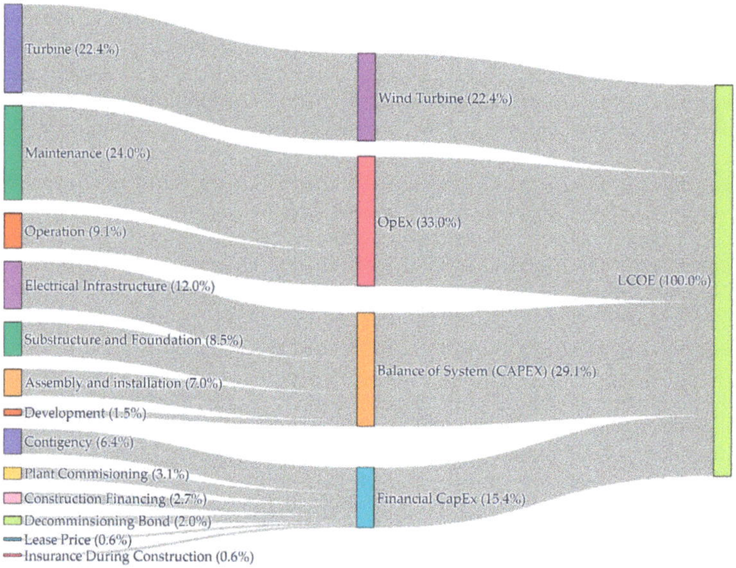

Figure 1. Sankey diagram depicting LCOE cost share for offshore wind turbines in 2021 (Data source: [7]).

1.3. Literature Review

1.3.1. Passive Control Concepts

Liebst [9], in 1986, studied the minimization of loads due to wind gusts on a curved wind turbine blade rotor. They observed that when a gust occurs, the curved wind blade changes its geometry, significantly reducing the pitch angle and therefore reducing aerodynamic loads on the blade.

In 1990, Christakis [10] proposed a WECS with sails instead of blades. The sails were controlled passively with the aid of elastic fibers. He also proposed a 9 m blade designed for twist–bending coupling.

Serra and Van Schoor [11], in 1995, presented an investigation of a self-regulating pitch control blade mechanism. The blade pitch results from the equilibrium between the bending moment of the aerodynamic load and the inverse torque generated by a spring. The spring was selected so that the blade operates in a range near maximum efficiency. The self-regulating pitch control blade mechanism exhibited improvements of up to 40% compared to a reference fixed blade. Serra and Van Schoor validated their mathematical model via experiments in a wind tunnel.

Eggers et al. [12], in 1996, demonstrated load reductions by linking a pitch control system to flapwise blade loads using simple integral control. The results indicated the potential to reduce the root-mean-square (rms) blade bending response to turbulent winds by about half.

Finally, although a departure from other work presented here for passive control of wind turbines, Tran, in 2010 [13], presented a methodology that obtained the optimal main characteristics (geometric and energetic features) of the Permanent Magnet Synchronous Generator (PMSG) for passive wind turbines, which allowed the design of a system without active electronic parts (power and control).

1.3.2. Passive Twist Coupled Control Concepts

Infield and Feuchtwang [14] in 1995 and [15] 1999 showed how small turbines can have improved speed regulation with extension twist coupling. They proposed and tested a "bend–twist-coupled" blade developed to control the rotor in a runaway scenario. Their composite blade was fabricated using a helical layup with layers of glass and carbon fibers. Measured twist coupling agreed well with predictions, and measured runaway speeds were actually less than predicted.

Lobitz and Veers [16], in 1996, studied the effects of twist–bend coupling on the annual energy production of a nominally 26 m diameter stall regulated HAWT. Using the generic utility-sized rotor as a test case, two blade twist configurations in conjunction with three twisting schedules were investigated to determine the benefits of blades that twist towards stall with applied loading. In all cases, the annual energy was increased by 10-15% for a maximum blade twist of two degrees and 5–7% for a one-degree maximum twist. In 1998, Lobitz and Veers [17] studied the stability effects of coupled rotors and addressed two of the most common stability constraints, namely classical flutter and divergence.

Eisler and Veers [18], in 1998, examined the performance gains of adaptive blades that twist under the action of centrifugal loads installed on a 26 m diameter variable speed rotor. The ability of bending–twist-coupled blades to attenuate (or exacerbate) cyclic loading was investigated by Lobitz and Laino [19], in 1999, and Lobitz, Veers, and Laino [20], in 2000, for a 33 m diameter rotor employing three different control strategies: constant speed stall-controlled, variable speed stall-controlled, and variable speed pitch-controlled. Results for the constant speed stall-controlled case indicate that twist coupling towards the stall produces significant increases in fatigue damage, and for a range of wind speeds in the stall regime, apparent stall flutter behavior is observed.

Christakis and Condaxakis [21,22], in 1998 and 1999, presented a passive pitch regulation flexible blade that constantly changed its geometry under bending. The initial wind turbine blade tests exhibited good performance in low-turbulence wind.

In 2002, Zuteck [23] conducted a study on a 30 m curved planform wind turbine blade constructed from a combination of glass and carbon fibers. He introduced the concept of bend–twist coupling for blades that allowed for passive control of the rotor. This design enabled the blades to self-adjust their geometry in response to aerodynamic loads. Zuteck found that decreasing the torsional stiffness was essential for achieving sufficient twisting of the blade. Furthermore, he suggested that the reduced aerodynamic load could allow for an increase in rotor diameter, thereby reducing the energy cost.

Continuing this line of research, Larwood and Zuteck [24] in 2006 explored a swept-blade concept, which was found to enhance energy capture without increasing turbine loads. Their study compared a 28 m backward-swept radius blade (STAR6) with a conventional 25 m straight–rigid blade rotor. The results indicated that the swept blade rotor achieved a 5–10% increase in energy capture while maintaining the same load envelope.

Building on these findings, researchers at Sandia National Laboratories [25] investigated the sweep–twist adaptive rotor (STAR) technology. This innovation aimed to reduce operating loads, thereby facilitating the development of larger, more efficient rotors. The design incorporated passive blade twisting at the outer sections to limit maximum rotor thrust. Experimental data showed that a 54 m backward-swept prototype blade, termed the STAR blade, increased average energy capture by 10–12% compared to a 48 m straight–rigid–bladed rotor without any increase in blade root bending moments.

M. Masoudi and K. Pope [26] investigated two swept blades with different bend depths for power enhancement potential of NREL Phase VI rotor via a curved planform geometry. Fluid–structure interaction analysis was performed at different wind speeds, and the CFD results were verified using experimental data. The induced elastic twist of the blade at different sections was reported, and a maximum elastic twist of 0.67° was predicted from a bend depth of 1.5 m for a wind speed of 15 m/s. A blade with a box spar was developed to investigate the effects of a closed-section spar on the elastic twist. The study predicts that a blade with a box spar experiences 18% less elastic twists compared to a blade with an L spar. A 1.89% improvement is estimated due to elastic twist at 10 m/s wind speed.

Marcus Wiens et al. [27] investigated the effect of bend–twist coupling in the design of large wind turbine rotors for passive load reduction. Starting from a straight rotor blade of the IWT 7.5MW-164, designed without sweep, he proposed a swept rotor blade and a new control method that exploits bend–twist coupling to mitigate gust effects. The feedforward of the blade tip twist rate establishes fast gust detection, and the pitch angle can be adjusted accordingly. The simulations demonstrated that this novel control strategy can lead to a reduction in maximum loads during gust events. In addition to regular closed-loop pitch control, the twist rate is used to adjust the collective pitch angle using feedforward control. Extreme loads for flapwise bending moment can be reduced on average by up to 6%, and tower acceleration is reduced by 15%.

Jorge Mario Tamayo-Avendaño et al. [28] explored the effects of bend–twist coupling on the performance of a small wind turbine rotor. He proposes the development of a small wind turbine blade using appropriate composite material layering to achieve passive control of the rotor. A comparative numerical analysis with a reference counterpart is performed using a fluid–structure interaction commercial software, which resulted in a 3% increase in the annual energy yield due to the bend–twist coupling, used as a passive pitch mechanism and an increment of 0.2% and 0.3% for the blade root flapwise moment and the rotor thrust force, respectively, when considering parked conditions.

R. Riva et al. [29] investigated the stability implications of a bend–twist coupling design for a wind turbine rotor. They perform a comparative stability analysis of a very large wind turbine, firstly on an isolated blade and then on the complete rotor, where the bend–twist coupling at the new non-conventional blade is obtained by rotating the fibers of the spar caps inside the blade. The analysis results show that the bend–twist coupling has a minor effect on the response of the isolated blade in the case it is dominated by stall-induced vibrations, and for the complete rotor, the edgewise whirling modes and the tower side–side mode characteristics are not affected by the bend–twist coupling.

1.3.3. Passive Control Numerical Codes and Algorithms Studies

In 1999 Voutsinas, Belesis and Rados [30], proposed a nonlinear, completely aero elastic 3D computer code for wind turbines, which is a useful tool for passive pitch control wind turbine blade studies. Capuzzi in 2015 [31] proposed an aero-elastically tailored blade, analyzed by using finite element models with structural stability and strength constraints imposed under realistic load cases.

MacPhee in 2011 [32] presented a methodology for developing a flexible turbine blade for passive blade pitch control in wind turbines, using a robust and accurate fluid–structure interaction routine, and introduced two morphing scenarios: one where rigid and flexible blades are identical when unloaded, and one where they are identical at the stall angle.

Larwood's paper [33] in 2014 described a parametric study of swept blade design parameters for a 750 kW machine. The amount of tip sweep had the largest effect on the energy production and blade loads; other parameters had less impact. The authors then conducted a design study to implement a swept design on 1.5 MW, 3 MW, and 5 MW turbines. An aeroelastic code, described in the paper, was developed to model the behavior and determine the loads of the swept blade. The design's goal was to increase annual energy production by 5% over the straight–rigid blade without increasing blade loads.

Xin Shen in 2015 [34] described a multi-objective optimization method for the design of horizontal axis wind turbines using the lifting surface method as the performance prediction model. The purpose of the optimization method is to achieve the best trade-off of the following objectives: maximum annual energy production and minimum blade loads, including thrust and blade rood flapwise moment. The result shows that the optimization models can provide more efficient designs.

M.G. Khalafallah [35] presented a CFD simulation methodology of a swept blade HAWT for a 0.9 m diameter wind turbine model. Four different scenarios were simulated: a backward sweep, a forward sweep, an upstream sweep, and a downstream sweep blade compared to a straight–rigid blade. The results show that the downstream swept blades are able to generate more output power than the straight–rigid blades but with a relatively higher axial thrust force. So, the blade deflection and tower clearance should be considered during the design phase of swept blade rotors.

Chen et al. [36,37], in 2014 and 2016, presented a passive pitch control mechanism for small horizontal axis wind turbines (HAWT), addressing the challenges of startup at low wind speeds and safe operation at high wind speeds. They developed a centrifugal force-based system to adjust the blade's pitch angle automatically as the wind speed varies, enhancing turbine performance across different wind conditions. The design, patented by their laboratory, was validated via wind tunnel experiments, demonstrating its effectiveness in both high and low wind speed scenarios.

In 2017, Jeh Chu [38] developed a biomimetic horizontal axis wind turbine (HAWT) blade modeled after the Dryobalanops aromatica seed, utilizing OpenFOAM® for computational fluid dynamics (CFD) analysis. Results indicate that the biomimetic turbine generates higher torque and demonstrates better self-start capabilities, with a power coefficient (C_P) of 0.386 at a tip speed ratio (TSR) of 1.5. they reported that it outperforms conventional blades [39] in low wind conditions and mitigates blade root bending stress via centrifugal forces.

1.4. Scope

This article aims to present preliminary results from a study on a novel flexible blade concept aimed to be part of a passively controlled wind turbine rotor system. It emphasizes the aerodynamic performance and behavior of a flexible wind turbine blade in comparison with a conventional straight blade. The primary focus is on a flexible blade designed to self-regulate its performance via intrinsic mechanical responses to varying wind speeds. This blade is characterized by its ability to twist around its longitudinal axis under bending loads, an attribute facilitated by its unique elasticity and geometric properties. These properties allow the blade to adjust its pitch angle dynamically in response to aerodynamic loads, thus optimizing power output and enhancing operational safety.

The scope of this research includes a detailed comparison of the aerodynamic performance of two small-scale model blades—one conventional straight (and relatively rigid) and one flexible curved—under controlled conditions in a wind tunnel. The performance metrics compared include rotational velocity, power output, and efficiency across various wind speeds and braking conditions. This comparison helps to illuminate the distinct advantages and potential drawbacks of the flexible blade design, particularly its ability to start generating power at lower wind speeds and to maintain controlled power output at higher speeds, akin to systems with active control mechanisms.

Furthermore, the study also reports on the alignment—or lack thereof—between the designed performance characteristics and the actual operational outcomes of the flexible curved blades. The findings indicate a misalignment in the activation of passive control mechanisms at designated wind speeds, pointing to areas requiring further design refinement.

The article also sets the stage for subsequent investigations as part of a series, with future work aiming to build on these findings by exploring advanced material composites and further optimizing blade design to fully realize the potential of passively controlled wind turbine blades. This ongoing research is intended to contribute significantly to the development of more efficient, safer, and economically viable wind energy solutions.

2. Methodology

2.1. Overview

As stated previously, the objective of this study is to present preliminary results from an innovative passively controlled wind blade concept, highlighting its distinctions from a traditional rigid system. To achieve this, the paper compares two directly comparable wind blades: one rigid straight and one flexible curved, both fabricated from isotropic materials using CNC machining stations. In some instances, the terms 'rigid' and 'straight', as well as 'flexible' and 'curved', may be used interchangeably in the text for the sake of clarity and conciseness. The comparisons will focus on the results from power versus wind speed tests, and from the coefficient of power (Cp) versus wind tip ratio (λ), with the goal of underscoring the differences between the two designs.

The methodology section of this paper will cover the following areas:

- A brief overview of the key points of the Blade Geometry Algorithm, which will not be detailed extensively in this paper as it is the subject of another study. This will include an introduction to the concept and the fundamental steps involved.
- A detailed presentation of the test campaign conducted for all three blades, highlighting experimental setups and findings.
- An explanation of the measurement apparatus and the configuration of the wind tunnel used during the tests.

2.2. Blade Geometry Algorithm

Figure 2 provides a comparative illustration of the straight and curved blades to introduce the reader to the flexible curved blade concept. The flexible curved blade can be conceptually divided into three distinct sections. The first section, mounted on the hub, constitutes about a quarter of the total blade length and is engineered for high rigidity to ensure stability at the blade's root. The second section, making up approximately half of the blade's length, is known as the control section. This part is designed and manufactured to possess the necessary elasticity to facilitate the desired aerodynamic control. The final quarter of the blade is further optimized for enhanced elasticity, enabling it to respond swiftly and effectively to rapid changes in wind velocity. This segmentation ensures that each part of the blade performs optimally under varying operational conditions.

The design of the flexible curved blade utilizes the same aerodynamic cross-sectional profiles as the straight blade, applied at corresponding radii. However, for each radius, the profile is offset by specific angles at different radii to introduce significant torsional deformation without altering the chord length or other basic geometric characteristics. As a result, despite the curved blade cross-sectional profiles exhibiting eccentricity compared to the straight blade (which maintains zero eccentricity), they remain tangent to the same radius. This configuration maintains fundamental aerodynamic properties consistent across both blade types, while the adjusted eccentricity offset induces a bend–twist coupling when aerodynamic bending loads are applied to the blade, which affects the torsional response of the flexible curved blade.

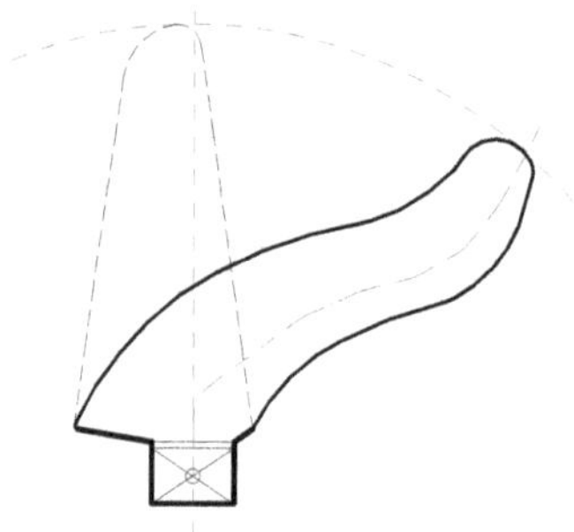

Figure 2. Comparison of straight and curved blade concept.

Figure 3 offers a cross-sectional view of the blade perpendicular to the plane of rotation, enhancing our understanding via two distinct operational scenarios: one at low wind speed with no rotation (bottom) and another at high wind speed (V2) with rotation (top). In both depictions, the wind (V1 for low speed and V2 for high speed) is assumed to originate from the bottom of the image, perpendicular to the plane of rotation, with the velocity of the cross-sectional profile (ω·R) shown horizontally.

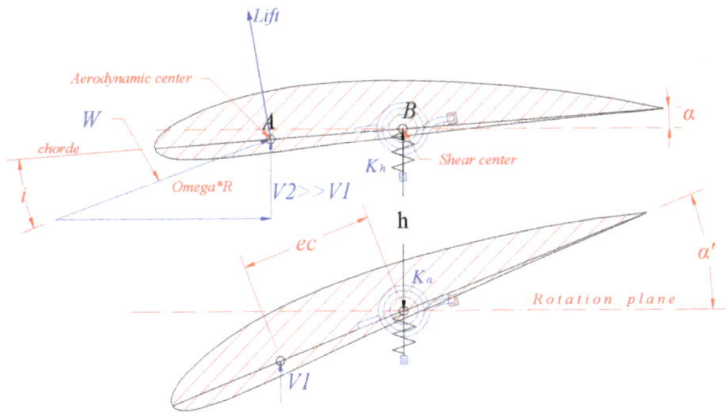

Figure 3. A: axis of aerodynamic centers, B: elastic axis, α: pitch angle, α-α′: pitch angle increase due to elastic deformation, w: bending deformation, ec: elastic–aerodynamic eccentricity.

In the lower section of the figure, the blade remains undeformed under low wind conditions (V1), resulting in no loading. Conversely, the upper image captures the blade under higher wind speeds (V2), where the offset between the aerodynamic center and the shear center induces bend–twist coupling. This coupling modifies the pitch angle of the blade, illustrating the dynamic response of the blade structure under varying wind forces.

It is crucial to note that the profile of the cross-section is a standard NACA4415. This profile was chosen because it is well-researched, with documented coefficients of lift (C_L) and drag (C_D), although it can be altered to meet specific user requirements. In this context, what we refer to as a "blade segment" in the paper is essentially a tangential cross-section of the blade at a specified radius from the center of rotation and follows a specific aerodynamic profile (e.g., NACA).

As depicted in Figure 3, the resultant wind speed $\left(\vec{W} = \vec{V} + \vec{\omega} \times \vec{R}\right)$ forms an angle of attack (i) with the chord, which is greater in the high-speed scenario (V2). The pitch angle difference (α′-α) shown in the upper sketch also varies, leading to both bending (h) and twisting deformations of the blade. At high wind speeds, the increase in the angle of attack (i) causes the blade to enter the aerodynamic stall region, highlighting the critical interplay between blade geometry and operational wind conditions. A challenging aspect of this design is accurately validating and calculating the eccentricities to ensure that they meet the design criteria. The design criteria defined for this project are as follows:

- Quick start-up of the blade to ensure rapid response to wind conditions.
- Achievement of nominal velocity at nominal wind speeds (8–10 m/s).
- Maintenance of a stable power yield plateau from nominal speed up to the cut-off speed.
- Initiation of stalling in the wind blade when wind speeds exceed the cutoff threshold.

This approach ensures that the blade's performance is optimized across a range of operational conditions, highlighting the sophisticated interplay between blade geometry and dynamic wind forces.

Figure 4 provides a comprehensive overview of the curved blade geometry derivation algorithm and the various steps involved. Initially, a straight–rigid blade is designed using geometry derived from an aerodynamic code. This code incorporates the Larsen, Frandsen, Soeresen, and Courtney [40] theoretical framework for aerodynamic behavior modeling. Additionally, it utilizes enhancements based on Glauert's theory for airfoils and airscrews as described in Désiré Le Gouriérès's "Les Éoliennes" [41], and incorporates aspects of the Hansen Blade Element Momentum method [42]. The code is not within the scope of this article and is planned to be presented extensively in another article upon maturity.

The parameters required for the aerodynamic code include the nominal power output, nominal wind speed, radial distribution of the chord length, and radial distribution of the pitch angle (with radial distribution referring to variations at different radii). Another parameter is the number of segments along the radius, typically set at 10 but adjustable to enhance algorithm precision. For a straight–rigid blade, aerodynamic loads are primarily aligned near the blade's elastic axis across all spanwise sections. The output from this phase includes the chord length and pitch angle at various radial segments.

The flexible curved blade geometry builds on the straight–rigid blade's parameters (chord length and pitch angle) with the additional parameter of eccentricity (usually expressed in degrees). This eccentricity, which shifts the elastic shear center of the profile, increases the torsional load due to aerodynamic lift and drag components, leading to more pronounced torsional rotation. The process begins with an initial eccentricity assumption, which modifies the profile segments. This adjustment and the corresponding aerodynamic loads are then input into a finite element code to calculate the bending–torsional deformation along the blade. This deformation, influenced by the blade's curve and spanwise loading distribution, alters the pitch angle. The blade is designed to self-adjust and bend and twist under aerodynamic forces, thereby keeping the pitch angle within an optimal range for maximum lift across various aerodynamic load conditions, ensuring optimal energy production over a broad range of wind speeds.

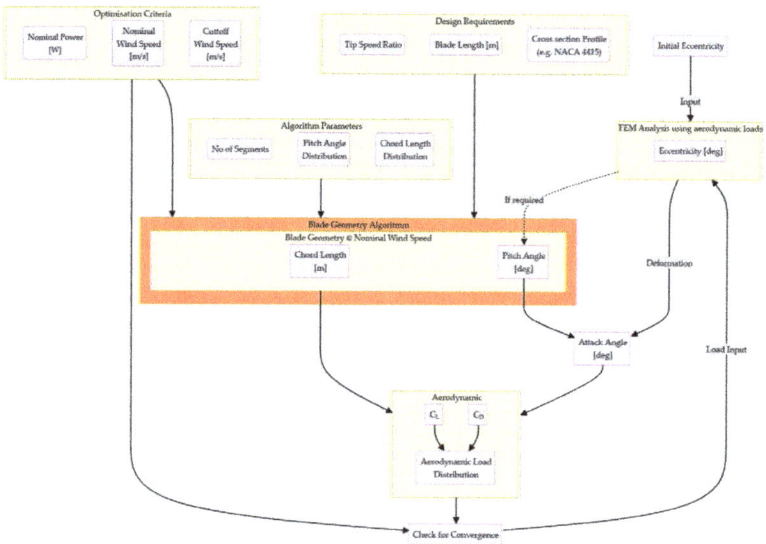

Figure 4. Flexible curved blade algorithm overview.

The flexible curved blade geometry has been designed in the Dassault Solidworks CAD system and optimized the deformation using its Simulation package for structural analysis. A static analysis was used where aerodynamic loads were imposed on the structure, and the deformation was measured. In terms of degrees of freedom (dof) in the simulation, the blade edge attached to the hub is fixed in all dof, while the opposite edge remains entirely free without any rotational or translational constraints. The load modeling reflects the dynamic pressure of the wind, adopting an inverse triangular thrust distribution. This load distribution assumes zero pressure load at the blade root and maximum pressure load at the tip, which is representative of the typical operational thrust distribution near rated power.

2.3. Blade Materials and Construction

2.3.1. Blade Material

A very important parameter for the blades is the construction material. High-Density Polyethylene (HDPE) material was selected because of its mechanical properties (800 N/mm^2

of elastic modulus and 30 N/mm^2 of tensile strength). For the construction of the blades, a two-axis CNC machine was used in order to obtain a solid blade structure with the required tolerances. After the construction of the blades, a hub was designed and constructed in a 3D printer using ABS material, and afterwards, the rotor was assembled.

2.3.2. Post-Construction Procedure

Upon completion of the rotor construction, several critical post-construction procedures were undertaken to ensure the structural integrity and functionality of the blades. Initially, the blade surfaces were meticulously sanded to enhance their aerodynamic smoothness, preparing them for further testing.

The blades then underwent static testing where weights, simulating aerodynamic loads, were applied. This testing was crucial for two primary purposes: measuring the displacements and angles along the blade to ensure that the empirical data aligned with the values predicted by our simulation models and confirming the structural strength of the blades under simulated operational stresses.

The results from these tests confirmed that the behavior of both the straight and curved blades was consistent with the expectations based on their design specifications. Furthermore, the method of static control in bending and torsion with distributed loads was carefully executed to validate the correlation with the simulation models, ensuring that the physical prototypes behaved under load as anticipated by theoretical analyses.

Finally, after assembly to the hub, the blades were subjected to balance testing on a horizontal shaft that was free to rotate. This test was essential to detect any preferential tilting or imbalance in the setup that could potentially affect performance and safety. These comprehensive post-construction tests are vital to guarantee that the rotor operates safely and efficiently, bridging the gap between theoretical design and practical application.

2.4. Measurement Campaign

2.4.1. Types of Measurements

Two distinct types of measurements were employed during the testing phase, categorized based on the wind velocity conditions: constant velocity (internally denoted as "RR") and increasing velocity ("VV").

Constant Wind Velocity Test (RR): In the constant wind velocity test, the wind blade is exposed to a steady wind until the revolutions per minute (rpm) of the blade stabilize at an equilibrium. Following this stabilization, an external load is gradually applied to the wind blade shaft, either mechanically (or electrically if the rotor is connected to a generator). This test format is commonly used to characterize wind energy blade performance, particularly for calculating the power coefficient (Cp) versus tip speed ratio (TSR or λ). It is important to note that the magnitude of the load significantly influences the power coefficient. Specifically, an excessively high load can stall the blade, while an insufficiently light load may prevent the blade from slowing sufficiently to observe the desired effects. During these tests, the braking load is gradually increased to a point where the blade decelerates enough to eventually stop.

Varying/Increasing Velocity Test (VV): Contrastingly, the increasing velocity test is designed to characterize the performance of the wind blade across a broader range of wind speeds and may include the application of braking loads. Although not as typical for standard characterization, these tests are crucial in this context as they highlight the benefits of the passively controlled wind energy converter system. This approach allows for an assessment of the blade's adaptability and performance under varying operational conditions.

2.4.2. Measurement Sets and Designation

The measurement campaign for isotropic blades involved testing blades with the same radius, curved and straight. Table 1 presents the constant velocity measurements for the isotropic blades. For the flexible curved blade, measurements were systematically recorded from wind speeds of 5 to 19 m/s at 1 m/s intervals, as per readings from the

Pitot–Prandtl tube. In contrast, the rigid blade was tested within a narrower range of 6 to 14 m/s, with intervals of 2 m/s. A critical consideration for not testing the straight blade at higher wind speeds is related to the methodology of the constant velocity tests, which necessitates allowing the blade to stabilize at an equilibrium rpm before applying a brake. Without a control system, the rigid blade can reach dangerously high rpms, posing a risk of damaging the testing setup—indeed, such damage has occurred during testing. Consequently, it was decided to limit the wind speed to safer, lower rpms for the rigid blade. Notably, the flexible curved blade did not exhibit this problem, underscoring one of the key advantages of its design highlighted in this study.

A final note regarding Table 1 is that because of the high number of constant velocity measurements, in order to avoid duplication, it was thought better to use the dataset name and suffix the velocity, e.g., for measurement at 5 m/s of the isotropic curved blade to use the designation "ISO.Curved.RR.05.y" (where y denotes the repletion, if any).

Table 1. Test parameters for isotropic wind blade RR measurements.

Dataset Name	Blade Type	Radius [m]	Wind Speed Velocity	Load
ISO.Curved.RR.<xx>.<y>	Isotropic curved	0.160	5, 6, 7, 8, 9, 10, 11, 12, 13, 14, 15, 16, 17, 18, 19	Ramp
ISO.Straight.RR.<xx>.<y>	Isotropic straight	0.160	6, 8, 10, 12, 14	Ramp

Table 2 presents the parameters for the measurement datasets for the increasing wind tunnel velocity test. The main parameter that changed in this is the braking load (the setting is a value between 0 and 50, with 0 being no brake and 50 the maximum braking). In this table, the parameters of the test are presented (blade type, radius, type of measurement and load) and also the wind speed range velocity and the max rpm during testing. It can be easily observed that the wind speed maximum measurement speed for the straight blade remained at 16 m/s while the wind velocity for the flexible curved blade was set to 19 m/s. The reason, as explained earlier, is the characteristic of the flexible curved blade to act like it has an embedded control system that regulates the power yield and rpm. Another interesting feature is that the flexible curved blade's rpm remained below 3000 rpm despite being exposed to higher wind speeds. Out of the 15 VV measurements, 4 are related to the flexible curved blade, and the remaining 11 are for the straight blade.

Table 2. Test parameters for isotropic wind blade VV measurements.

Dataset Name	Exp. UID	Blade Type	Radius [m]	Type of Meas.	Load	Wind Speed Velocity Range	Max rpm during Test
VV. 0907.23	23	straight	0.16	VV	15	6–15	6300
VV. 0907.24	24	straight	0.16	VV	15	6–15	6300
VV. 0908.25	25	straight	0.16	VV	15	6–15	6400
VV. 0908.26	26	straight	0.16	VV	0	6–15	6900
VV. 0908.27	27	straight	0.16	VV	0	5–15	6700
VV. 0908.28	28	straight	0.16	VV	0	7–15	6700
VV. 0908.29	29	straight	0.16	VV	15	6–15	6300
VV. 0908.30	30	straight	0.16	VV	15	6–15	6200
VV. 0908.31	31	curved	0.16	VV	0	4–19	2900
VV. 0908.32	32	straight	0.16	VV	0	6–15	6900
VV. 0908.33	33	curved	0.16	VV	20	12–19	2800
VV. 0908.34	34	straight	0.16	VV	20	9–15	6300
VV. 0908.35	35	curved	0.16	VV	20	10–19	2800
VV. 0908.36	36	curved	0.16	VV	30	15–19	2700
VV. 0908.37	37	straight	0.16	VV	30	12–15	5800

2.5. Experimental Apparatus

The rotor tests were conducted in the Wind Energy Lab's 600 × 600 mm Wind Tunnel at the Hellenic Mediterranean University. The testing apparatus was equipped to measure various parameters such as wind speed, rotor thrust (T), rotor torque (Q), and rotor speed (N), as illustrated in Figure 5. Table 3 presents the measurement devices, the model names and the manufacturing companies.

Table 3. Measurement devices.

Measurement	Device	Model Name	Company, Country
Wind speed Pitot	differential pressure transducer	HD408T	Delta Ohm (now Senseca), Italy
Torque	Rotating Torquemeter	No DR2112L	SCAIME, France
Rotational Velocity	Rotating Torquemeter	No DR2112L	SCAIME, France
Drag	Load Cell	model No SP4MC6MR	HBM (now HBK), Germany
DAQ	Multifunction Data Acquisition card	NI-USB-6353	National Instruments, USA

Wind speed at the tunnel outlet was recorded using a Pitot–Prandtl tube coupled with a Delta Ohm HD408T differential pressure transducer (Senseca, Italy). To ensure accuracy, the Pitot–Prandtl system was calibrated against an externally calibrated hotwire anemometer.

The rotor's axis was connected to a SCAIME DR2112L Rotating Torquemeter (France), which provided measurements for torque and revolutions per minute (rpm). Thrust (or drag) was measured using an HBM SP4MC6MR load cell (Germany) interfaced with an ADAM 3016 (Advantech Taiwan) isolated strain gauge input module.

Additionally, a small mechanical brake, activated by a spring-loaded actuator driven by a multiturn servo rotor connected to a screw, was employed to control the braking of the rotor. While the rotor rotates the screw, the spring is loaded. Thus, a relatively constant force is applied to the mechanical brake. One of the shortcomings of this configuration—which is presented in Figure 5—is that the numerical value of the brake setting does not correspond linearly to a braking force.

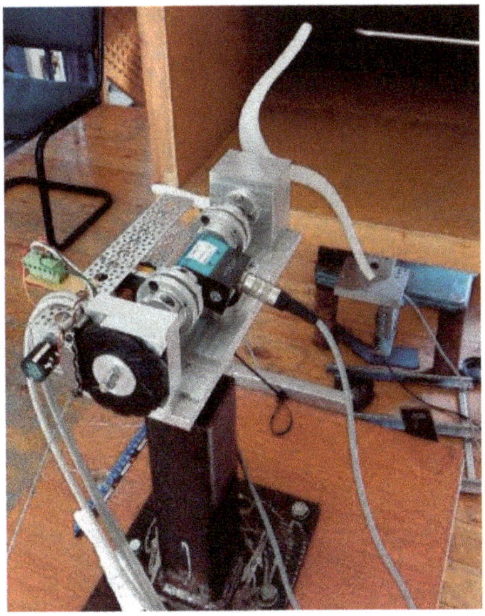

Figure 5. Test configuration with the flexible curved blade.

Data acquisition was handled by a National Instruments NI-USB-6353 card, utilizing the LabVIEW 2014 Development System for setup. A custom LabVIEW application was developed to capture all relevant signals at a sampling rate of 1000 Hz for each channel. These channels included measurements for two wind speeds: drag, torque, and rpm. Data from each channel was segmented into 0.1-s intervals, with recorded values representing the average across each interval.

To ensure accuracy, each measurement device underwent annual calibration. The calibration procedures involved the following:

- An externally calibrated hotwire anemometer for wind speeds (already mentioned before).
- Known weights for calibrating the load cell used for measuring drag.
- Standard weights for torque calibration.
- A tachometer to verify the rpm measurements.

3. Results for Isotropic Blades

3.1. Geometry for Isotropic Blades

Table 4 presents the geometrical characteristics of the straight isotropic blade and the curved isotropic blade. As mentioned before, the difference lies only in the eccentricity angle, which offsets the elastic axis. Another comment is that for this particular setup, the first two segments correspond to radii, which are technically part of the rotor hub (not part of the actual aerodynamic blade).

Table 4. Geometric characteristics of a straight blade and the eccentricity of the curved isotropic blade.

Seg. No	Radius [m]	Chord [m]	Pitch Angle [deg]	Eccentricity [deg]
1	0.016	0.0238	27.095	0.00
2	0.032	0.0242	19.599	0.00
3	0.048	0.0227	14.057	2.51
4	0.064	0.0202	10.436	8.71
5	0.080	0.0177	8.009	17.22
6	0.096	0.0156	6.310	25.11
7	0.112	0.0138	5.073	32.37
8	0.128	0.0123	4.141	37.44
9	0.144	0.0111	3.418	40.73
10	0.160	0.0101	2.846	42.18

The curved blade geometry is presented in Figure 6, and it is expected to develop a twist of 8 degrees at the tip under a wind speed of 12 m/s.

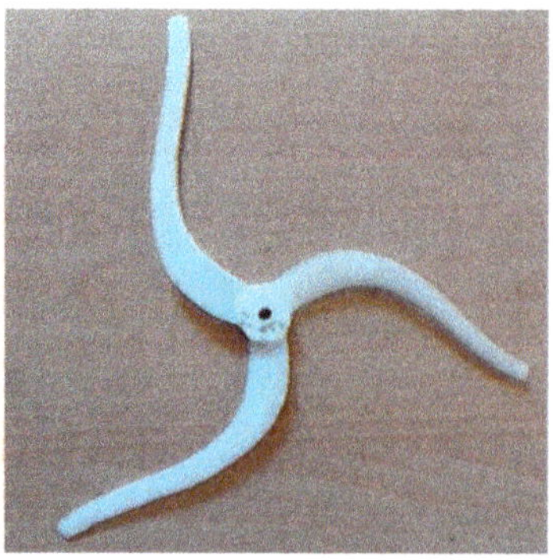

Figure 6. Eccentricity at each section of a flexible curved blade.

3.2. Straight Blade Isotropic Material

3.2.1. Straight Blade Measurements with Increasing Velocity

Figure 7 features an automated plot summary from a measurement session with increasing velocity utilized for preliminary data validation. The first plot at the top displays wind tunnel velocity, as measured by the Pitot–Prandtl Tube, which ranged from 6 to

15 m/s throughout this particular test. The second graph illustrates two variables over time: revolutions per minute (rpm), shown in blue on the primary axis, and mechanical power in red on the secondary axis. In this test, rpm starts to increase once a specific velocity threshold is reached, with mechanical power consistently rising for the straight blade. The third graph offers a detailed view by plotting each power data point against the corresponding wind speeds measured by the Pitot–Prandtl tube, providing insights into the blade's mechanical performance. The final graph in the sequence presents the power coefficient (Cp) against the calculated tip speed ratio (TSR) derived from the Pitot–Prandtl measurements. The maximum TSR observed is between 7 and 8, which is consistent with established practices in wind blade performance.

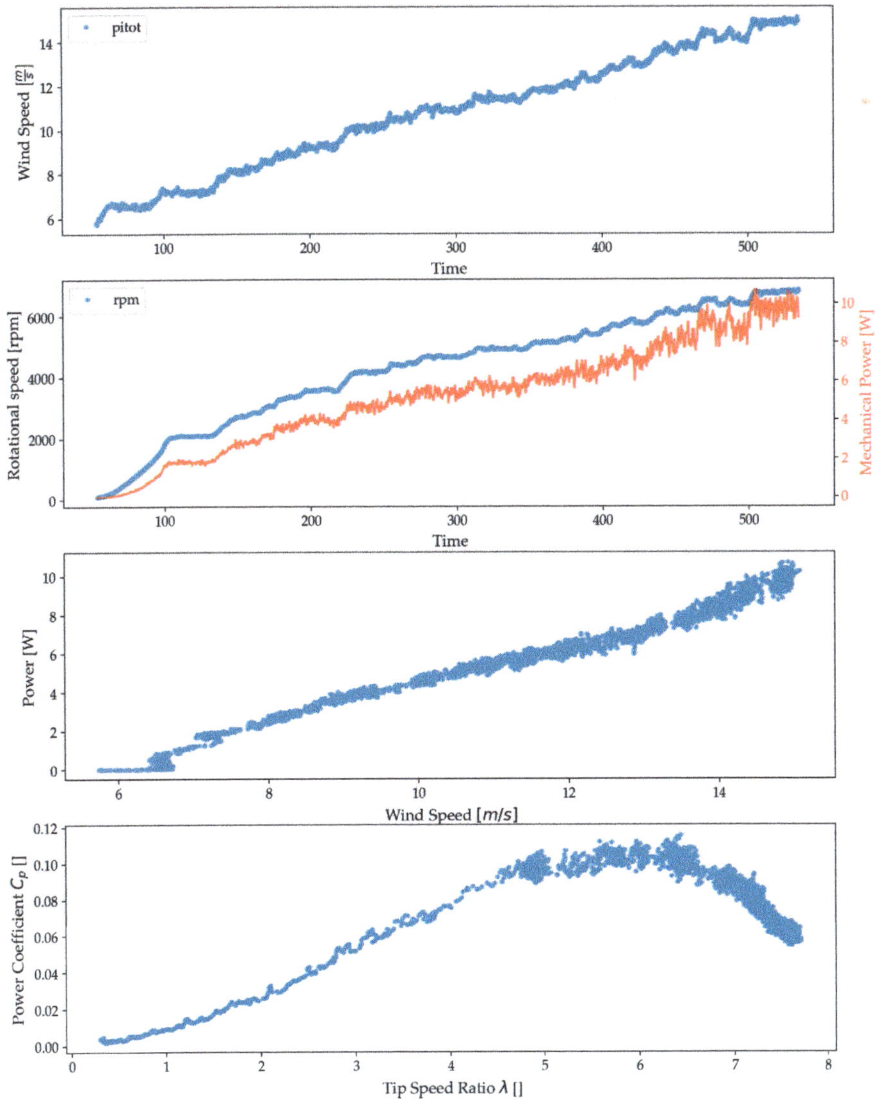

Figure 7. Example of automated plot summary for exp_id 26 (straight blade—isotropic material—VV type—brake setting: 15).

Figure 8 presents mechanical power measurement against the wind speed velocity from all increasing velocity tests of the straight blade, with data points color-coded by brake setting and distinct shapes representing different experiment IDs. The plot demonstrates good repeatability among tests with identical brake settings, underscoring the consistency of the experimental setup. Notably, experiments with higher brake settings commence at elevated wind speeds; for instance, a brake setting of 30 corresponds to starting velocities above 12 m/s. This pattern suggests that the blade requires some time to initiate rotation under higher load conditions. Additionally, as anticipated, the straight blade's performance increases monotonically with the wind speed, reflecting its predictable operational characteristics under varying aerodynamic loads.

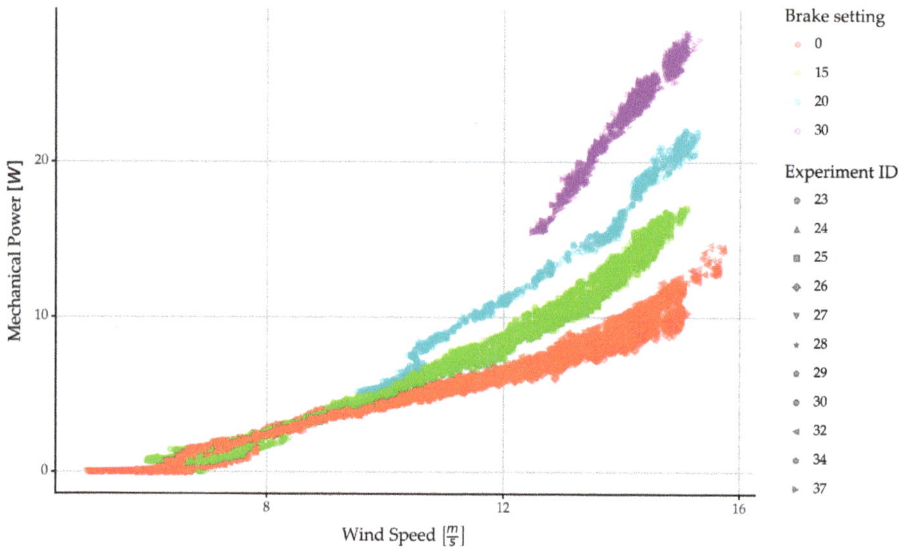

Figure 8. Mechanical power of straight blade vs. the wind speed (with and without brake) for tests with increasing wind speed.

Figure 9 illustrates the power coefficient (Cp) versus tip speed ratio (TSR) for all increasing velocity tests of the straight blade. Data points are color-coded by brake setting, with distinct shapes representing different experiment UIDs. The plot demonstrates good repeatability among tests with identical brake settings, highlighting the consistency of the experimental setup. Notably, experiments with higher brake settings tend to exhibit higher maximum power coefficients. This observation underscores the impact of mechanical loading on the turbine's output, suggesting the presence of an optimal loading for each condition, analogous to Maximum Power Point Tracking (MPPT) used in solar energy systems. This concept is similar to impedance matching in electrical engineering, where optimal power transfer is achieved by matching the 'impedance'—in this case, the aerodynamic load on the turbine blades—to the mechanical resistance provided by the generator and braking systems. Such matching ensures that the turbine operates efficiently under varying wind conditions and load settings. It is important to note that the authors acknowledge that the measurements are not at the optimal brake setting to ensure optimal power coefficient (C_P) (it was beyond the scope of this study).

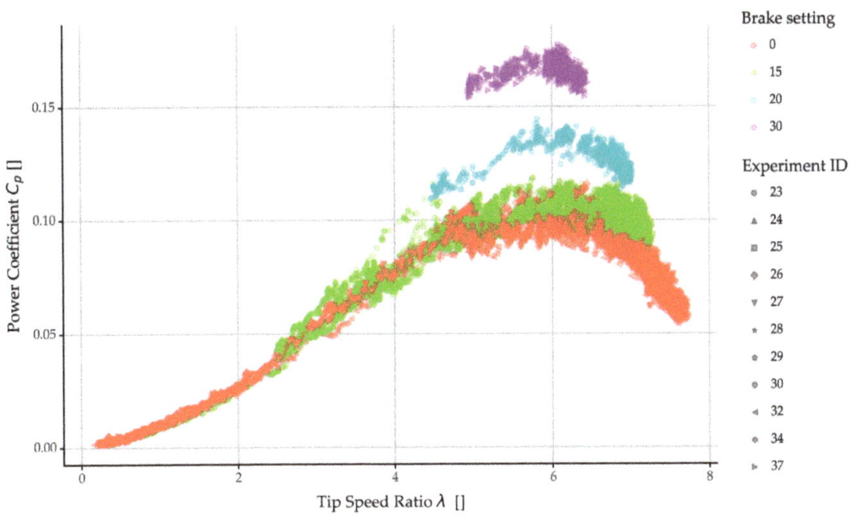

Figure 9. Coefficient of power with respect to tip speed ratio for straight blade measurements.

3.2.2. Straight Blade Measurements with Constant Velocity

Figure 10 showcases an automated plot summary from a constant velocity test on a straight blade, complemented by the application of a load, primarily serving for preliminary data validation and offering insights into test parameters. The first plot at the top records the wind tunnel velocity as measured by the Pitot–Prandtl Tube, maintaining a narrow range between 9.875 and 10.050 m/s throughout this test. The second graph displays revolutions per minute (rpm) in blue on the primary axis and mechanical power in red on the secondary axis. Initially, both rpm and power remain constant until approximately 15 s into the test, when the mechanical brake engages, increasing the torque on the rotor shaft, thereby elevating the power. Subsequently, as the rotational velocity begins to decline, it causes a corresponding decrease in power.

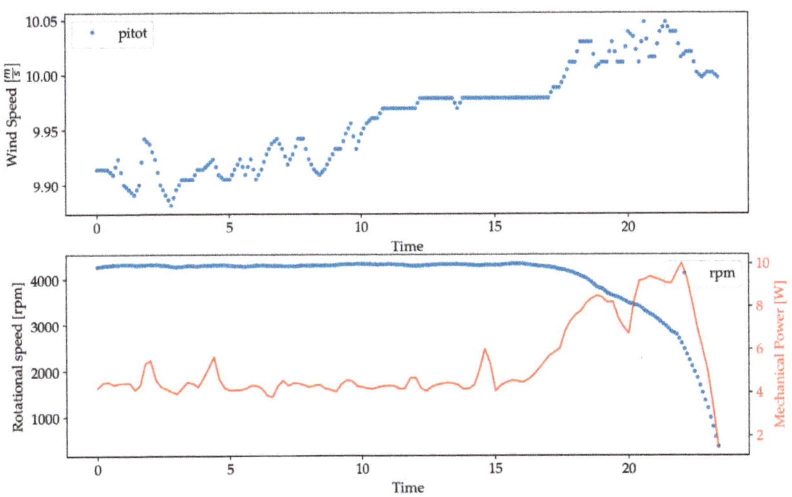

Figure 10. *Cont.*

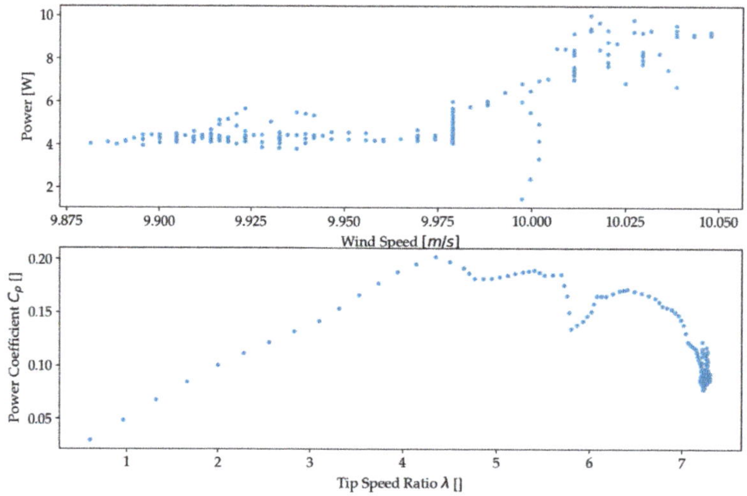

Figure 10. Example of automated plot summary for exp_id 20 (straight blade—isotropic material—RR type—10 m/s).

The third graph, plotting each power data point against the corresponding wind speeds measured by the Pitot–Prandtl tube, turns out to be less informative for this particular test due to the minor, stochastic variations in wind speed, rendering this data visualization not very useful. This is common with all constant velocity (RR) tests in this work.

The final graph in the sequence illustrates the power coefficient (Cp) against the calculated tip speed ratio (TSR) based on the Pitot–Prandtl measurements. The maximum TSR observed falls between 7 and 8, aligning with established norms for wind blade performance.

Figure 11 illustrates the power coefficient (Cp) versus tip speed ratio (TSR) for all constant velocity tests of the straight blade. Data points are color-coded by nominal velocity. Notably, experiments with higher nominal velocities tend to exhibit higher maximum power coefficients suggesting that the blade is operating better at those velocities.

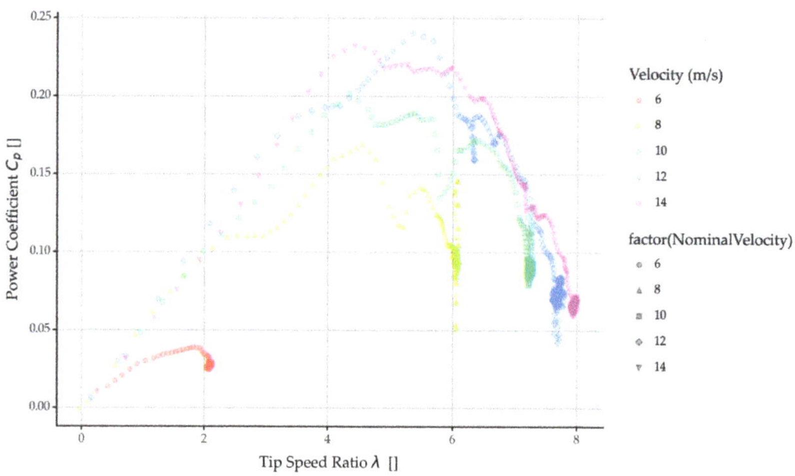

Figure 11. Power coefficient vs. tip speed ratio for different velocities of the straight blade.

3.3. *Flexible Curved Blade: Isotropic Material*

3.3.1. Flexible Curved Blade Measurements with Increasing Velocity

Figure 12 presents an automated plot summary from a flexible curved blade measurement with increasing velocity, serving as a tool for preliminary data validation. The first graph at the top displays wind tunnel velocity, measured by the Pitot–Prandtl Tube, which varied from 5 to 18 m/s during this test. The second graph tracks two variables over time: revolutions per minute (rpm), shown in blue on the primary axis, and mechanical power, depicted in red on the secondary axis. The rpm begins to increase at 5 m/s and continues rising until 15 m/s. Correspondingly, the mechanical power, calculated as the product of mechanical torque and angular velocity, also increases in tandem with the rpm. The constancy of the mechanical power suggests that the torque remains stable as both rpm and power plateau.

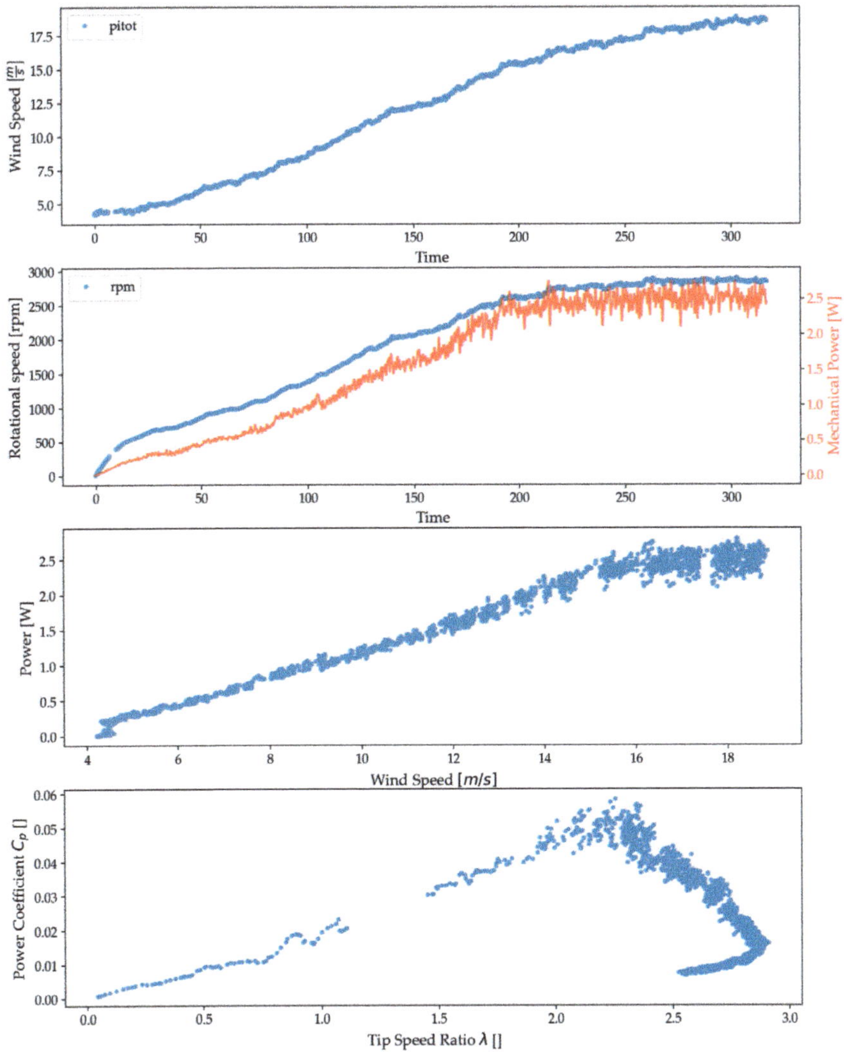

Figure 12. Example of automated plot summary for experiment id 31 (flexible curved blade—isotropic material—VV type—brake setting: 0).

The third graph provides a detailed analysis by plotting each power data point against the corresponding wind speeds measured by the Pitot–Prandtl tube, offering insights into the blade's mechanical performance. Notably, beyond 15 m/s, the mechanical power appears to plateau, indicating activation of the passive control features of the flexible curved blade concept.

The final graph illustrates the power coefficient (Cp) against the calculated tip speed ratio (TSR) based on Pitot–Prandtl measurements. This graph reveals several key aspects: the TSR remains below 3, significantly lower than the optimal TSR. The shape of the Cp curve initially increases with TSR, reaching a peak, and then decreases, reminiscent of the straight blade's performance profile. However, it uniquely features a 'hook-like' tail beyond 15 m/s, marking the onset of control mechanisms that moderate the blade's response to higher wind speeds.

Figure 13 presents mechanical power measurement against the wind speed velocity from all increasing velocity tests of the flexible curved blade, with data points color-coded by brake setting and distinct shapes representing different experiment IDs. In a similar manner to the straight blade, in experiments with higher brake settings, the rotation of the blade commenced at elevated wind speeds; for instance, a brake setting of 30 corresponds to starting velocities above 16 m/s. This pattern suggests that the blade requires some time to initiate rotation under higher load conditions. Finally, although for no braking, the blade shows that there is a power plateau at high speeds, the settings with higher brake settings do not exhibit this behavior.

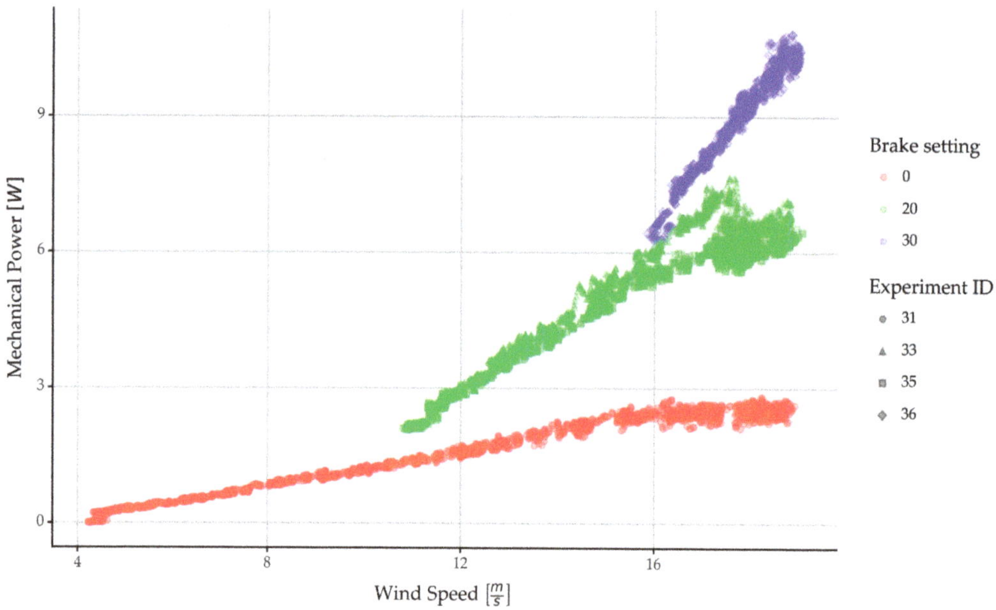

Figure 13. Mechanical power of curved blade vs. the wind speed (with and without brake) with increasing speed.

Figure 14 presents the rotational velocity (measured in rpm) of the flexible curved blade vs. the wind speed. This plot is presented to discuss a finding in Figure 13, which showed that for higher brake settings, the power increased monotonically. This plot shows that the rpm of the flexible curved blade at different settings seems to converge to 3000 rpm. This shows that the wind blade does not increase the rpm uncontrollably but is able to obtain higher Torque values. In the discussion stage, this will be contrasted more widely.

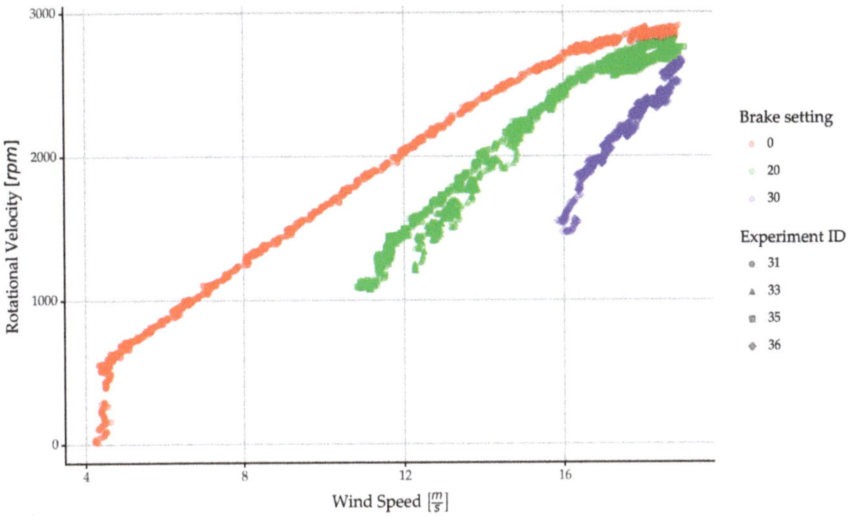

Figure 14. Flexible blade's rotational velocity of blade vs. the wind speed during the test (with and without brake).

Figure 15 illustrates the power coefficient (Cp) versus tip speed ratio (TSR) for all increasing velocity tests of the flexible curved blade. Data points are color-coded by brake setting and marked with distinct shapes to represent different experiment UIDs. Notably, experiments with higher brake settings generally exhibit lower maximum power coefficients, although these maximum values appear to converge quickly. This behavior starkly contrasts with that observed in straight blade measurements. Additionally, a hook-like feature is consistently present across all flexible curved data measurements. The lower Cp values are attributed to the control features of the wind turbine blade. Despite higher wind speeds providing potentially more energy, the blade's design and control mechanisms result in it harnessing less energy than possible, leading to lower observed Cp values.

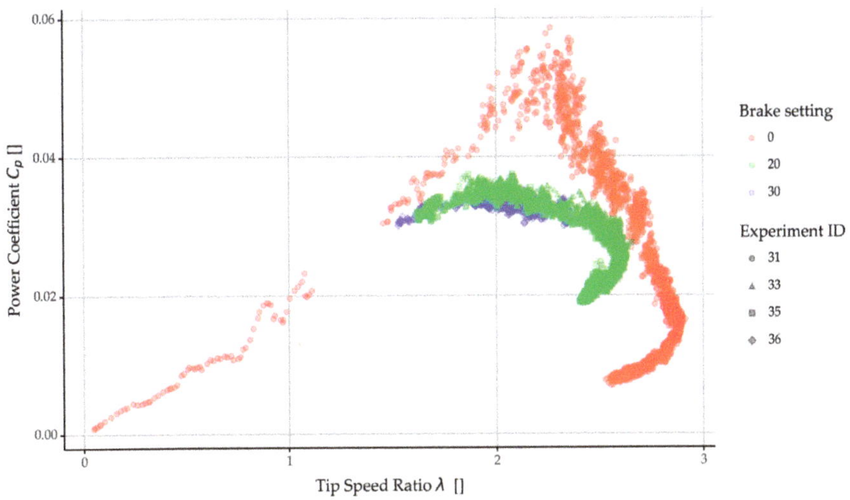

Figure 15. Coefficient of power with respect to tip speed ratio for flexible curved blade measurements.

3.3.2. Flexible Curved Blade Measurements with Constant Velocity

Figure 16 presents an automated plot summary from a measurement used for preliminary data validation, featuring four plots. The first plot displays the wind tunnel velocity from the Pitot–Prandtl Tube, consistently around 10 m/s, a benchmark value for this test. The second graph shows the rpm on the primary axis and the mechanical power on the secondary axis. As the brake increasingly engages, rpm begins to drop, while mechanical power rises until reaching a critical point where both power and rpm start to decline. The third graph, plotting each data point against different wind speeds, is similar to the straight blade's tests and provides limited informative value due to the small range of wind speeds. The final graph depicts the power coefficient (Cp) against the calculated tip speed ratio (TSR), showing a TSR range from 0 to 3. This range is significantly different from that of the rigid blade and will be discussed further in subsequent sections.

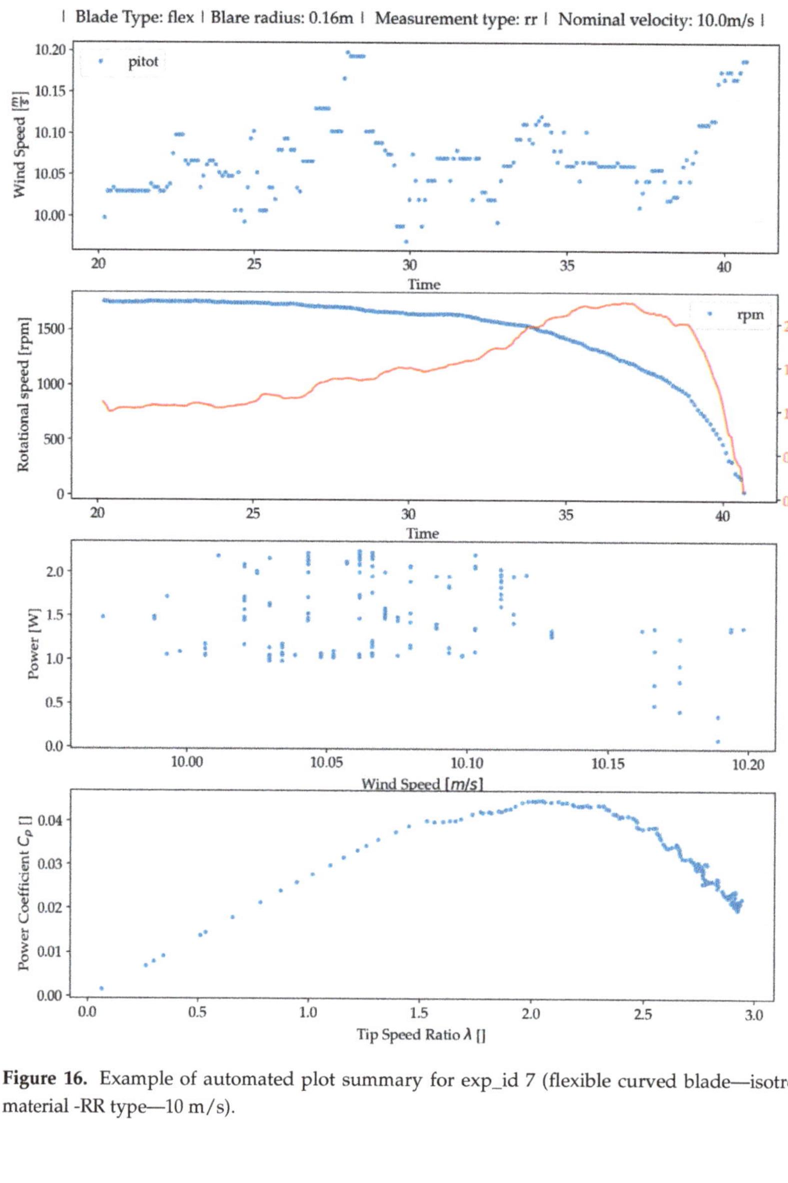

Figure 16. Example of automated plot summary for exp_id 7 (flexible curved blade—isotropic material -RR type—10 m/s).

Figure 17 illustrates the power coefficient (Cp) versus tip speed ratio (TSR) for all constant velocity tests of the flexible curved blade, with data points color-coded by nominal velocity. Contrary to the behavior observed in straight blades, experiments with higher nominal velocities, in this case, tend to show lower maximum power coefficients. This suggests that the flexible curved blade extracts less energy from the wind at these higher speeds. However, this reduced energy extraction can be attributed to the passive control features inherent in the flexible curved blade's design, which adjusts the blade's response dynamically to maintain structural integrity and operational efficiency under varying wind conditions.

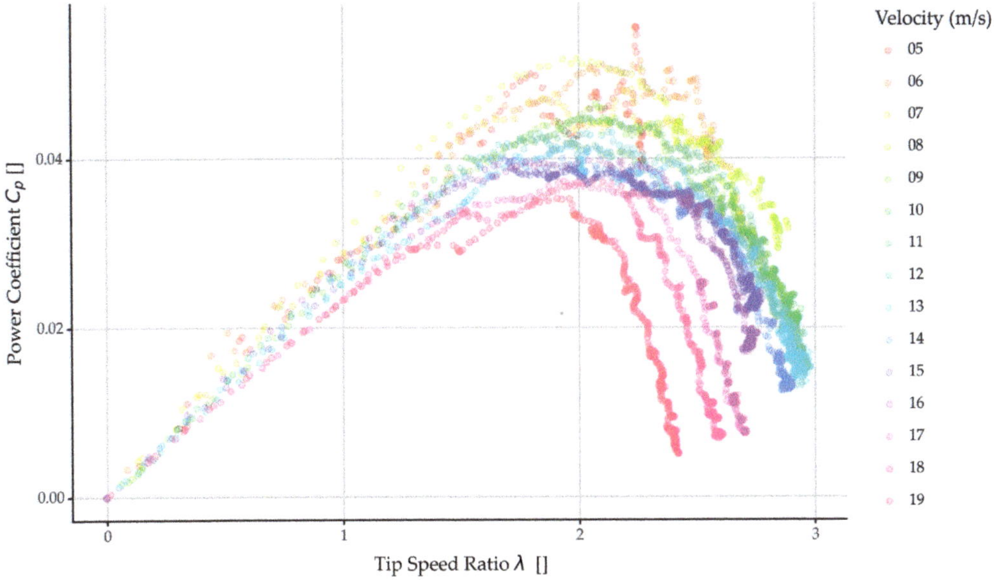

Figure 17. Power coefficient vs. tip speed ratio for different velocities of the flexible curved blade.

It is important to note that a rigid blade in a wind turbine would also show a similar reduction in performance. This can be easily observed in the power curve of any commercial wind turbine: power output increases up to the nominal power at the nominal wind speed, beyond which the power remains flat and constant and then drops off at the cutoff speed. Therefore, although the flexible curved blade may extract less energy when tested in isolation, as part of a functional wind energy conversion system (WECS), it would result in comparable overall production levels to the rigid blade."

4. Discussion

4.1. Self-Regulation of Wind Turbine Blade

In this part of the article, the performance of the flexible curved wind turbine blade is discussed, and more specifically the passive control traits. A conventional straight blade made from a similar material with the same diameter and cross-sectional profiles, is used as a reference for the comparison.

4.1.1. Rotational Velocity

Figure 18 provides a comparative illustration of the rotational velocity against wind speed for straight and curved blades in different subgraphs, with the data points color-coded by brake settings. To facilitate direct comparison, the axes of the subgraphs are uniformly scaled. A notable observation is that the flexible curved blade exhibits a consistent maximum rotational velocity of 3000 rpm across all brake settings when wind speeds exceed 18 m/s, significantly lower than the straight blade, which accelerates more rapidly, reaching and often surpassing 5000 rpm at just 15 m/s.

The relationship between wind speed and rotational velocity reveals notable differences between blade types. For the straight blade, this relationship is almost linear, indicating a steady increase in rpm with wind speed. In contrast, the flexible curved blade demonstrates a marked non-linearity, displaying an asymptotic behavior where the rpm approaches 3000 rpm irrespective of the brake setting. This suggests that the flexible curved blade design effectively achieves self-regulation, maintaining a controlled rotational velocity despite increasing wind speeds.

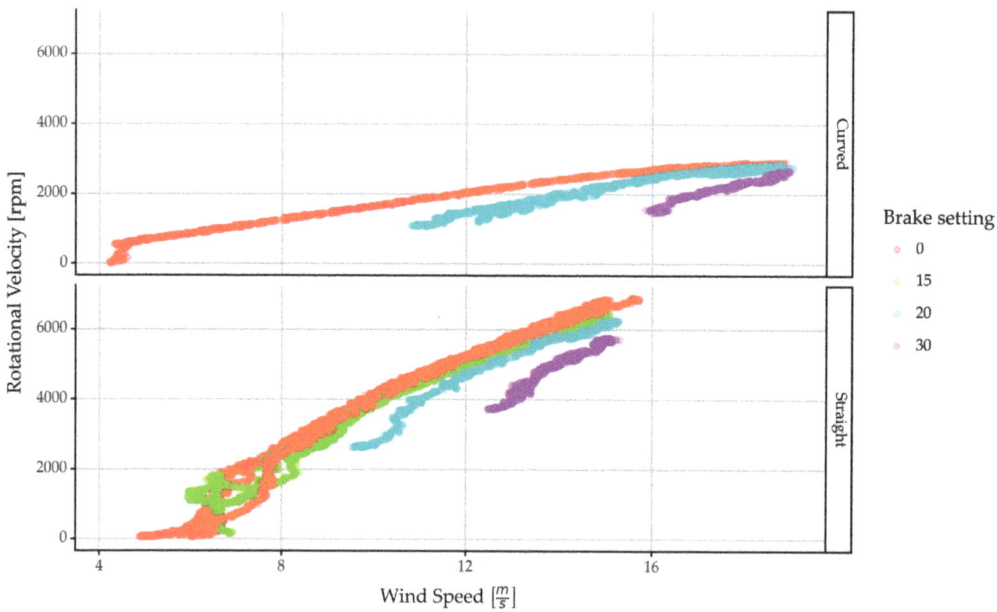

Figure 18. Comparison of Rpm vs. wind speed for different blade types and brake settings.

Figure 19 zooms in on this phenomenon by presenting a direct comparison of the 'no brake' settings, with an adjusted aspect ratio favoring the y-axis to highlight the behavior of the flexible curved blade more clearly. This figure substantiates the previous observations by illustrating that up to 15 m/s, the rpm increases linearly. Beyond this point, the passive control mechanisms activate, asymptotically stabilizing the rpm at 3000, effectively modulating the rotational acceleration as wind speed continues to increase.

An additional yet significant observation from Figure 19 is that the passive control of the flexible curved blade activates at higher wind speeds than initially designed. This delayed activation suggests that while the blade does achieve self-regulation, the onset of passive control mechanisms does not align perfectly with the intended design parameters, indicating that the design objectives were not fully met.

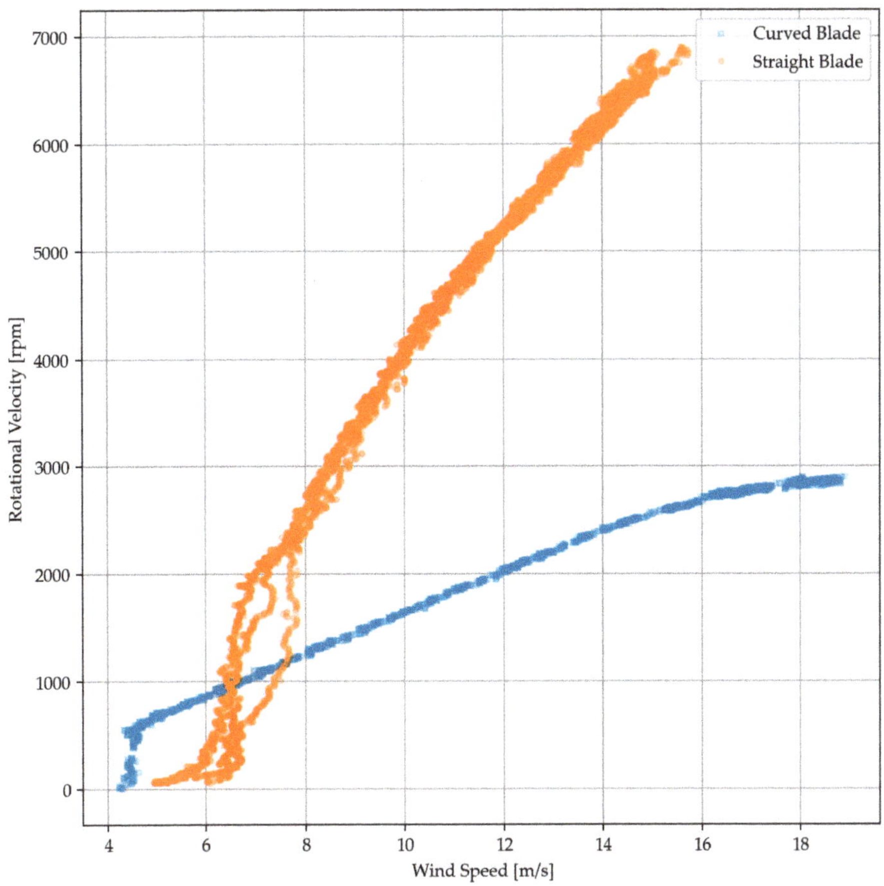

Figure 19. Comparison of Rpm vs. wind speed for different blade types for the case of no brake.

4.1.2. Power vs. Wind Speed

Similar effects are observed in the case of power versus wind speed. Figure 20 provides a comparative illustration of the mechanical power (obtained by torque multiplied by angular velocity) against wind speed for straight and flexible curved blades in different subgraphs, with the data points color-coded by brake settings. To facilitate direct comparison, the axes of the subgraphs are uniformly scaled. The data in the figure are from the tests with gradually increasing velocity.

A notable observation is that the flexible curved blade exhibits significantly lower mechanical power values compared to the straight blade. Additionally, the rate at which the power climbs with increasing wind speed is significantly greater.

Figure 21 provides a detailed examination of the 'no brake' settings by employing a log10 scale on the y-axis, which represents the logarithm of mechanical power, to accentuate the pronounced differences in behavior between the straight and flexible curved blades. An initial observation is that the flexible curved blade commences power production at a lower wind speed of 4.5 m/s, compared to 6 m/s for the straight blade, thus demonstrating an enhanced capability to generate energy at lower wind velocities.

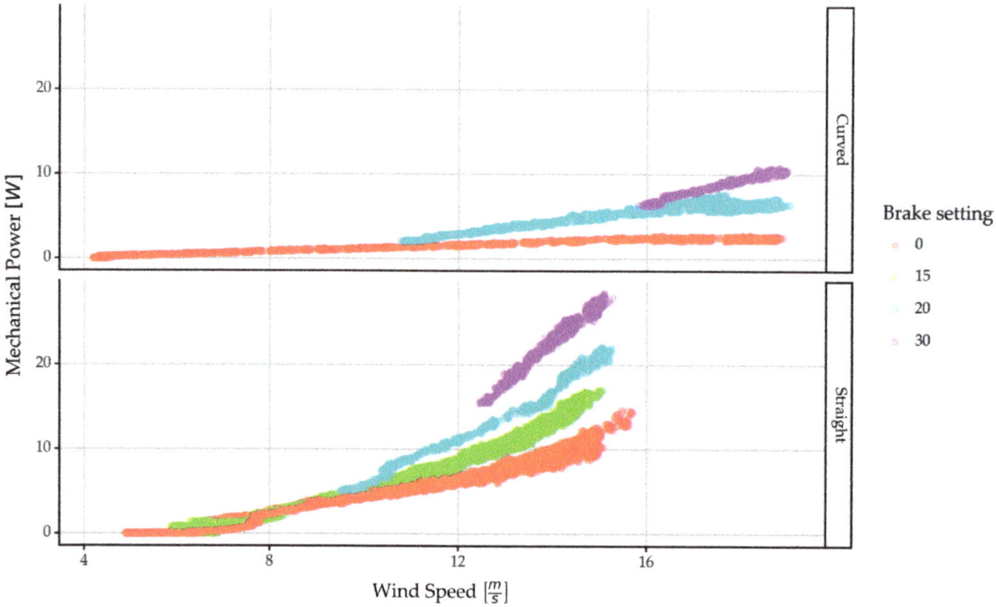

Figure 20. Power vs. wind speed for a passive controlled rotor.

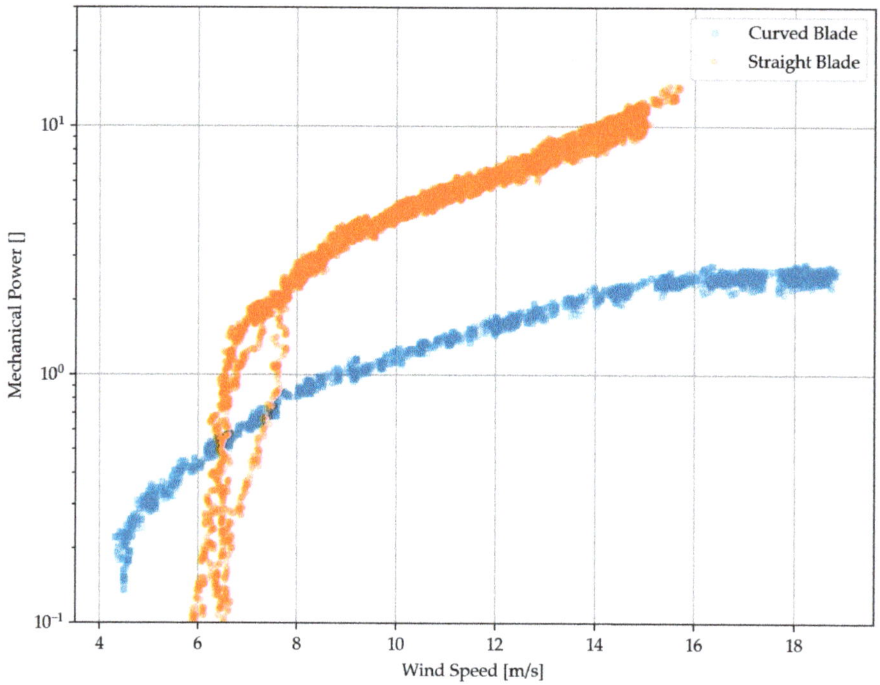

Figure 21. Log scale of power vs. wind speed comparison between a curved and straight blade, with no brake.

During the initial phase, both blades demonstrate an increase in power output with rising wind speeds. However, on the log scale, the straight blade exhibits a linear relationship between 8 m/s and 15 m/s, suggesting that its power output is proportional to the power of velocity. In contrast, the flexible curved blade shows an initial increase in power between 6 m/s and 16 m/s, but the rate of increase diminishes with higher wind speeds, as observed in the log scale. This behavior indicates an early onset of self-regulation in power harvesting. At higher speeds above 16 m/s, the power output of the flexible curved blade stabilizes, remaining constant and effectively demonstrating the passive control capabilities of this blade design.

Additionally, a critical observation from Figure 21 (and is consistent with the findings for the rotational velocity from Figure 19) is that the maximum power yield does not occur at the intended velocities of between 8 and 10 m/s but at significantly higher velocities. This indicates that while passive control is present and operational to a degree in the flexible curved blade, it may not be optimally effective, as the power regulation does not align with the designed target velocities.

4.2. Power Coefficient vs. Tip Speed Ratio

Figure 22 provides a detailed comparison of the power coefficient (Cp) versus tip speed ratio (TSR) for both straight and curved blades, with each subgraph's data points color-coded according to brake settings. This comparison highlights significant performance and behavioral differences attributable to the curved blade design.

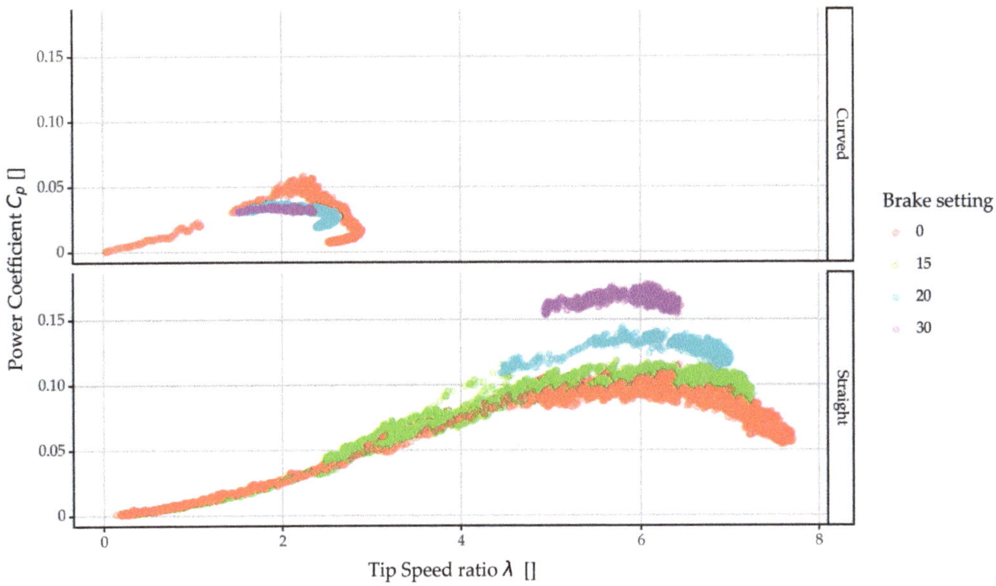

Figure 22. Power coefficient with respect to tip speed ratio for different blade types using the increasing velocity speed data.

A notable distinction is observed in the range of TSR. The flexible curved blade operates within a TSR range from 0 to approximately 3, substantially lower than the straight blade, which ranges up to almost 7—closer to the optimal TSR value. Interestingly, greater braking settings correlate with narrower TSR ranges for both blade types, as the mechanical brake impedes startup and restricts the increased rate of rotational speed, affecting the maximum TSR achievable.

Another key observation concerns the maximum power coefficient values. For cases without braking, the maximum Cp is nearly double compared to those with braking.

Intriguingly, increasing the brake setting decreases the Cp for the flexible curved blade, while the opposite trend is observed for the straight blade, where higher braking values enhance the Cp. This divergence highlights the passive control mechanism inherent to the flexible curved blade, which, unlike the straight blade that maximizes power output with increased wind force, regulates itself to harvest only the nominal power at higher velocities—similar to an actively pitch-controlled wind turbine.

Regarding the influence of braking on C_P, it is critical to recognize that load settings can significantly affect C_P measurements, potentially leading to misleading conclusions if not properly considered. This is particularly relevant in comparative analyses of blades with different design goals: a straight blade aimed at maximizing power and a flexible curved blade designed for safe and controlled power production. Ideally, measurements would be made using an MPPT inverter [43,44], as a load to align more closely with operational conditions. However, given the limitations of the current measurement configuration setup in the Power Synthesis Laboratory, which lacks such a system, the measurement results should be interpreted qualitatively.

Lastly, a consistent feature in flexible curved blade measurements is a hook-like pattern observed in Figure 15. This downward-facing hook pattern demonstrates lower C_P values at the same TSR at higher velocities, attributed to the passive control mechanisms of the flexible curved blade. As wind velocity increases, the rotational velocity remains capped at 3000 rpm, thus reducing the TSR. Concurrently, the C_P decreases because while the available energy (the denominator) increases, the power output (the numerator) remains capped by the blade's control features.

The findings from the analysis of Figure 22 provide insights into the operational dynamics of straight versus curved blades under various braking conditions. These observations not only reinforce the claim of passive control in the flexible curved blade concept but also underscore the necessity for appropriate load settings in power coefficient measurements. The ongoing development of tailored testing systems, including an MPPT inverter for more accurate load simulation, is crucial for advancing our understanding of blade performance across different scales. This endeavor will enhance the measurement capability at the Power Synthesis Laboratory to design blades that optimally balance power production with dynamic control.

5. Conclusions

This work aimed to present preliminary results on the aerodynamic performance and behavior of a novel flexible curved wind turbine blade concept in comparison with a conventional straight blade. The flexible curved blade concept was designed to enhance operational safety and energy efficiency by incorporating sufficient elasticity to allow for bend–twist deformation around its longitudinal axis under bending loads. This passive control mechanism is intended to self-regulate the blade's response to varying wind speeds, thereby tailoring its power profile.

The study rigorously examined the performance characteristics of the preliminary results for these flexible curved blades, utilizing a series of tests to highlight differences in rotational velocity, power output, and efficiency under varying wind speeds and braking conditions. Key findings revealed that the flexible curved blades, due to their innovative design, managed to demonstrate a capping on the maximum rotational speed and mechanical power output at high speeds compared to the straight blades, which generally displayed higher operational metrics under similar conditions. Additionally, the flexible curved blades were able to start generating power at lower wind speeds (flexible curved 4 m/s, while straight started at 6 m/s) and exhibited a degree of power stabilization (beyond 16 m/s)—traits indicative of the effective passive control integrated into their design.

Despite these advantages, the passive control mechanisms did not activate at the designed wind speeds (8 to 10 m/s), indicating a misalignment between performance expectations and actual outcomes. This misalignment suggests areas for further refinement in the blade's design and control settings. Moreover, while the flexible curved blades maintained a stable power output at higher wind speeds, similar to actively controlled systems, this capability marks a significant advancement in blade technology, aligning with sustainability goals by potentially reducing active control systems costs and possibly mitigating potential turbine damage.

Additionally, the comparison of the Power coefficient (Cp) versus tip speed ratio (TSR) demonstrated that flexible curved blades operate within a significantly lower TSR range. Under no braking conditions, their performance at lower TSR was superior, challenging conventional expectations about blade dynamics under passive control. Additionally, the flexible curved blade demonstrated a 'hook-like' feature in Cp behavior under high wind conditions (over 16 m/s), which indicated that the flexible curved blades effectively cap power in a controlled manner.

In conclusion, this study has provided valuable insights into the capabilities and limitations of flexible curved wind turbine blades and test configurations, affirming the feasibility of designing and manufacturing passively controlled wind turbine blades tailored for specific performance goal criteria. The flexible curved blade concept, which allows for self-regulation without the need for electrical or hydraulic control systems, shows promise. However, further adjustments and optimizations are required to fully realize their potential and ensure alignment with design specifications. Some of these optimizations will be explored in more detail in the next article of this series.

Future Work

Building on the insights provided in this study, future research will focus on enhancing the design, efficiency, and operational reliability of flexible curved wind turbine blades. Key areas for development include refining design parameters and simulation codes to align more closely with real-world performance, and materials for improved adaptability and resilience. Additionally, field testing should be extended to validate the long-term behavior of these blades under diverse environmental conditions.

Additionally, future studies will explore the scalability and economic viability of these designs, from small-scale models to potential applications (which is probably a very niche market), ensuring their commercial feasibility.

A particular emphasis for the Power Synthesis Laboratory will be placed on developing and adapting testing equipment, such as MPPT inverters or similar, to improve the accuracy of load simulations and power measurements. These efforts aim to advance our understanding of wind turbine technology and contribute to the broader field of renewable energy, supporting the transition towards more sustainable energy systems globally.

Author Contributions: Conceptualization, C.C. and N.P.; methodology, N.P. and C.C.; software, N.P.; validation, N.P.; investigation, N.P. and C.C.; resources; data curation, N.P.; writing—original draft preparation, N.P.; writing—review and editing, N.P.; visualization, N.P.; supervision, N.P.; project administration, C.C.; funding acquisition, No funding. All authors have read and agreed to the published version of the manuscript.

Funding: This research received no external funding.

Data Availability Statement: The original contributions presented in the study are included in the article, further inquiries can be directed to the corresponding authors.

Conflicts of Interest: The authors declare no conflict of interest.

Abbreviations:

C_p	Power Coefficient
MPPT	Maximum Power Point Tracking
TSR	Tip Speed Ratio
WECS	Wind Energy Converter Systems

References

1. Doukas, H. Energy Citizenship—A New Dimension in Energy Policy Making. Available online: https://www.e-mc2.gr/news/energy-citizenship-new-dimension-energy-policy-making (accessed on 6 February 2024).
2. Roesch, R.; Taylor, M.; Al-Zoghoul, S.; Ralon, P.; Sorokina, O. *IRENA Report—Renewable Power Generation Costs in 2022*; International Renewable Energy Agency: Abu Dhabi, United Arab Emirates, 2022; ISBN 978-92-9260-544-5.
3. IEA. *World Energy Outlook 2013*; IEA: Paris, France, 2013.
4. International Energy Agency (IEA). *Global Electricity Generation by Technology, 2015, 2021 and 2027*; IEA: Paris, France, 2022.
5. European Union Electricity and Heat Statistics. Available online: https://ec.europa.eu/eurostat/statistics-explained/index.php?title=Electricity_and_heat_statistics (accessed on 3 May 2023).
6. Frequently Asked Questions (FAQs)—U.S. Energy Information Administration (EIA). Available online: https://www.eia.gov/tools/faqs/faq.php (accessed on 9 February 2024).
7. Hutchinson, M.; Zhao, F. *Global Wind Report 2022*; Global Wind Energy Council: Brussels, Belgium, 2022.
8. Ponta, F.L.; Otero, A.D.; Rajan, A.; Lago, L.I. The Adaptive-Blade Concept in Wind-Power Applications. *Energy Sustain. Dev.* **2014**, *22*, 3–12. [CrossRef]
9. Liebst, B.S. Wind Turbine Gust Load Alleviation Utilizing Curved Blades. *J. Propuls. Power* **1986**, *2*, 371–377. [CrossRef]
10. Christakis, D. The Development of Wind Energy Systems in Crete. In Proceedings of the Seminar for the New Trends on Wind Energy, Chanea, Greece; 1991.
11. Serra, M.F.G.; van Schoor, M.C. Aeroelastic Tailoring of a Horizontal Axis Wind Turbine. *Wind Eng.* **1995**, *19*, 193–207.
12. Eggers, A.J., Jr.; Ashley, H.; Rock, S.M.; Chaney, K.; Digumarthi, R. Effects of Blade Bending on Aerodynamic Control of Fluctuating Loads on Teetered HAWT Rotors. *J. Sol. Energy Eng.* **1996**, *118*, 239–245. [CrossRef]
13. Tran, D.-H.; Sareni, B.; Roboam, X.; Espanet, C. Integrated Optimal Design of a Passive Wind Turbine System: An Experimental Validation. *IEEE Trans. Sustain. Energy* **2010**, *1*, 48–56. [CrossRef]
14. Infield, D.G.; Feuchtwang, J.B. Design Criteria for Passive Pitch Control of Wind Turbines Using Self-Twisting Blades. *Int. J. Ambient. Energy* **1995**, *16*, 139–146. [CrossRef]
15. Infield, D.G.; Feuchtwang, J.B.; Fitches, P. Development and Testing of a Novel Self-Twisting Wind Turbine Rotor. In *1999 European Wind Energy Conference*; Routledge: London, UK, 1999; ISBN 978-1-315-07433-7.
16. Lobitz, D.W.; Veers, P.S.; Migliore, P.G. Enhanced Performance of HAWTs Using Adaptive Blades. In Proceedings of the Wind 1996 ASME Wind Energy Symposium, Houston, TX, USA, January 29–2 February 1996; Report Number: CONF-960154. American Society of Mechanical Engineers: New York, NY, USA; Sandia National Laboratories (SNL): Albuquerque, NM, USA; Livermore, CA, USA, 1996.
17. Lobitz, D.; Veers, P. Aeroelastic Behavior of Twist-Coupled HAWT Blades. In *1998 ASME Wind Energy Symposium*; Aerospace Sciences Meetings; American Institute of Aeronautics and Astronautics: Reston, VA, USA, 1998.
18. Eisler, G.R.; Veers, P.S. *Parameter Optimization Applied to Use of Adaptive Blades on a Variable Speed Wind Turbine*; Sandia National Laboratories: Albuquerque, NM, USA, 1998.
19. Lobitz, D.; Laino, D. Load Mitigation with Twist-Coupled HAWT Blades. In *37th Aerospace Sciences Meeting and Exhibit*; Aerospace Sciences Meetings; American Institute of Aeronautics and Astronautics: Reston, VA, USA, 1999.
20. Lobitz, D.; Veers, P.; Laino, D. Performance of Twist-Coupled Blades on Variable Speed Rotors. In *2000 ASME Wind Energy Symposium*; American Institute of Aeronautics and Astronautics: Reston, VA, USA, 2000.
21. Condaxakis, C.G.; Christakis, D.G.; Frandsen, S.T.; Eboueya, M. Passive Controlled Wind Turbine Blades. In Proceedings of the 1999 European Wind Energy Conference, Nice, France, 1 March 1999; James & James Ltd.: Nice, France, 1999.
22. Christakis, D.G.; Condaxakis, C.G.; Eboueya, M.; Granier, P. A Contribution to the Passive Controlled Optimised Wind Motor Design. In Proceedings of the Computational Engineering in Systems Applications CESA, Nabeul-Hamamet, Tunisia, 1 April 1998; IEEE: Nabeul-Hamamet, Tunisia, 1998.
23. Zuteck, M. *Adaptive Blade Concept Assessment: Curved Platform Induced Twist Investigation*; Sandia National Lab. (SNL-NM): Albuquerque, NM, USA; Livermore, CA, USA, 2002.
24. Larwood, S.; Zutek, M. Swept Wind Turbine Blade Aeroelastic Modeling for Loads and Dynamic Behavior. In Proceedings of the WINDPOWER 2006, Pittsburgh, PA, USA, 4–7 June 2006.
25. Ashwill, T.D. *Sweep-Twist Adaptive Rotor Blade: Final Project Report*; Sandia National Laboratories (SNL): Albuquerque, NM, USA; Livermore, CA, USA, 2010.
26. Masoudi, M.; Pope, K. Numerical Predictions on Fluid-Structure Bend-Twist Coupling of Wind Turbine Blades. *Int. J. Mech. Eng. Robot. Res.* **2019**, *8*, 506–510. [CrossRef]
27. Wiens, M.; Meyer, T.; Wenske, J. Exploiting Bend-Twist Coupling in Wind Turbine Control for Load Reduction*. *IFAC-Pap.* **2020**, *53*, 12139–12144. [CrossRef]
28. Tamayo-Avendaño, J.M.; Patiño-Arcila, I.D.; Nieto-Londoño, C.; Sierra-Pérez, J. Fluid–Structure Interaction Analysis of a Wind Turbine Blade with Passive Control by Bend–Twist Coupling. *Energies* **2023**, *16*, 6619. [CrossRef]
29. Riva, R.; Spinelli, M.; Sartori, L.; Cacciola, S.; Croce, A. Stability Analysis of Wind Turbines with Bend-Twist Coupled Blades. *J. Phys. Conf. Ser.* **2018**, *1037*, 062014. [CrossRef]

30. Voutsinas, S.G.; Belessis, M.A.; Rados, K.G. Investigation of the Yawed Operation of Wind Turbines by Means of a Vortex Particle Method. In *Proceedings of the AGARD Conference Proceedings AGARD CP*; Advisory Group for Aerospace Research and Development (AGARD): Neuilly sur Seine, France, 1995.
31. Capuzzi, M.; Pirrera, A.; Weaver, P.M. Structural Design of a Novel Aeroelastically Tailored Wind Turbine Blade. *Thin-Walled Struct.* **2015**, *95*, 7–15. [CrossRef]
32. MacPhee, D.; Beyene, A. A Flexible Turbine Blade for Passive Blade Pitch Control in Wind Turbines. In Proceedings of the 2011 IEEE Power Engineering and Automation Conference, Wuhan, China, 8–9 September 2011; Volume 1, pp. 196–199.
33. Larwood, S.; van Dam, C.P.; Schow, D. Design Studies of Swept Wind Turbine Blades. *Renew. Energy* **2014**, *71*, 563–571. [CrossRef]
34. Shen, X.; Chen, J.-G.; Zhu, X.-C.; Liu, P.-Y.; Du, Z.-H. Multi-Objective Optimization of Wind Turbine Blades Using Lifting Surface Method. *Energy* **2015**, *90*, 1111–1121. [CrossRef]
35. Khalafallah, M.G.; Ahmed, A.M.; Emam, M.K. CFD Study of Some Factors Affecting Performance of HAWT with Swept Blades. *Int. J. Sustain. Energy* **2017**, *36*, 489–501. [CrossRef]
36. Chen, Y.-J.; Tsai, Y.-F.; Huang, C.-C.; Li, M.-H.; Hsiao, F.-B. The Design and Analysis of Passive Pitch Control for Horizontal Axis Wind Turbine. *Energy Procedia* **2014**, *61*, 683–686. [CrossRef]
37. Chen, Y.-J.; Shiah, Y.C. Experiments on the Performance of Small Horizontal Axis Wind Turbine with Passive Pitch Control by Disk Pulley. *Energies* **2016**, *9*, 353. [CrossRef]
38. Chu, Y.-J.; Chong, W.-T. A Biomimetic Wind Turbine Inspired by *Dryobalanops Aromatica* Seed: Numerical Prediction of Rigid Rotor Blade Performance with OpenFOAM®. *Comput. Fluids* **2017**, *159*, 295–315. [CrossRef]
39. Krogstad, P.-Å.; Lund, J.A. An Experimental and Numerical Study of the Performance of a Model Turbine. *Wind Energy* **2012**, *15*, 443–457. [CrossRef]
40. Larsen, G.C.; Frandsen, S.T.; Sørensen, P.E.; Courtney, M. *Design Basis for Horizontal-Axis Wind Turbines. Theoretical Background*; Risoe National Lab.: Roskilde, Denmark, 1990.
41. Le Gouriérès Désiré; Le Gouriérès, D. *Energie Éolienne: Théorie, Conception et Calcul Pratique Des Installations/Par Désiré Le Gourières, . . .*; Eyrolles: Paris, France, 1980.
42. Hansen, M. The Classical Blade Element Momentum Method. In *Aerodynamics of Wind Turbines*, 2nd ed.; Routledge: London, UK, 2007; ISBN 978-1-84977-040-8.
43. Zhang, X.; Jia, J.; Zheng, L.; Yi, W.; Zhang, Z. Maximum Power Point Tracking Algorithms for Wind Power Generation System: Review, Comparison and Analysis. *Energy Sci. Eng.* **2023**, *11*, 430–444. [CrossRef]
44. Kesraoui, M.; Korichi, N.; Belkadi, A. Maximum Power Point Tracker of Wind Energy Conversion System. *Renew. Energy* **2011**, *36*, 2655–2662. [CrossRef]

energies

MDPI

Article

An Experimental Performance Assessment of a Passively Controlled Wind Turbine Blade Concept: Part B—Material Oriented with Glass-Fiber-Reinforced Polymer

Nikolaos Papadakis * and Constantinos Condaxakis

Power Plant Synthesis Laboratory, Department of Mechanical Engineering, Hellenic Mediterranean University, 71410 Heraklion, Greece; condax@hmu.gr
* Correspondence: npap@hmu.gr; Tel.: +30-2810-379722

Abstract: This paper is the second in a two-part series presenting preliminary results on a passively controlled wind turbine rotor system using a flexible curved blade concept. Building on the initial findings, this segment explores the application of glass-fiber-reinforced polymer (GFRP) composites with strategically oriented layers to enhance blade flexibility and aerodynamic performance and ensure operational safety. Previously, we demonstrated that flexible blades fabricated from isotropic materials with an NACA4415 airfoil profile could self-regulate rotor RPM and power output in response to aerodynamic loads, offering a glimpse of controlled operational behavior, in contrast to straight blades of similar material geometry and aerodynamic characteristics. However, they did not fully meet the design objectives, particularly in achieving nominal power at the intended wind speeds and in safely halting operation at high wind speeds. The current study employs a GFRP blade with a simpler, flat geometry due to manufacturing constraints, diverging from traditional airfoil contours to focus on material behavior under aerodynamic loads. Despite these changes, the blade exhibited all desired operational characteristics: quick startup, stable power output across operational wind speeds, and effective shutdown mechanisms at high speeds. This success illustrates the potential of passively controlled blades designed with appropriately oriented composite layers. Challenges with load application methods—that were identified in the first installment—were addressed by adopting a generator connected to a rheostat, offering improved control over load variations compared to the mechanical brakes used previously. This advancement enabled more consistent data collection, particularly at lower Tip–Speed Ratio (TSR) values, although real-time control for maximum power point tracking was still out of reach. These findings not only confirm the effectiveness of the flexible blade concept but also highlight the need for further refinement in blade design and testing methodology to optimize performance and ease of manufacturing. Future work will continue to refine these designs and explore their scalability and economic viability for broader applications in wind energy technology and in particular to those of small Wind Energy Converter Systems (WECSs).

Keywords: passive control; aero elastic tailoring; adaptive blades; curved blades

1. Introduction

1.1. Findings from Part A: Isotropic Materials

In the first installment of this two-part series, we introduced a novel concept in wind turbine technology: a flexible wind turbine blade designed for passive control, aimed at enhancing aerodynamic performance and operational safety. The flexible blade concept is particularly suited for small Wind Energy Conversion Systems (WECSs) because their presence would reduce the cost of an active electromechanical control system [1].

This innovative approach utilizes inherent blade elasticity to allow for bend–twist deformation around its longitudinal axis under bending loads. This deformation is critical for the blade's ability to self-regulate in response to varying wind conditions, thereby optimizing its power profile without the need for electrical or hydraulic control systems.

The philosophy of passive control in wind energy systems is based on the turbine's ability to respond to excitation loads and the blade's capacity to function as a feedback mechanism, adapting to operational conditions to meet design criteria. The primary design objectives include maximizing power yield, minimizing operational loads, and optimizing control within the WECS. Implementing a passive pitch control philosophy can drastically reduce blade aerodynamic loads and overall structural stress. Turbines equipped with passive pitch-controlled blades could potentially eliminate the need for complex, moving parts found in actively controlled variable-pitch blades. Instead, these blades adjust their geometry in response to aerodynamic loads, combining the benefits of variable-pitch blades—such as responsiveness, maximization of energy yield, and eventually reduced energy cost—with the simplicity of manufacturing akin to fixed-pitch blades. An additional advantage of passive pitch control blades is their quicker response time, presenting a viable solution for enhancing the feasibility of small WECS.

Our initial findings, presented in the first paper, demonstrated several key advantages of this flexible blade design over conventional straight blades. The study highlighted that the flexible blades could cap their maximum rotational speed and mechanical power output at high wind speeds, exhibiting a controlled and sustainable energy production model. Notably, these blades began generating power at lower wind speeds and maintained a stable power output even as conditions became less favorable, showcasing their efficiency and the effectiveness of the passive control mechanisms integrated into the design.

However, the study also uncovered certain challenges. The passive control mechanisms did not activate at the intended wind speeds, revealing a significant gap between design expectations and actual performance. This misalignment, along with observations of a 'hook-like' feature in the Power Coefficient (C_P) versus Tip–Speed Ratio (TSR) behavior under high wind conditions, suggested areas for further refinement.

As we move into the second part of this series, our focus will shift to addressing these challenges through advanced material usage, design optimization, and enhanced testing procedures. We aim to build on the promising foundation laid by the initial study, refining our designs and methodologies to fully realize the potential of passive control in wind turbine blades. This next phase will explore in depth the opportunities from integration of GFPR-composite-layered orientation to improve blade flexibility and performance.

1.2. Literature Review

The literature on the development of wind turbine blades, particularly those crafted from fiber-reinforced polymers such as GFRP, highlights a significant focus on enhancing blade performance through passive control mechanisms. This review underscores the role of composite layup orientations in achieving desired aerodynamic responses and structural efficiencies, particularly through the implementation of bend–twist coupling.

Liebst [2] in 1986 pioneered studies in this area by examining the effects of wind gusts on curved blades. He noted that such blades could autonomously adjust their geometry, specifically by reducing their pitch angle, thus mitigating aerodynamic loads significantly. This early work set the stage for further exploration into blades that could passively adapt to varying wind conditions. Subsequently, Infield and Feuchtwang in 1995 [3] and in 1999 [4] introduced the concept of "stretch–twist coupled" blades, which were designed to control rotor behavior in runaway scenarios. Their innovative use of a helical layup incorporating both glass and carbon fibers proved effective, as demonstrated by their experimental results aligning closely with predictive models.

Furthering this research, Lobitz and Veers [5,6] conducted studies throughout the late 1990s, focusing on how twist–bend coupling could enhance annual energy production and ensure stability in utility-sized rotors. Their findings revealed that even minor blade twists could result in significant increases in energy output and provided insights into managing common stability issues like flutter and divergence. Similarly, Eisler and Veers [7] examined adaptive blades on a 26 m diameter variable speed rotor.

The ability of bending twist-coupled blades to attenuate (or exacerbate) the cyclic loading has been investigated by Lobitz and Laino [8], in 1999, and Lobitz, Veers, and Laino [9] in 2000 for a 33 m diameter rotor employing three different control strategies: constant speed stall-controlled, variable speed stall-controlled, and variable speed pitch-controlled. Results for the constant speed stall-controlled case indicate that twist-coupling toward stall produces significant increases in fatigue damage, and for a range of wind speeds in the stall regime apparent stall flutter behavior is observed.

The early 2000s saw continued advancements with researchers like Zuteck [10] exploring the passive control capabilities of bend–twist coupling in larger blades. His work emphasized the necessity of reducing torsional stiffness to enable sufficient blade twisting, suggesting that such designs could also support larger rotor diameters and thus lower energy costs. This line of inquiry was extended by Larwood and Zuteck [11], who compared backward-swept blades to traditional designs, finding improvements in energy capture without additional mechanical strain on the turbine.

Sandia National Laboratories [12] further built on these concepts with their sweep-twist adaptive rotor (STAR) technology, which aimed to enhance rotor efficiency by incorporating passive twisting in blade designs. Their findings demonstrated notable increases in energy capture without exacerbating blade root bending moments, marking a significant step forward in rotor design.

Moreover, the research community has persistently explored optimizing composite ply structures to maximize the mechanical coupling effects. Earlier studies [13,14] and more recent analyses [15] have investigated optimal configurations of bidirectional layups that balance structural integrity with dynamic performance. Tsai and Ong [16] specifically addressed the optimal angular orientation of fibers to maximize bend–twist coupling, pinpointing a 20-degree orientation relative to the span-wise axis as the most effective for certain blade cross-sections.

These collective efforts in the literature underscore a continuous strive towards refining wind turbine blade designs through innovative material use and structural strategies, setting a firm foundation for further advancements in this field. This historical context serves as a critical backdrop for the current study, which aims to push the boundaries of wind turbine blade technology by exploring the integration of GFRP in non-traditional blade configurations, assessing their viability and performance in modern wind energy applications.

1.3. Scope

This second installment of our study series delves deeper into the exploration of novel wind turbine blade designs by examining a curved blade with a rectangular cross-section made from layered Glass-Fiber-Reinforced Polymer (GFRP) in preferential orientation. This design is a departure from the first installment, which focused on a curved blade with a conventional NACA 4415 airfoil cross-section crafted from isotropic materials. The use of a simple, non-airfoil, rectangular shape cut directly from a sheet of composite material signifies a shift towards more practical and cost-effective manufacturing processes within a research laboratory setting.

The decision to utilize a rectangular cross-section arises from the necessity (recognized from the previous installment) to enhance blade flexibility and is driven by the limitations of our current simulation tools, which are not yet capable of accurately predicting the aerodynamic behavior of complex curved shapes with oriented GFRP layers. This simpler shape allows for investigation of the strategic orientation of GFRP layers without the need for creating multiple molds for shaping and curing, significantly reducing both the complexity and the cost of blade production. Such an approach facilitates rapid iterations and modifications, crucial for advancing blade technology through experimental research.

While adopting a non-airfoil shape inevitably leads to certain performance trade-offs, such as reduced aerodynamic efficiency and altered lift-to-drag ratios, this study primarily focuses on evaluating the mechanical performance of the blade under operational conditions. We aim to assess whether the increased mechanical flexibility, achieved through the preferential orientation of GFRP layers, can be tailored in order to meet specific design goals: achieving a tailored startup at lower wind speeds, maintaining stable power output between nominal and cutoff speeds, and safely stopping the turbine when wind speeds exceed operational thresholds.

The overarching goal of this research is to provide preliminary findings on the effectiveness of using layered GFRP composites in enhancing the flexibility of wind turbine blades. By exploring these non-traditional design and material choices, we hope to offer insights that could benefit small-scale wind turbines or other applications where the ease and cost of manufacturing are more critical than maximizing energy output at an affordable cost. Ultimately, the results from this investigation may open new avenues for future studies into unconventional blade shapes and materials, potentially broadening the scope of design strategies in small wind turbine converter systems and supporting the evolution towards more sustainable and adaptable energy solutions.

2. Methodology

2.1. Overview

In this study, we aim to explore the potential of tailoring the orientation of layered Glass-Fiber-Reinforced Polymer (GFRP) materials to influence the bend–twist deformation of a wind turbine blade, thereby tailoring its aerodynamic performance. This paper presents wind tunnel results for a single curved layered GFRP blade with a rectangular cross-section, selected as the optimal from a series of specimens. The focus of our analysis is on the power versus wind speed tests and the Coefficient of Power (C_P) versus Tip–Speed Ratio (λ). The objectives are threefold: to achieve nominal power output at nominal wind speed, maintain this output between nominal and cutoff speeds, and halt the blade operation when wind speeds exceed safe operational thresholds.

To this end, several blade specimens with varying layer counts and dimensions were fabricated. Of these, we present results from the specimen designated as number 010, identified as the optimal design based on preliminary tests. This specimen measures approximately 155 mm in length, and when mounted on the rotor hub—which has a diameter of approximately 55 mm—the tip's radius extends to 172.5 mm.

The methodology section of this paper will provide a comprehensive overview of the experimental approach undertaken to evaluate the aerodynamic performance of the optimized blade. Although a detailed discussion of the Blade Geometry Algorithm is beyond the scope of this paper, as it will be covered extensively in a subsequent study, we direct the interested reader to the first part of this series which provides an overview. This will set the stage for a deeper understanding of the blade design principles under investigation.

Furthermore, we will detail the test campaign executed for these blades, emphasizing the experimental setups and key findings. This includes a thorough description of the measurement apparatus and the configuration of the wind tunnel used during the tests. By presenting these methodologies, this paper aims to provide clear insights into the experimental procedures and the rationale behind the design choices, supporting the validity of the results obtained from this innovative study.

2.2. Blade Geometry Algorithm

Figure 1 introduces the core concept by comparing straight and curved blades. The curved blade is divided into three sections: the root section, the control section, and the tip section. The root section, which constitutes 25% of the blade length, is engineered for high rigidity to ensure stability. The control section, making up 50% of the blade length, is designed for necessary elasticity to facilitate aerodynamic control. The tip section, comprising the remaining 25% of the blade length, is optimized for enhanced elasticity to respond quickly to wind changes. This segmentation ensures that each part of the blade performs optimally under varying operational conditions.

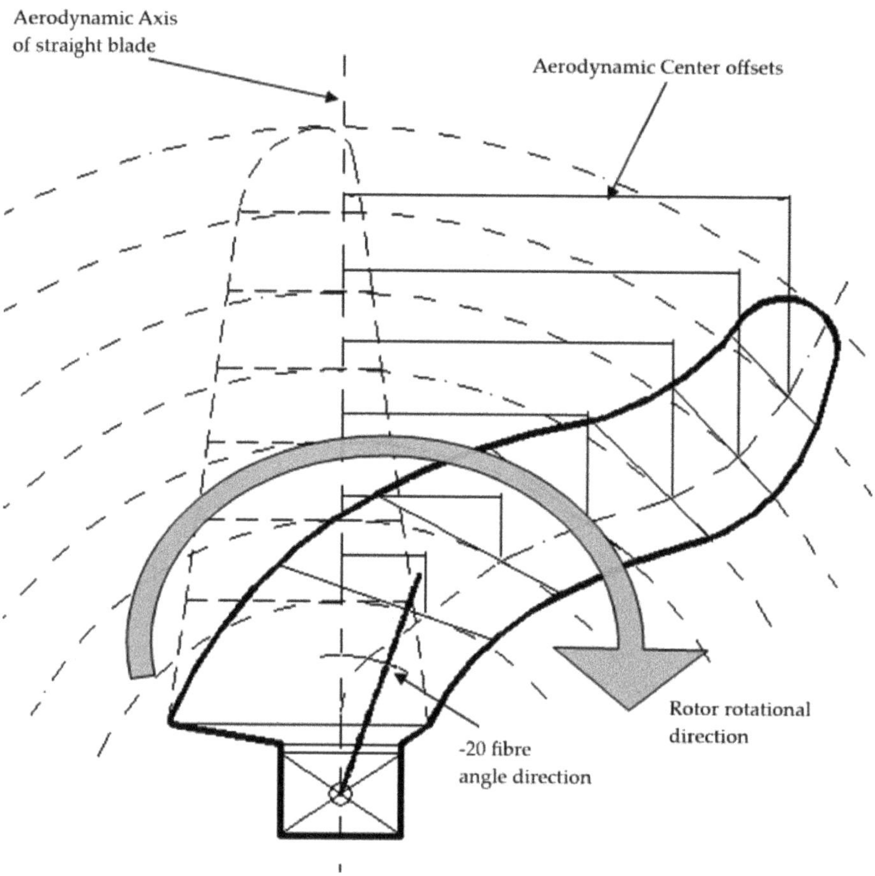

Figure 1. Eccentricity at each section of an early iteration of the curved blade.

The curved blade uses the same aerodynamic cross-sectional profiles as the straight blade, but the profiles for the curved blade are offset by specific angles at different radii. This introduces torsional deformation without altering the chord length or other geometric characteristics, maintaining consistent aerodynamic properties. The offset induces bend–twist coupling under aerodynamic loads, affecting the torsional response of the curved blade.

Part A of this series [17] provided a detailed examination of the aerodynamic behavior of wind turbine blades under various conditions utilizing the NACA4415 airfoil profile. It highlighted how blade geometry interacts with wind conditions to achieve specific aerodynamic responses. For instance, entering aerodynamic stall under high wind speeds helps prevent damage and maintain efficiency. Key design criteria include quick startup to respond to changing wind conditions, achieving nominal power output at nominal wind speeds, maintaining stable power output from nominal to cutoff speeds, and initiating stalling mechanisms at high wind speeds to prevent damage.

Figure 2 provides an overview of the curved blade geometry derivation algorithm and the various steps involved. Initially, a straight/rigid blade is designed using geometry derived from an aerodynamic code. This code incorporates the Larsen, Frandsen, Soeresen, and Courtney [18] theoretical framework for aerodynamic behavior modeling. Additionally, it utilizes enhancements based on Glauert's theory for airfoils and airscrews as described in Désiré Le Gouriérès's "Les Éoliennes" [19] and incorporates aspects of the Hansen

Blade Element Momentum method [20]. Key parameters include nominal power, wind speed, radial chord length distribution, and pitch angle distribution, with adjustments possible to enhance algorithm precision. The curved blade geometry builds on these parameters with the addition of eccentricity, which increases torsional loads and rotations. The process involves modifying profile segments based on initial eccentricity assumptions, calculating bending-torsional deformation using finite element analysis, and optimizing blade geometry in CAD software (Solidworks 2020).

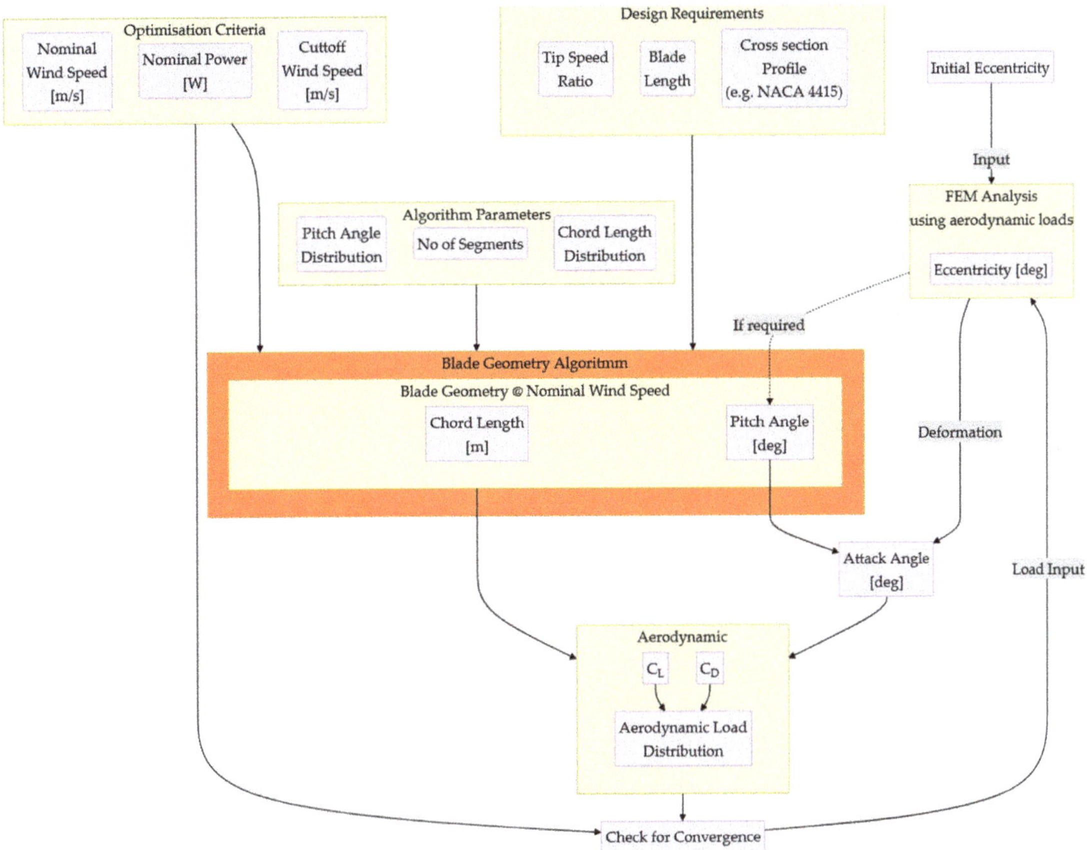

Figure 2. Curved blade design algorithm overview.

Structural analysis in Dassault Solidworks involves fixing the blade edge to the hub, applying aerodynamic loads, and measuring deformation. The blade edge attached to the hub is fixed in all degrees of freedom, while the opposite edge remains entirely free, without any rotational or translational constraints. Load modeling reflects the dynamic pressure of the wind, adopting an inverse triangular thrust distribution. This load distribution assumes zero pressure load at the blade root and maximum pressure load at the tip, representative of typical operational thrust distribution near rated power. The optimization of eccentricity and chord length is based on experience and expertise. Prototypes are manufactured and tested, followed by iterative improvements.

For a more detailed presentation of the wind blade design algorithm, the interested reader should refer to Part A of this series [17].

2.3. GFRP Blade Materials, Fabrication, and Rotor Assembly

In the development of Glass-Fiber-Reinforced Polymer (GFRP) blades, the design extends beyond just the curved planform geometry to also include the orientation of the composite material layers. Given the complexities of constructing a real model with varying fiber orientations across each section of the blade, a uniform fiber orientation was adopted across the span of the micro model blade. However, incorporating different angle orientations throughout the span could potentially provide additional flexibility tailoring.

The fiber orientation is optimized specifically to maximize torsional blade deformation. As illustrated in Figure 3, the graph demonstrates the relationship between the tip torsional deformation angle and the blade material layer fiber orientation angle. The fiber orientation angle is defined as the angle between the elastic axis of the uncurved blade and the fiber axis.

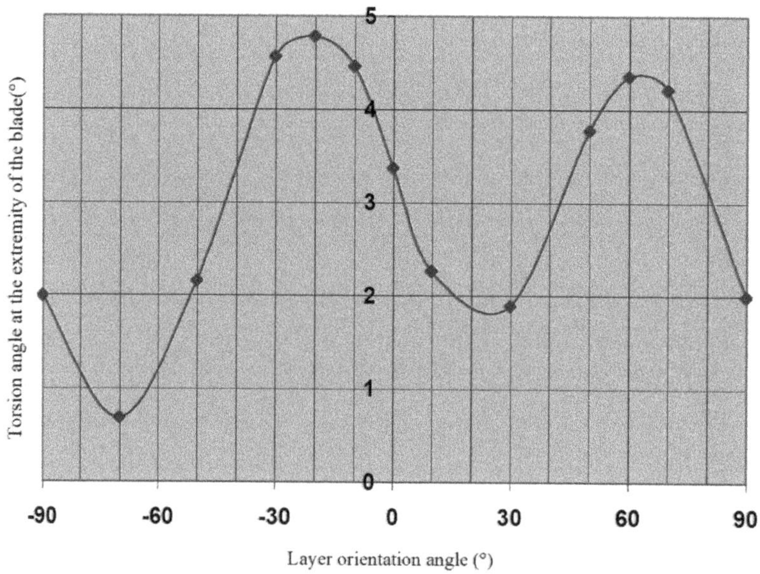

Figure 3. Torsional deformation angle versus the layer fibers orientation angle.

Figure 3 presents the torsional deformation angle versus the layer fiber orientation angle. Simulation results indicate that fiber orientation significantly affects torsional deformation. From −90° to −70°, as well as from −20° to +25° and from 60° to 90°, the torsion angle at the blade's extremity decreases as the layer orientation angle increases. Conversely, from −70° to −20° and from 25° to 60°, the torsion angle increases with an increase in the layer orientation angle. This results in two local minima at angles of −70° and 25° and two local maxima at angles of −20° and approximately 60°. The global maximum occurs at a layer orientation angle of −20°; at this angle, the torsion angle at the blade's extremity reaches 4.79°, correlating with aerodynamic loads at a wind speed of 12 m/s, a finding consistent with other studies [13,14].

For the construction of the GFRP blades in this study, the blades were fabricated from a balanced standard E-glass fabric with a warp/weft ratio of 60/40, weighing 202 gr/m^2, and bonded with an epoxy resin. The construction utilized three layers, though it is important to note that variations in the number of layers can significantly influence both the bending and torsional behavior of the blades. The layers were oriented at 20°, which aligns with the identified optimal orientation for achieving desired mechanical properties and torsional responses. This orientation results in the following orthotropic mechanical properties:

Young modulus: Ex = 23.5 GPa, Ey = 18.0 GPa, Ez = 18.0 GPa

$$\text{Poisson's ratio}: \ \nu_{xy} = 0.153, \ \nu_{yz} = 0.122, \ \nu_{xz} = 0.153$$

Shear modulus: Gxy = 4.81 GPa, Gyz = 8.02 GPa, Gxz = 4.81 GPa

To complete the rotor assembly, three blades were fabricated and mounted onto the rotor hub. Due to the manufacturing method—cutting the blades from a sheet of GFRP—it was not feasible to integrate a non-zero pitch angle directly into the blades themselves. Consequently, it was determined that the rotor hub would be designed to incorporate a constant 5-degree pitch angle, thereby ensuring that each blade maintained the necessary angle throughout its length. This design approach effectively compensated for the fabrication constraints, allowing for aerodynamic functionality despite the blades' flat profile.

2.4. Measurement Campaign

2.4.1. Types of Measurements

Two distinct types of measurements were employed during the testing phase, categorized based on the wind velocity conditions: constant velocity (internally denoted as "rr") and increasing velocity ("vv").

Constant Wind Velocity Test: In the constant wind velocity test, the wind blade is exposed to a steady wind until the revolutions per minute (rpm) of the blade stabilize at an equilibrium. Following this stabilization, an external load is gradually applied to the wind blade shaft, either mechanically (or electrically if the rotor is connected to a generator). This test format is commonly used to characterize wind energy blade performance, particularly for calculating the Power Coefficient (C_P) versus Tip–Speed Ratio (TSR or λ). It is important to note that the magnitude of the load significantly influences the Power Coefficient. Specifically, an excessively high load can stall the blade, while an insufficiently light load may prevent the blade from slowing sufficiently to observe the desired effects. During these tests, the braking load is gradually increased to a point where the blade decelerates enough to eventually stop.

Increasing Velocity Test: Contrastingly, the increasing velocity test is designed to characterize the performance of the wind blade across a broader range of wind speeds and may include the application of braking loads. Although not as typical for standard characterization, these tests are crucial in this context as they highlight the benefits of the passively controlled wind energy converter system. This approach allows for an assessment of the blade's adaptability and performance under varying operational conditions.

2.4.2. Measurement Campaign for GFPR Blades (Curved)

Table 1 outlines the key parameters for the measurements of the curved GFRP blade under conditions of increasing velocity. This work reports on four different load settings. Load settings higher than 10 Ohms were excluded from consideration because, under those conditions, the blade failed to initiate rotation. Each test scenario was conducted with a single repetition. During these tests, the wind tunnel velocity was manually adjusted by the operator at a rate of approximately 1 m/s every 10 s, allowing for controlled variations in wind speed.

Table 2 presents the salient parameters for the measurements conducted under constant velocity and constant load conditions using the curved GFRP blade. The tests spanned wind speeds from 4 m/s up to and including 24 m/s, in 2 m/s increments. Two distinct loading methods were employed during these tests.

Table 1. Test parameters for GFRP wind blade (010) VV measurements.

Experiment ID	Resistive Load
58	2
59	4
60	6
62	10

Table 2. Test parameters for GFRP wind blade RR measurements.

Nominal Wind Tunnel Velocity	Ramp Load Experiment ID	Step Load Experiment ID
4	47	66, 80, 106, 132
6	48	67, 81, 108, 134
8	49	68, 82, 110, 136,
10	50	69, 83, 112, 138
12	51	70, 84, 114, 140
14	52	71, 85, 116, 142
16	53	72, 86, 118, 144
18	54	73, 87, 120, 146
20	55	74, 88, 122, 148
22	56	75, 89, 90, 124, 150
24	57	126, 152

The first method involved a ramp load, where the resistive load started at zero and was continuously adjusted upward. This method is referred to as "ramp load", and the corresponding experiment IDs are listed in the second column of the table. The second method, known as "step load", involved incrementally increasing arbitrarily the resistive load in discrete steps and then allowing the rotor hub to stabilize at the new rpm equilibrium. This step-loading method was favored for repeated measurements as it provided more consistent and representative data, capturing the rotor's behavior as it settled into each new load condition.

2.5. Experimental Apparatus

The testing facility for this series of experiments was conducted at the Hellenic Mediterranean University's Wind Energy Lab in a 600 × 600 mm wind tunnel, presented in Figure 4, largely maintaining the setup used in the first part of this series. Despite the similarities in the papers of this two-part series, notable modifications were introduced to address limitations observed in the previously employed braking methods.

The wind tunnel facilitated measurements of various parameters, including wind speed, rotor thrust (T), rotor torque (Q), and rotor speed (N). Table 3 presents manufacturing companies and models of the Measurement apparatus. Wind speed measurements at the tunnel outlet were accurately captured using a Pitot–Prandtl tube coupled with a Delta Ohm HD408T differential pressure transducer, calibrated against an externally calibrated hotwire anemometer to ensure precision. The rotor's axis was connected to a SCAIME DR2112L Rotating Torquemeter, which measured torque and revolutions per minute (rpm). Thrust (or drag) data were collected via an HBM SP4MC6MR load cell interfaced with an ADAM 3016 isolated strain gauge input module. Data acquisition was managed through a National Instruments NI-USB-6353 card, using a custom LabVIEW application developed on the LabVIEW 2014 Development System. This setup allowed data capture at a sampling rate of 1000 Hz for each channel, including measurements for two wind speeds, drag, torque, and rpm, with data segmented into 0.1 s intervals and averaged across each interval.

Figure 4. Wind tunnel at Power Systems Synthesis Laboratory.

Table 3. Measurement Devices.

Measurement	Device	Model Name/No	Company, Country
Wind Speed Static	Differential Pressure Transducer	HD408T	Delta Ohm (Senseca) Padua, Italy
Wind Speed Pitot	Differential Pressure Transducer	HD408T	Delta Ohm (Senseca) Padua, Italy
Torque	Rotating Torquemeter	DR2112L	SCAIME, Haute-Savoie, France
Rotational Velocity	Rotating Torquemeter	DR2112L	SCAIME, Haute-Savoie, France
Drag	Load Cell	SP4MC6MR	HBM (now HBK), Darmstadt, Germany
DAQ	Multifunction Data Acquisition Card	NI-USB-6353	National Instruments, Austin, TX, USA

A significant change in the setup involved the rotor shaft's connection to a small generator, which was linked to a Vishay wirewound rheostat ranging from 0 to 1 k Ohm. This new configuration served as a load on the generator wires, effectively acting as an electromagnetic brake. This modification enabled easy adjustments of the braking load applied to the rotor, facilitating detailed investigations into the generator's performance under varied load conditions.

To ensure that our experimental setup remained robust and reliable and the measurements accurate, each measurement device underwent annual calibration. The calibration procedures involved:

- An externally calibrated hotwire anemometer for wind speeds (already mentioned before).
- Known weights for calibrating the load cell used for measuring drag.
- Standard weights for torque calibration.
- A tachometer to verify the rpm measurements.

3. Results for GFRP Blades

3.1. Geometry for GFRP Blades

Table 4 presents the geometrical characteristics of the GFRP blade (iteration no. 010). As mentioned before, the design starts from a standard straight blade, and an initial eccentricity is applied to the aerodynamical centers of the blade. Then, the aerodynamic loads are applied and a structural analysis is performed in order to determine the torsional angle. Then, adjustments are made on the blade's chord length and eccentricity. Note that the pitch angle that is reported is due to the way it is installed on the rotor (using a wedge with the appropriate), and also note that the radius of the segments is reported from the center of the rotor. The geometry adjustments/optimization procedure currently are human driven using engineering intuition. After several iteration steps (optimization, manufacturing, and testing), the geometry that was tested is presented in Table 4.

The curved blade geometry is presented in Figure 5 (front view and side view) and Figure 6 (side view), and it was expected to develop a twist of 8 degrees at the tip under a wind speed of 12 m/s. In Figure 6, the eccentricity of the blade and the chord lengths can be seen.

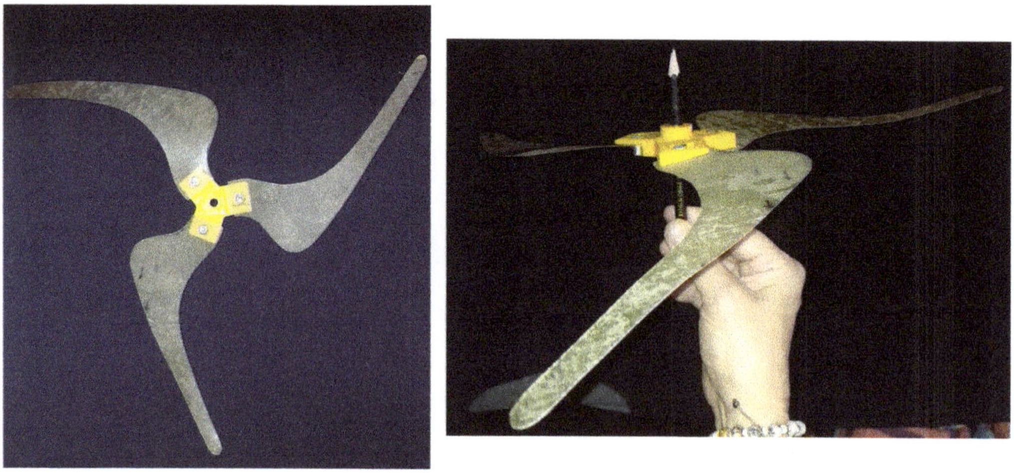

Figure 5. Curved blades assembled on rotor (number 010): front view (**left**) and side view (**right**).

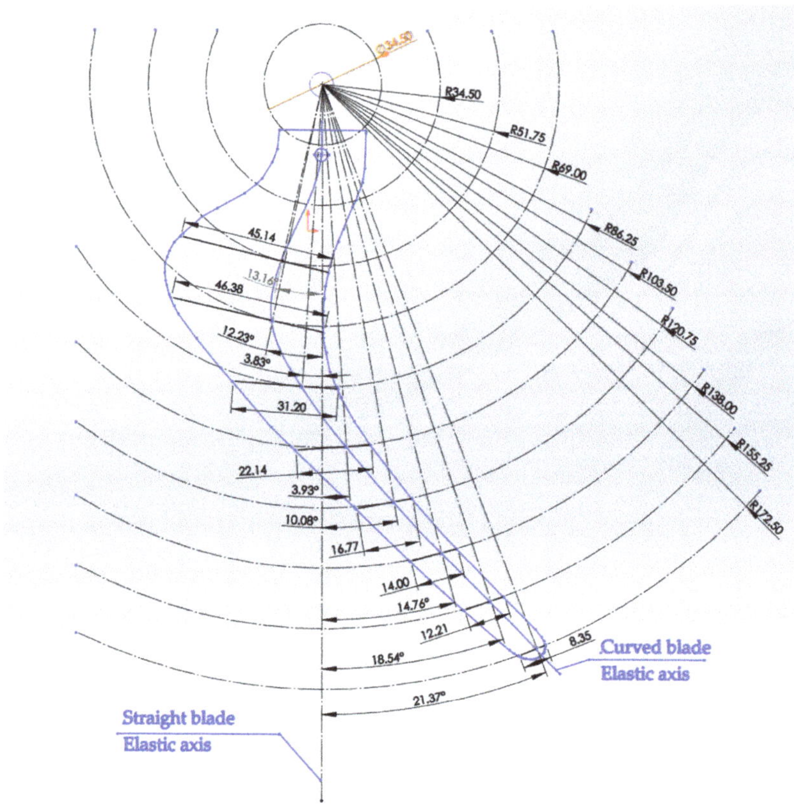

Figure 6. Curved blade geometry (Mk010) with eccentricity details.

Table 4. Geometric characteristics of GFRP blade 010 (center is the rotor hub center).

Seg. No.	Radius [mm]	Chord [mm]	Eccentricity from Elastic Axis [mm]	Eccentricity from Rotor Center [deg]	Pitch Angle [deg]
1	17.25				5
2	34.50				5
3	51.75	45.14	−11.78	−13.2	5
4	69.00	46.38	−14.62	−12.2	5
5	86.25	31.20	−5.61	−3.8	5
6	103.50	22.08	7.09	3.9	5
7	120.75	16.77	21.14	10.1	5
8	138.00	14.00	35.17	14.8	5
9	155.25	12.22	49.37	18.5	5
10	172.50	8.35	62.85	21.4	5

3.2. *Increasing Velocity (VV) Measurements*

3.2.1. Typical Behavior

Figure 7 presents an automated plot summary from a curved GFRP blade measurement with increasing velocity, used for preliminary data validation. The top graph displays wind tunnel velocity, as measured by the Pitot–Prandtl Tube, which varied from 5 to 24 m/s during this test. The second graph illustrates two variables over time: revolutions per minute (rpm), represented in blue on the primary axis, and mechanical power, depicted in red on the secondary axis. The rpms begin to increase abruptly at 5 m/s, reaching nearly 80% of the maximum value, and continues to rise slowly until the cutoff speed at 22 m/s, where it then drops sharply to zero. Correspondingly, mechanical power, calculated as the product of mechanical torque and angular velocity, increases in tandem with the rpm. The constancy of mechanical power between 5 and 22 m/s suggests that the torque remains approximately stable, as both rpm and power plateau.

The third graph provides a detailed analysis by plotting each power data point against the corresponding wind speeds measured by the Pitot–Prandtl tube, offering insights into the blade's mechanical performance. Notably, this graph shows that beyond 15 m/s, the mechanical power appears to plateau completely, indicating the activation of the passive control features of the flexible curved blade concept.

The final graph displays the Power Coefficient (C_p) against the calculated Tip–Speed Ratio (TSR), derived from wind speed measurements taken by the Pitot–Prandtl tube This graph reveals that the TSR extends beyond 8, and the C_p curve increases non-linearly, following a power-law relationship with TSR. Notably, the higher values of TSR are linked to low wind speeds, arising because the curved GFRP blade starts rotating from rest and rapidly accelerates to 80% of its maximum rpm within a narrow range of wind speeds, leading to observable variability at these high TSR values. It is important to acknowledge that while comparisons of C_p against TSR are typically useful for evaluating different wind blade designs, the results obtained from this study using a blade with a rectangular cross-section cannot be quantitatively compared to those from blades with aerodynamically efficient profiles. Therefore, these plots should be primarily utilized for qualitative analysis.

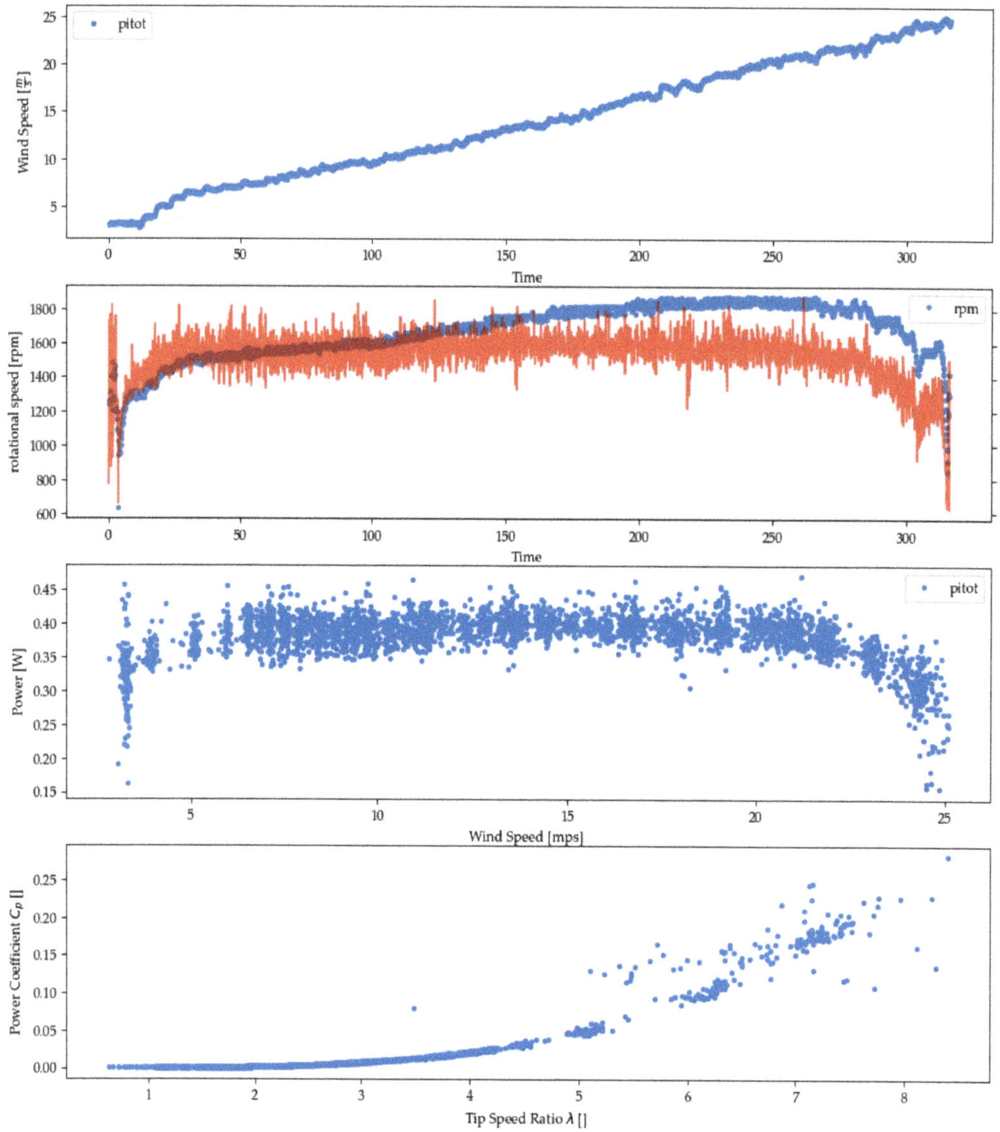

Figure 7. Example of automated plot summary for experiment id 58 (curved blade–GFRP material–vv type–brake resistive load: 4 Ohm).

3.2.2. Power Output Control

Figure 8 displays the mechanical power measurements against wind speed velocity for all increasing velocity tests of the curved GFRP blades, with data points color-coded by brake settings (as resistive loads) and distinct shapes representing different experiment IDs. This figure is pivotal as it illustrates the behavior of the rotor across a range of wind speeds, providing a crucial assessment of how well the curved blade adheres to the established design criteria.

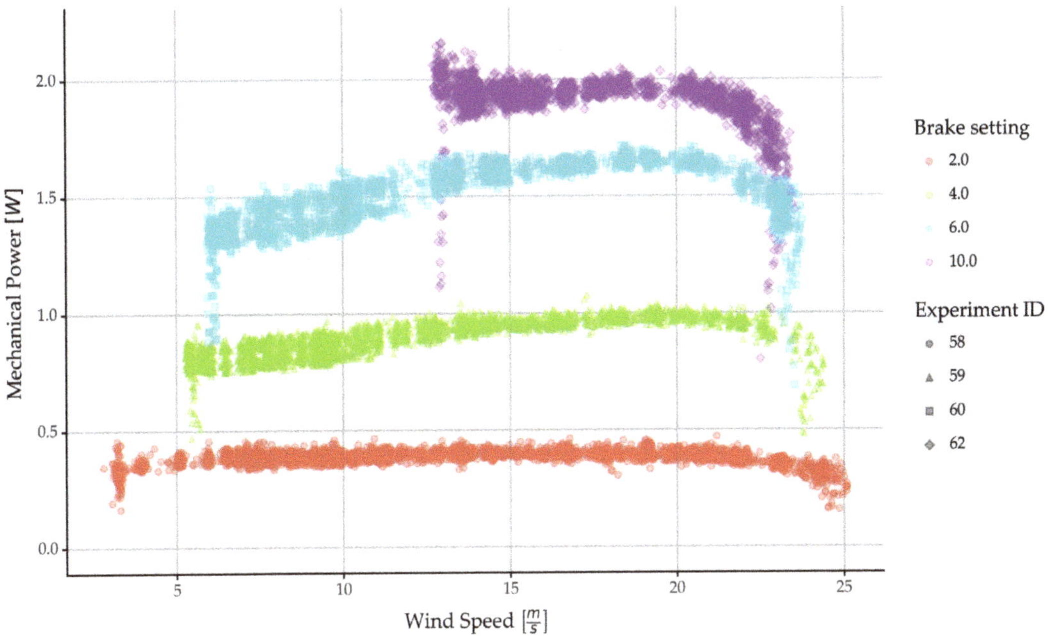

Figure 8. Mechanical power of curved GFRP blade vs. the wind speed (with different load settings).

The graph shows that experiments with higher brake settings result in the commencement of blade rotation at higher wind speeds, which is consistent with observations made with both straight and curved isotropic rotors in the first part of this series. For example, a brake setting of 10 correlates with starting velocities above 13 m/s. Notably, once the rotor starts rotating, a sharp increase in power is typically observed. This behavior indicates that the rotor equipped with the flexible curved blade concept requires overcoming some initial inertia under high load conditions but rapidly achieves nominal power (and rpm) levels once rotation begins, supporting the design criterion for quick startup.

Additionally, the figure demonstrates that with higher brake settings, higher power outputs are observed, suggesting that the rotor can effectively reach and maintain its power output under varying load conditions. Between high speeds, and specifically between the nominal and the cutoff speeds, power output ranges from 70% to 100% of the maximum recorded, indicating that the blade's power output remains stable and predictable, meeting another key design criterion.

A critical observation related to safety is how the blade behaves at cutoff speeds. The graph indicates that beyond certain wind speed thresholds, the rotor's rpm decrease sharply, demonstrating the curved GFRP blade's ability to halt operation under excessive wind conditions to prevent structural damage, thus fulfilling the design criterion for initiating stalling mechanisms at safe operational thresholds. Additionally, it is observed that the cutoff speed increases with decreasing load, likely due to the lower torque values developed by the blade under these conditions.

Overall, Figure 8 is essential for verifying the curved GFRP blade's performance against design expectations, demonstrating its capability to respond effectively to varying wind conditions, manage power output efficiently, and ensure operational safety through appropriate stalling mechanisms.

3.2.3. Rotational Velocity Control

Figure 9 presents the rotational velocity (measured in rpm) of the curved blade against wind speed. This plot demonstrates that the rpm of the curved GFRP blade follow a consistent pattern across all brake settings, with only minor reductions in maximum rpm as the brake load increases, slightly altering the startup and stopping behaviors.

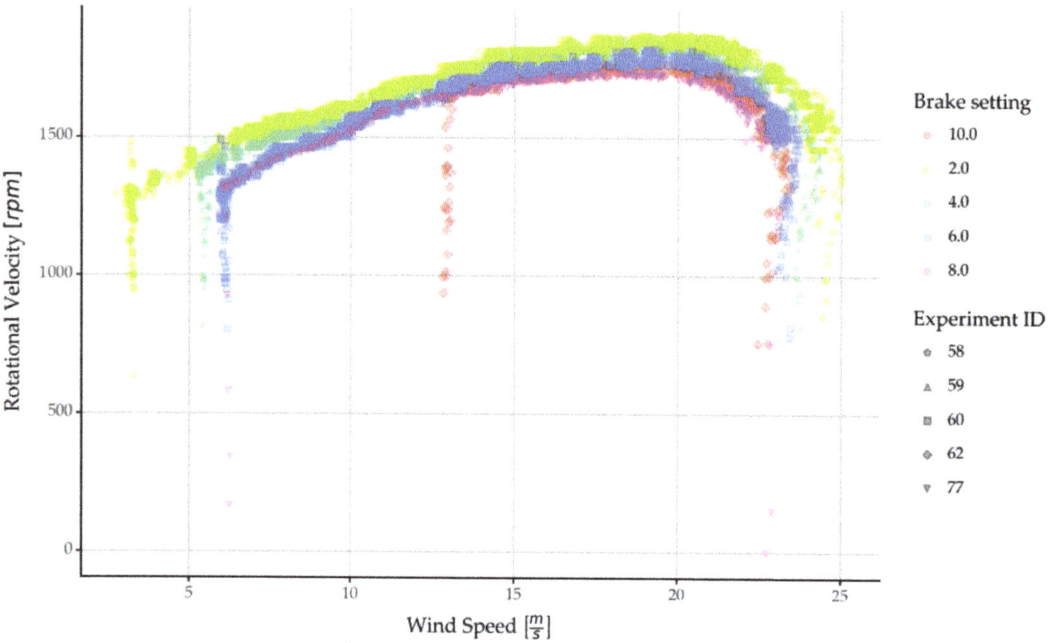

Figure 9. Rotation velocity of blade vs. the wind speed during test (with and without brake).

More specifically, for brake settings ranging from 2 to 8, the rpm increase sharply between wind speeds of 4 and 6 m/s. At the highest brake setting of 10, the rpm do not begin to increase until the wind speed reaches 13 m/s (which also explains why higher brake settings were not utilized).

Additionally, the maximum rpm decrease as the brake load decreases, asymptotically approaching a minimum. This minimum value of maximum rpm is observed at wind speeds of 8 and 10 m/s.

Overall, this data indicate that the rotor equipped with the curved GFRP blade behaves in a controlled and predictable manner regarding rpm, regardless of the brake setting within a certain range of wind speeds. This consistent performance suggests that the desired control features of a WECS can be effectively achieved through the curved GFRP blade design.

Figure 10 depicts the Power Coefficient (C_p) versus Tip–Speed Ratio (TSR) for all increasing velocity tests of the curved blade across different brake settings. Data points are color-coded by brake setting and are differentiated by distinct shapes representing various experiment UIDs. While all brake settings exhibit a trend of increasing C_p with increasing TSR, significant differences are evident among the settings.

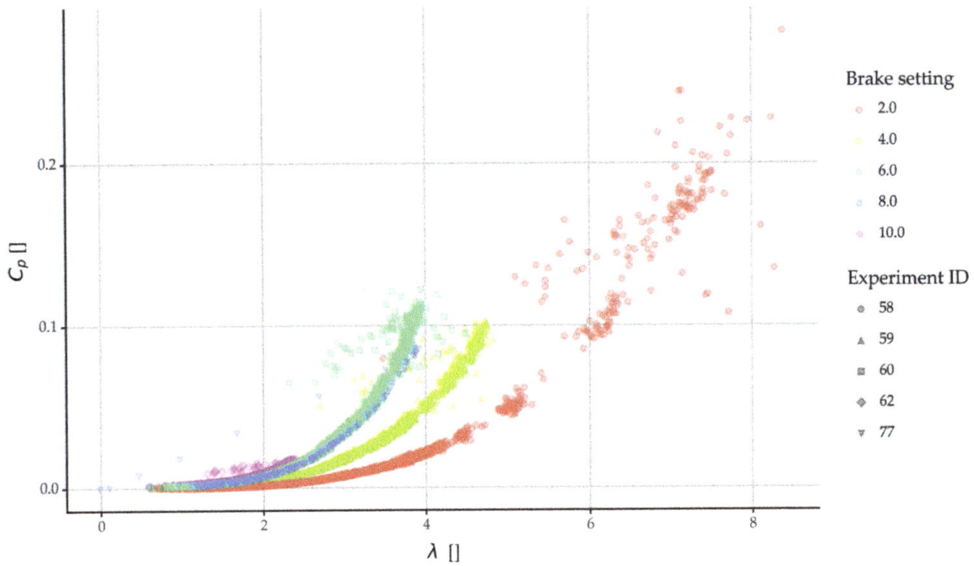

Figure 10. Coefficient of Power with respect to Tip Speed Ratio for curved blade measurements.

Specifically, higher brake settings correspond to lower TSR values and decreased C_p values. This correlation can be attributed to the rapid increase in rpm during startup, which achieves a high proportion of the maximum rpm at very low wind speeds. Due to the rpm increasing quickly within a short timeframe—and thus within a narrow range of wind speeds—the Tip Speed Ratio, defined as $\left(\lambda = \frac{rotational\ velocity}{wind\ speed}\right)$, experiences a fast-changing numerator (which grows to values of 80% of the max rotational velocity) with a relatively stable denominator (which is in the order of 20% of the maximum wind speed). Consequently, TSR values are considerably higher compared to those seen in conventional wind turbines.

Additionally, lower braking settings are associated with higher maximum TSR and C_p values. This is because, under lower loads, the rotor starts to rotate at lower wind speeds, enabling a quicker ramp-up in rotational velocity.

It is important to note that the findings from this plot should primarily be interpreted qualitatively rather than quantitatively. This advisement stems from the fact that such plots are generally more informative under constant velocity conditions, which will be discussed in further detail later in the analysis. This context is crucial for understanding the behavior of the curved blade under variable operational conditions.

In conclusion, the data from this graph further suggest that the curved GFRP blade concept being tested exhibits performance characteristics similar to those found in Wind Energy Conversion Systems (WECSs) with active control systems. Specifically, the rotor equipped with this curved blade design demonstrates a quick startup and maintains relatively constant rpm across a broad range of wind speeds. These findings underscore the potential of the curved blade design to effectively mimic the dynamic response and control typical of actively managed systems, providing a strong foundation for further exploration and development of this technology.

3.3. Constant Velocity (RR) Measurements

3.3.1. Typical Behavior of Constant Velocity (RR) Measurements

Figures 11 and 12 display two automated plot summaries side by side from constant velocity (RR) measurements for experiments with IDs 50 and 69. The key distinction between these plots is that the first graph (exp_id: 50) uses a ramp increase in resistive load, while the right graph (exp_id: 69) uses a step increase in resistive load. These plots serve as tools for preliminary data validation and are used here to underscore the differences in behavior between ramp load and step load configurations.

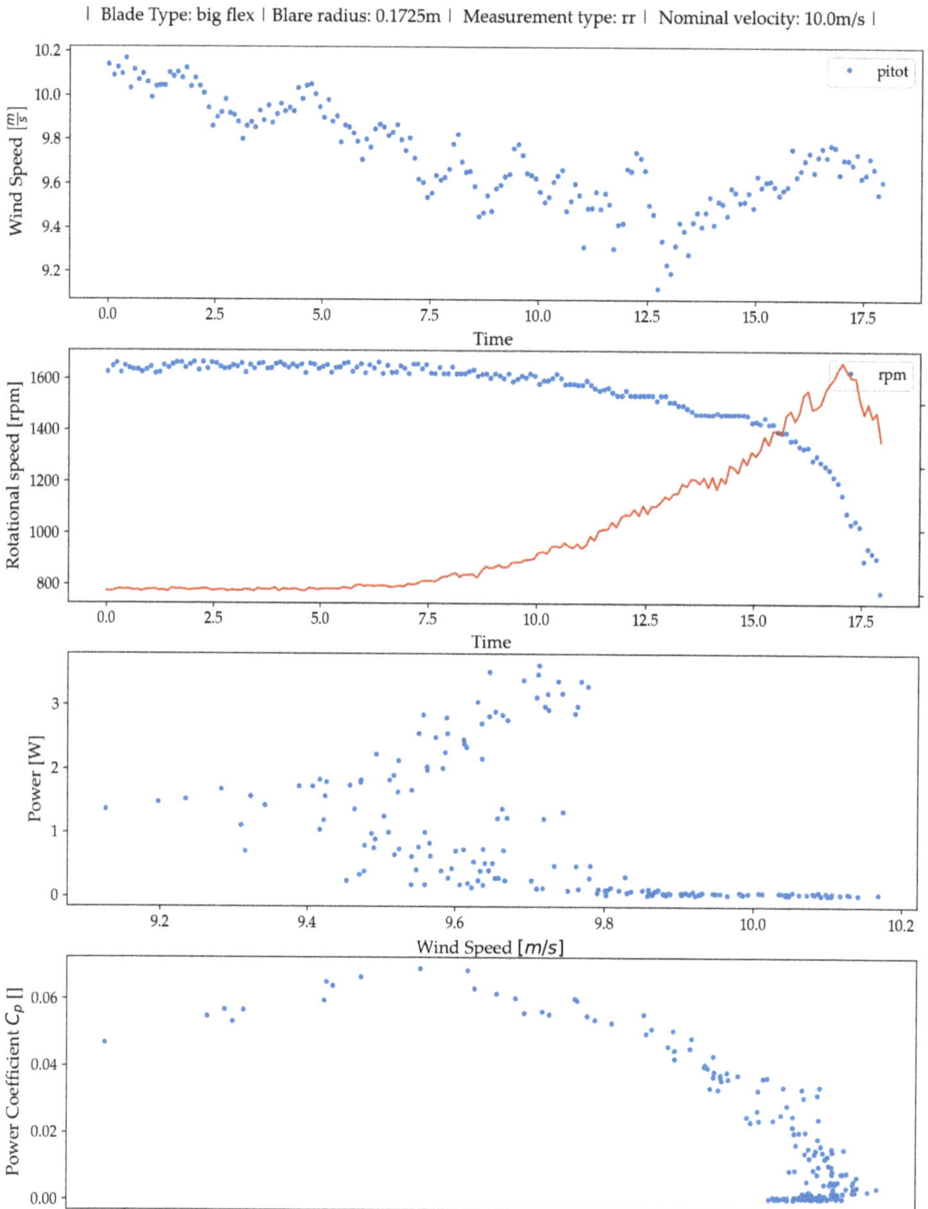

Figure 11. Example of automated plot summary for ramp load: exp_id 50 (flexible blade–GFRP material–rr type–10 m/s–ramp load).

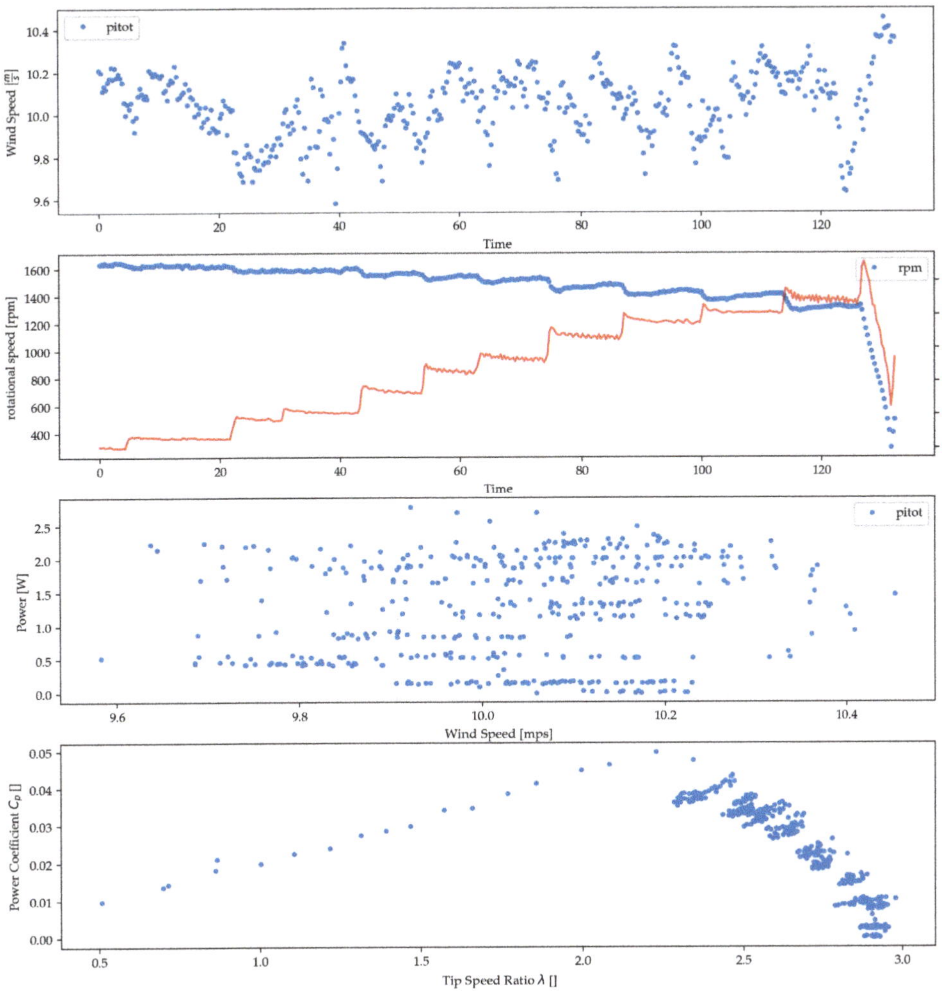

Figure 12. Example of automated plot summary step load for exp_id 69 (curved blade–GFRP material–rr type–10 m/s–step load).

Starting from the top plot, both graphs display wind tunnel velocity from the Pitot–Prandtl tube, which remains around 10 m/s throughout the tests. This velocity shows the typical variability inherent in this type of testing, with no significant differences between the two experiments.

The second graph from the top illustrates the rotational velocity (rpm) on the primary axis and mechanical power on the secondary axis. Here, the impact of the different loading strategies becomes apparent. In the left graph (ramp load), as the brake load gradually increases, the rotational velocity begins to drop, while mechanical power rises until it reaches a critical point where both power and rpm start to decline. In contrast, the right graph (step load) shows that the mechanical power increases in steps, with rpm exhibiting a corresponding stepped decrease.

The third graph, which plots each mechanical power data point against different wind speeds, offers limited informational value due to the narrow range of wind speeds tested.

The final graph at the bottom of the figure depicts the Power Coefficient (C_P) against the calculated Tip–Speed Ratio (TSR), with both scenarios showing a TSR range from 0 to 3. In this graph, the differences between the ramp and step methods are more pronounced. Specifically, in the ramp load scenario, C_P values appear to drop from the maximum towards zero in a less structured manner, whereas in the step load scenario, a pattern resembling inclined lines is observable. These lines illustrate the "settling" of rpm following an abrupt step change in the load. This visualization provides clearer insight into the blade's response dynamics, particularly demonstrating that while the two methods show similar results in the latter half of the TSR range (from the maximum C_p value to the maximum TSR), the step load approach offers more descriptive results in the initial half of the TSR range.

3.3.2. Power Coefficient vs. Tip–Speed Ratio Plots

Figure 13 shows the Power Coefficient (C_p) versus Tip–Speed Ratio (TSR) for all constant velocity tests of the curved blade, with data points color-coded by nominal velocity and grouped by the load type (ramp or step). The black dashed lines are automatically calculated through the Locally Weighted Scatterplot Smoothing (LOWESS) local regression algorithm.

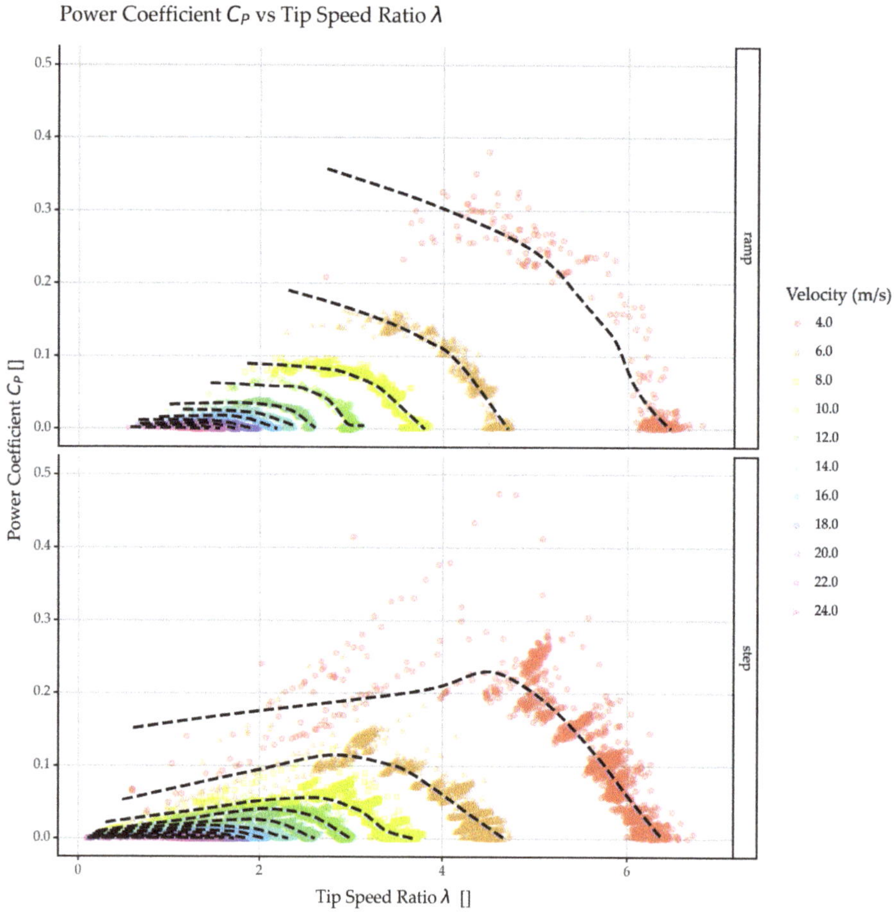

Figure 13. Power Coefficient vs. Tip–Speed Ratio for different velocities of the curved GFRP blade.

A significant observation from this figure is the distinct difference between the ramp and step responses across the TSR range from zero to the maximum value. The step response, which aligns more closely with expected C_P vs. TSR diagrams, was preferred for its repeatability and clearer data patterns. Beyond the maximum TSR value, the responses between the two loading types appear similar. This difference in behavior can be attributed to the stepped response providing more time for the rotor hub to settle into a new rpm equilibrium and stabilize, allowing for more reliable data collection. Conversely, in the ramped load scenario, the rotational velocity decrease is too rapid, resulting in insufficient data capture.

Another prominent feature is that although the maximum C_P values are low compared to conventional wind turbine blade designs, they reach appreciably high levels of about 0.2, particularly at lower wind speeds, which is slightly over one-third of the Betz Limit. The decrease in C_P values with increasing wind speed indicates that less energy is being harvested relative to the available energy in the wind. This pattern, similar to observations in previous sections of this work, underscores the curved GFRP blade concept's ability to initiate rotor hub rotation early and to control and stabilize power output at a nominal level. This performance is akin to that of actively pitch-controlled Wind Energy Conversion Systems (WECSs), suggesting that the curved GFRP blade concept effectively meets the design criteria. This capability demonstrates the potential of the curved blade design to function efficiently in variable wind conditions, emulating some advantages of active control systems while employing passive mechanisms.

3.3.3. Measurement Repeatability

Figure 14 presents displays the Power Coefficient (C_P) versus Tip–Speed Ratio (TSR) for all experiments at a nominal wind speed of 10 m/s, grouped by load type. This graph gives crucial insights for the evaluation of the measurement repeatability. The measurements depicted involve a stepped increase in load, manually adjusted by an operator, which could potentially introduce variability. Despite these challenges, the graph demonstrates that the measurements for the stepped load increases display commendable repeatability across different days, even without a strictly defined protocol. This consistency lends credibility to the reliability of the results.

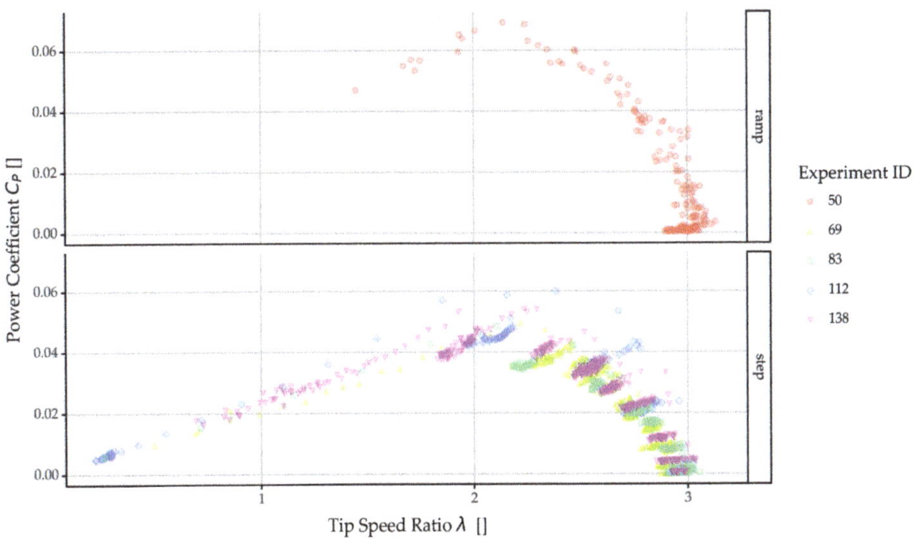

Figure 14. Curved GFPR blade's Power Coefficient vs. Tip–Speed Ratio curve at 10 m/s wind speed for different measurements.

When comparing these results to those obtained under ramped load conditions, the step case typically shows slightly lower C_p values for the same TSR values in the higher TSR range. However, in the lower TSR value range, due to the greater number of measurements taken in the step load case, it provides a more detailed depiction of behavior. This enhanced data collection under step load conditions better illustrates the turbine's performance across a broader range of operational scenarios, underscoring the effectiveness of this measurement approach in capturing the blade performance.

4. Discussion

4.1. General Performance of the GFRP Blade Concept

The performance of the curved, passively controlled GFRP blades developed in this study under a wide range of wind speeds is presented in Figure 15. Three distinct operational regions can be observed: (a) Initial (Buildup) Region, (b) the Second (Plateau) Region, and (c) the Tertiary (Braking) Region.

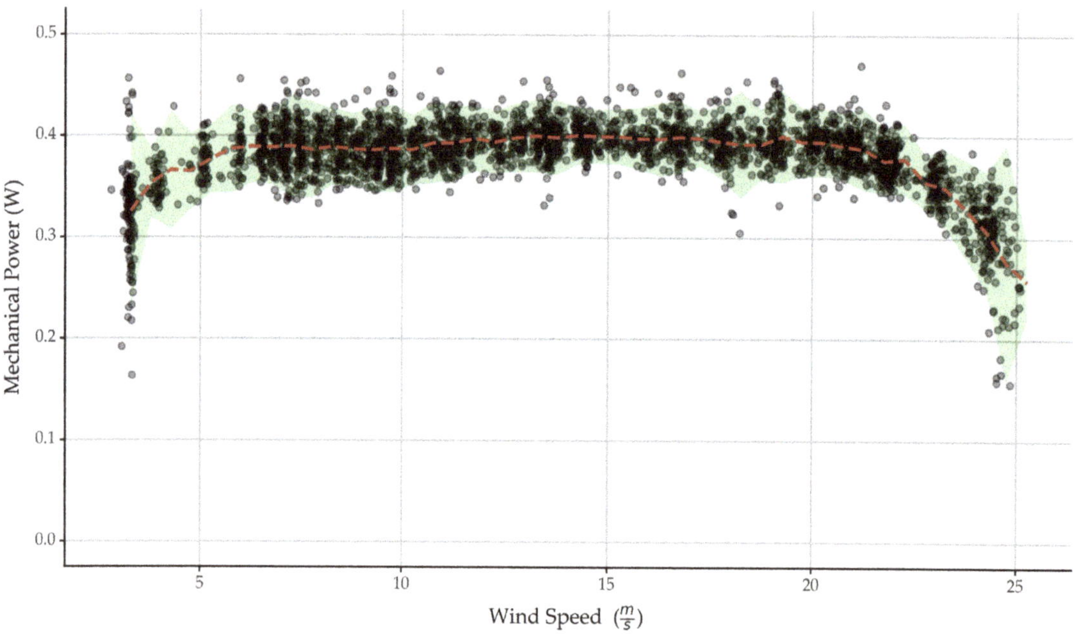

Figure 15. Experiment ID 58 power output vs. wind speed curve at 10 m/s for the passive rotor with diameter 345 mm, 3 layers GFRE, with $\pm 1\sigma$ confidence intervals.

The Initial (Buildup) Region as depicted in Figure 15 spans from 0 to 4 m/s. The curve showing power output versus wind speed begins at zero and rises sharply. The minimal twisting deformations during the early phase are crucial for maximizing torque, which aids in initiating blade rotation. As the curve approaches nominal wind speeds (around 5 m/s in this test), there is a noticeable stabilization in power output. This corresponds to the blade's design feature where the twisting deformation starts to self-regulate. The gradual change in pitch, influenced by the material's elastic properties and blade geometry, modulates the power effectively, preventing any abrupt spikes or drops in energy production as the blade transitions to higher operational speeds. This illustrates how the curved GFRP material and blade design contribute to a controlled and efficient ramp-up in power, aligning with the design criteria for quick startup and efficient energy capture at nominal speeds.

Second (Plateau) Region: This region extends from the nominal wind velocity (5 m/s) to near the cutoff wind velocity (20 m/s). Here, the pitch angle adjusts in response to increases in wind velocity. As the wind velocity exceeds a desirable level, the pitch angle decreases to deepen the blade's entry into the stall region, thereby reducing power yield. If the power output falls below the nominal level, the pitch angle increases to raise the incidence angle, allowing the blade to gradually exit the stall region. This mechanism is designed to maintain a stable, flat power output curve, which can be observed in the figure.

Tertiary (Braking) Region: Beyond the cutoff wind velocity, this region aims to safely halt the turbine rotation. When the wind velocity exceeds the cutoff threshold, the pitch angle changes significantly, forcing the blade into a deep stall. This action effectively reduces the turbine's rotation until it stops completely. Conversely, if the wind velocity drops below the cutoff threshold, the pitch angle increases, which raises the incidence angle and transitions the blade back into the Second (Plateau) Region to optimize power output.

It should be noted that the actual power output values are dependent on the load setting, as can be seen in Figure 8.

The analysis confirmed that the curved GFRP blade design, despite its simple planar geometry, successfully met all the established design criteria: quick start-up and rapid achievement of nominal power, maintenance of stable power output between nominal and cutoff velocities, and secure shutdown beyond cutoff speeds. This demonstration affirms the curved concept's full capability to meet the functional criteria outlined earlier. This success underscores the viability of designing and manufacturing passively controlled wind turbine blades that could potentially incorporate all the advantages of active pitch and stall control systems found in Wind Energy Converter Systems but with the added benefits of simplicity and cost efficiency inherent in passive systems. While the initial results are promising, there are several considerations for scaling up, which could affect the dynamics and forces involved.

4.2. Resistive Load Setup

To address the limitations identified with the mechanical brake system in the initial study—specifically, its non-linear response characteristics and slow adjustment times which hindered effective power optimization—this study introduced a new approach for applying load to the wind turbine rotor. By integrating a generator connected to a resistive load via a rheostat, we aimed to refine the wind turbine blade testing methodology. The mechanical brake previously used had settings that did not correspond to measurable physical quantities that could be adjusted rapidly, making it challenging to track the optimal power point effectively. The shift to a resistive load was driven by the need for a method that offered more linear and swiftly adjustable load application. Although this innovative approach enhanced the clarity of load-response relationships, the setup still encountered several challenges that affected its capacity to precisely achieve optimal power tracking.

Firstly, the resistive load setup allowed for quick adjustments but did not support real-time digital control, crucial for tracing optimal power values dynamically as wind conditions change. This limitation persisted despite the shift from a mechanical braking system, indicating that further enhancements are necessary to fully capitalize on the potential of this new method.

The load was applied using two distinct methods: a continuous, gradual increase (ramp) and discrete, incremental steps (step). The step-wise method proved particularly effective, as it facilitated more consistent stabilization of the rotor hub's rotational velocity, enhancing data reliability, especially at lower Tip–Speed Ratio (TSR) values. This method also enabled more precise control over experimental conditions, allowing for detailed performance analysis across a broad operational range.

Despite these advancements, the method of manually adjusting the stepped increases by an operator introduced potential variability. However, the results demonstrated commendable repeatability across different testing days, which reinforces the reliability of the findings despite the lack of a strictly defined protocol. Comparatively, the step load conditions typically showed slightly lower C_p values for the same TSR values in the higher

TSR range (see Figure 14); however, it offered a richer depiction of behavior in the lower TSR range due to a greater number of measurements.

Moreover, while the C_p values obtained under ramped load conditions were generally higher, the clarity and control provided by the step approach made it more favorable for consistent data collection. This suggests that while the new loading method has significantly improved the experimental setup, ongoing adjustments and development are required to optimize this system for real-time power optimization and to further enhance the fidelity of data collection for wind turbine performance assessments.

5. Conclusions

This paper is the second installment in a two-part series investigating a novel flexible curved blade concept for wind turbines. In the first part, we fabricated and tested straight and curved blades with NACA4415 airfoil profiles using isotropic polymeric materials. Although the curved isotropic blades demonstrated some desirable control characteristics like fast starting, keeping power output from increasing uncontrollably and ultimately a power plateau beyond critical wind speeds, they fell short of meeting all design criteria. Notably, the nominal power was reached well beyond the nominal velocity, and the blades failed to safely shutdown at or beyond the cutoff velocity. Furthermore, in the first part of the series, the influence of load settings on power measurements and the mechanical brake's inability to track the maximum power point were identified as significant issues.

Building on these findings, this work aimed to explore the potential of adjusting the GFRP orientation to enhance the blade's flexibility, tailor its aerodynamic performance, and meet specific design criteria: quick start-up and rise to nominal power, stable power output between nominal and cutoff velocities, and safe shutdown beyond cutoff speeds. A secondary goal was to explore a different measurement setup that might enable optimal power yield to be obtained at different operating conditions.

To this end, identical GFRP blades with three layers oriented at 20 degrees relative to the blade's longitudinal axis were fabricated from flat sheets of resin-impregnated three-layered GFRP. Each blade measured approximately 155 mm in length and was assembled onto a rotor hub with a final diameter of 355 mm, set at a fixed pitch angle of 5 degrees. This setup was then tested in a wind tunnel.

The methodology employed in this study to arrive at the optimal geometry and layup was labor-intensive and largely intuitive, reflecting the developmental stage of our simulation tools. The process was iterative, with each test cycle refining the blade design and setup based on the previous results. These experiences underline the need for further development of our tools and methods, which will be covered in subsequent publications.

Among various GFRP blades with different geometries and curvatures that were fabricated and tested, results are reported here for blade number 010. The analysis demonstrated that the curved GFRP blade design successfully met all the prescribed design criteria, showcasing that the curved concept is capable of achieving in full the functional criteria that were reported earlier. This success underscores the feasibility of designing and manufacturing passively controlled wind turbine blades that could potentially offer all the benefits of active pitch and stall control Wind Energy Converter Systems but with the simplicity and cost-efficiency of passive systems.

Despite the positive outcomes, this work primarily involved planar geometry as the blade was flat, being cut from a GFRP sheet. The achievement of design criteria with this simple geometry suggests that similar results might be feasible with more aerodynamically efficient cross-sectional profiles, warranting further investigation.

Moreover, a new approach to load application was explored by connecting the rotor shaft to a generator and employing a resistive load via a rheostat. This method provided a clearer and more linear relationship to load application compared to the mechanical brake used previously. Although this approach facilitated quick load adjustments, it did not allow (at least in the first iteration of our setup) for real-time digital control necessary for optimal power point tracking.

Two methods for applying load were explored in this study: a ramp method, where load is increased continuously and gradually, and a step method, where load is applied in discrete increments. Notably, the step-wise method was found to be more effective than the ramped method. It allowed the rotor hub's rotational velocity to stabilize more consistently, which significantly improved data collection, particularly at lower Tip–Speed Ratio (TSR) values. Finally, the Power Coefficient (C_P) values obtained from the ramped approach were generally slightly higher than those from the step approach.

Future Work

Further work in this study will focus on several key areas to enhance the understanding and application of curved wind turbine blade technology. First, there is a need to develop and refine the computational code used for designing and simulating the blades. More detailed presentations of the code will be provided, alongside comparisons between predicted and actual performance metrics. Consideration will also be given to making the code open source, which would facilitate broader collaboration and innovation in the field.

The investigation will continue with an emphasis on the flexible curved blade concept using Glass-Fiber-Reinforced Polymer (GFRP)-oriented materials. Future studies will detail the manufacturing processes, material orientation techniques, and resultant aerodynamic performances of these blades. Comprehensive results from these investigations will be reported to assess the viability and efficiency of the GFRP blades in practical applications.

Another significant aspect of upcoming research will involve a thorough analysis of the thrust/drag forces and the dynamic loading on both the rotor and the blade. There is also an imperative to quantify the flutter and dynamic behavior of the blades under various operational conditions. Although preliminary data are available, detailed studies are required to fully understand these dynamics and implement necessary design adjustments.

Scaling up the design from laboratory or small-scale models to larger, potentially commercial-sized applications is another critical area of focus. This scaling will help in understanding the challenges and adjustments needed for larger turbine implementations.

Additionally, improvements in the experimental setup are planned, particularly the inclusion of a Maximum Power Point Tracking (MPPT) system. Integrating MPPT technology will enhance the capability to precisely find the optimal operational conditions for the turbine systems, thereby maximizing efficiency and output under varying environmental conditions.

Together, these efforts will not only deepen the understanding of flexible wind turbine blades but also push forward the boundaries of renewable energy technology, making it more adaptable, efficient, and applicable on a larger scale.

Author Contributions: Conceptualization, C.C.; methodology, N.P.; software, N.P.; validation, N.P.; investigation, C.C.; resources, C.C.; data curation, N.P.; writing—original draft preparation, N.P.; writing—review and editing, N.P.; visualization, N.P.; supervision, C.C.; project administration, C.C. All authors have read and agreed to the published version of the manuscript.

Funding: This research received no external funding.

Data Availability Statement: Dataset available on request from the authors.

Conflicts of Interest: The authors declare no conflict of interest.

Abbreviations

C_P	Power Coefficient
MPPT	Maximum Power Point Tracking
rpm	Revolutions Per Minute
TSR	Tip–Speed Ratio
WECS	Wind Energy Converter Systems

References

1. Ponta, F.L.; Otero, A.D.; Rajan, A.; Lago, L.I. The Adaptive-Blade Concept in Wind-Power Applications. *Energy Sustain. Dev.* **2014**, *22*, 3–12. [CrossRef]
2. Liebst, B.S. Wind Turbine Gust Load Alleviation Utilizing Curved Blades. *J. Propuls. Power* **1986**, *2*, 371–377. [CrossRef]
3. Infield, D.G.; Feuchtwang, J.B. Design Criteria for Passive Pitch Control of Wind Turbines Using Self-Twisting Blades. *Int. J. Ambient Energy* **1995**, *16*, 139–146. [CrossRef]
4. Infield, D.G.; Feuchtwang, J.B.; Fitches, P. Development and Testing of a Novel Self-Twisting Wind Turbine Rotor. In *1999 European Wind Energy Conference*; Routledge: Abingdon, UK, 1999; ISBN 978-1-315-07433-7.
5. Lobitz, D.W.; Veers, P.S.; Migliore, P.G. Enhanced Performance of HAWTs Using Adaptive Blades. In Proceedings of the Wind 1996 ASME Wind Energy Symposium, Houston, TX, USA, 29 January–2 February 1996.
6. Lobitz, D.; Veers, P. Aeroelastic Behavior of Twist-Coupled HAWT Blades. In *1998 ASME Wind Energy Symposium*; Aerospace Sciences Meetings; American Institute of Aeronautics and Astronautics: Reston, VA, USA, 1998.
7. Eisler, G.R.; Veers, P.S. *Parameter Optimization Applied to Use of Adaptive Blades on a Variable Speed Wind Turbine*; Sandia National Laboratories: Albuquerque, NM, USA, 1998.
8. Lobitz, D.; Laino, D. Load Mitigation with Twist-Coupled HAWT Blades. In *37th Aerospace Sciences Meeting and Exhibit*; Aerospace Sciences Meetings; American Institute of Aeronautics and Astronautics: Reston, VA, USA, 1999.
9. Lobitz, D.; Veers, P.; Laino, D. Performance of Twist-Coupled Blades on Variable Speed Rotors. In Proceedings of the 2000 ASME Wind Energy Symposium, Reston, VA, USA, 10–13 January 2000; American Institute of Aeronautics and Astronautics: Reston, VA, USA.
10. Zuteck, M. *Adaptive Blade Concept Assessment: Curved Platform Induced Twist Investigation*; Sandia National Lab. (SNL-NM): Albuquerque, NM, USA; Sandia National Lab. (SNL-CA): Livermore, CA, USA, 2002.
11. Larwood, S.; Zutek, M. Swept Wind Turbine Blade Aeroelastic Modeling for Loads and Dynamic Behavior. In Proceedings of the WINDPOWER 2006, Pittsburgh, PA, USA, 4–7 June 2006.
12. Ashwill, T.D. *Sweep-Twist Adaptive Rotor Blade: Final Project Report*; Sandia National Laboratories (SNL): Albuquerque, NM, USA; Livermore, CA, USA, 2010.
13. Karaolis, N.M.; Mussgrove, P.J.; Jeronimidis, G. Active and Passive Aeroelastic Power Control Using Asymmetric Fibre Reinforced Laminates for Wind Turbine Blades. In Proceedings of the 10th British Wind Energy Conference, London, UK, 22 March 1988; Milbrow, D.J., Ed.
14. Karaolis, N.M.; Jeronimidis, G.; Mussgrove, P.J. Composite Wind Turbine Blades: Coupling Effects and Rotor Aerodynamic Performance. In Proceedings of the European Wind Energy Conference, EWEC'89, Glasgow, Scotland, 10–13 July 1989.
15. Chen, J.; Shen, X.; Zhu, X.; Du, Z. Study on Composite Bend-Twist Coupled Wind Turbine Blade for Passive Load Mitigation. *Compos. Struct.* **2019**, *213*, 173–189. [CrossRef]
16. Ong, C.-H.; Tsai, S.W. Design, Manufacture and Testing of A Bend-Twist D-Spar. In Proceedings of the 37th Aerospace Sciences Meeting and Exhibit, Reno, NV, USA, 11–14 January 1999.
17. Papadakis, N.; Condaxakis, C. An Experimental Performance Assessment of a Passively Controlled Wind Turbine Blade Concept: Part A—Isotropic Materials. *Energies* **2024**, *17*, 3049. [CrossRef]
18. Larsen, G.C.; Frandsen, S.T.; Sørensen, P.E.; Courtney, M. *Design Basis for Horizontal-Axis Wind Turbines. Theoretical Background*; U.S. Department of Energy: Washington, DC, USA, 1990.
19. Le Gouriérès, D. *Energie Éolienne: Théorie, Conception et Calcul Pratique Des Installations.*; Eyrolles: Paris, France, 1980.
20. Hansen, M. The Classical Blade Element Momentum Method. In *Aerodynamics of Wind Turbines*, 2nd ed.; Routledge: Abingdon, UK, 2007; ISBN 978-1-84977-040-8.

energies

Article

The Problem of Power Variations in Wind Turbines Operating under Variable Wind Speeds over Time and the Need for Wind Energy Storage Systems

Cristian Paul Chioncel [1], Elisabeta Spunei [1,*] and Gelu-Ovidiu Tirian [2]

1 Department of Engineering Sciences, Faculty of Engineering, Babes-Bolyai University, 400028 Cluj-Napoca, Romania; cristian.chioncel@ubbcluj.ro
2 Faculty of Engineering Hunedoara, Polytechnic University of Timisoara, 300006 Timisoara, Romania; ovidiu.tirian@fih.upt.ro
* Correspondence: elisabeta.spunei@ubbcluj.ro; Tel.: +40-0751-194693

Abstract: One of the most important and efficient sources of green electricity is catching air currents through wind turbine technology. Wind power plants are located in areas where the energy potential of the wind is high but it varies. The time variation of the wind generates fluctuations in the power produced by the wind farms that is injected into the grid. This elevates, depending on the intensity, problems of network stability and the need for balancing energy, thus raising both technical and cost issues. The present paper analyzes the behavior of a wind turbine (WT) over time in varying wind speed conditions, highlighting that without automation algorithms, a WT is far from the operation at the maximum power point (MPP). However, even when it is brought to operate at MPP, there are still significant variations in the power injected into the network. These power variations can be compensated if the wind system has energy storage facilities for the captured wind. All of these assumptions are analyzed using improved mathematical models and processed in simulations, with experimental data used as input from a wind turbine with an installed power of 2.5 [MW] in operation on the Romanian Black Sea coastal area. Consequently, the paper demonstrates that during an operation in the optimal area, from an energy perspective, the wind turbine's maximum power point requires a storage system for the captured wind energy.

Keywords: wind system; wind turbine; time-varying wind speeds; peak power points and power gaps; wind energy storage

1. Introduction

Regarding the energy crisis and socio-economic development correlated with climate change, as well as energy security in a geopolitical context, many countries have initiated the transition from an energy system predominantly dependent on fossil fuels to a diversified and sustainable energy mix [1], with an increasing share of green energy [2].

The purpose of this paper is to analyze the necessity of installing captured wind energy storage facilities in wind systems that operate at variable wind speeds over time [3]. This will ensure that the electroenergetic system continues to operate as intended and that the wind turbine operates at its maximum power point (MPP), at the optimal mechanical angular speed (MAS), and at the optimal power point. Energy storage systems can be categorized based on their storage configuration as local, distributed, or centralized [4]. The optimal configuration of the storage systems must be carried out according to different algorithms so that they ensure the reliability of the system and contribute to the improvement of the use of energy from renewable sources within the energy system [5]. At the same time, the storage facility must be properly sized [6] so as to cover the energy gaps, as well as to minimize costs [7]. Storage pumping plants with large storage capacities are used as storage facilities [8].

In existing systems, the load on the electric generator (P_{EG}) is adjusted based on the wind speed value [9,10] and the optimal speed's dependency on wind speed. The actual wind speed may be different from the speed measured by the sensor on the nacelle, which is why different methods are used to determine the actual wind speed value [11]. In all cases, wind systems must operate optimally from an energy point of view at time-varying wind speeds [12,13]. The wind turbine (WT) must run at the optimal MAS, ω_{OPTIM}, or the optimal rotational speed, n_{OPTIM}, in order to accomplish a maximum capture of wind energy at a given wind speed [14,15]. Thus, the provision of auxiliary services can also be achieved by regulating the frequency [16]. Due to the significant mechanical inertia of the wind system, operating the WT at the MPP and at the MAS, ω_{OPTIM}, is a complicated challenge at time-varying wind speeds [17]. Different control strategies are used with different types of algorithms, such as PI to control the power output [18] and optimal operation according to wind speed [19,20]. The adjustment of the electric generator's load in accordance with the wind speed, v, is the basis for having the WT operating at the ideal mechanical angular speed and the MPP [12].

The function $P_{WT}(\omega)$, which represents the WT power characteristic, reaches its maximum MPP at the optimum mechanical angular speed (ω_{OPTIM})—the reference quantity for the control system [21,22]. When operating at the MPP, the time variation of the wind speed imposes the time variation of the mechanical angular speed. At optimal energy control of the wind system, the time variation of the wind speed must be identical to the time variation of the mechanical angular velocity ω.

Thus, the problems treated in the paper consider the operation of wind turbines at the MPP at variable wind speeds over time. Major difficulties appear in wind systems operating at variable wind speeds, while the specialized literature has addressed these aspects mainly at constant wind speeds. The novelties brought about are: the wind speed being variable over time means that the MPP operating point is not fixed, and in this sense, the authors of this article analyze these aspects; the operation at the MPP is achieved by prescribing the value of the power at the electric generator, and this paper presents the fact that, at wind speeds increasing over time, the power value at the electric generator must decrease, and at wind speeds decreasing over time, the power value at the electric generator must grow.

The main sections of this paper are an overview of the main implemented storage facilities for wind power, an analysis of the wind and mechanical speeds at the generator, the MPP of the operating WT on the power curve at variable wind speeds, the determination of the mathematical model for the WT, the development of an algorithm for determining power losses, discussion, results, perspectives, and the references used.

2. Maximum Capture of Wind Energy and the Need for Its Storage

Obtaining electricity through the extensive growth of wind systems poses major obstacles for the electroenergetic system in the sense that [11]:

1. With time-varying wind speeds, the electrical power provided by wind systems in the national energy system is variable, depending on the wind speed;
2. Due to the high-value equivalent moment of inertia, J, it is necessary to adjust the power of the electric generator in order to ensure that the mechanical angular speed, ω, equals the ω_{OPTIM}, in order to achieve an operation of the WT at the maximum power point and at time-varying wind speeds, in addition to the significantly time-varying WT power [23,24]:

$$\omega = \omega_{OPTIM} = k_V \cdot v \quad (1)$$

At high values of the J and at significant variations in wind speed, as well as at high values of its derivative, the value of the power at the electric generator can even become negative, which requires a switch to a motor regime, in which case there are power gaps in the network.

3. When a WT operates at its maximum power point, electric power fluctuations are produced by:
   - The power generated by WTs is directly proportional to the cube of the wind speed;
   - Inertial power, $P_{INERTIAL}$, is dependent on the wind speed and its derivative:

$$P_{INERTIAL} = J\cdot(d\omega/dt)\cdot\omega = J\cdot(k_V)^3\cdot(dv/dt)\cdot v \quad (2)$$

In cases where the variations in the time of the optimal rotational speeds (optimal mechanical angular speeds) are identical to the variation in the wind speed in time, a maximum wind energy will be captured.

In the utilization of wind systems, the fluctuation of the power injected into the grid is a key consideration. From the perspective of the stability of the network, it is desired that the value of the injected power be as constant as possible [25], even if the value of the wind speed varies over time, which is an apparently unsolvable contradiction. However, wind energy storage facilities can be implemented to solve this issue of wind systems.

Two significant valid storage solutions were identified in the specialized literature [26,27] of the field: storage in batteries and storage in terms of hydraulic power.

In [11], a hydraulic energy storage system is presented, which involves the use of hydro-towers and an asynchronous generator with a wound rotor. The control of the ω is accomplished by adjusting the rotor power of the generator. Consequently, the converter that connects the rotor of the generator to the pumping station operates at a low power level. The nacelle's support pillar is constructed in the form of an upper tank, while the lower tank serves as a reservoir, a lake, or the sea with a significant capacity. The doubly fed induction generator's rotor transfers power to the hydraulic storage system through a permanent magnet synchronous machine, functioning as a motor during pumping and as a generator when converting hydro energy into electricity. The power flow in the rotor circuit is two-way. During periods of high wind power, water is pumped from the lake water source to the hydro tower, where it stores the excess energy. When required, this stored energy is then released back into the lake or ocean.

In [10], a wind system with synchronous generator and pumped storage is presented on the Greek island of Ikaria. The hybrid power plant consists of three hydroelectric units (H1: 1.05 MW, H2-3: 2·1.55 MW) all equipped with Pelton turbines, a pumping station, 8 fixed speed pumps, and 4 variable speed pumps with a variable speed of 0.25 MW each and a 3·0.9 MW wind farm.

In small standalone systems, frequency and voltage control as well as power system stability are difficult problems at significantly time-varying wind speeds. Currently, the pumping station is controlled to track electricity production, compensating for the variations in wind power that can disrupt frequency control. The latter is provided by diesel units, which must be replaced by hydroelectric turbines if a high penetration of renewable energy sources is desired. However, this introduces significant challenges for frequency regulation due to the slow response of hydroelectric units, water column inertia, and wind turbines.

In [8], a classic model of a wind turbine and pumped storage hydropower plant is presented in Tunisia, where the wind power plant has an installed capacity of 200 MW. It is stated that a high penetration of energy from wind sources raises the problem of instability of the Tunisian energy system, caused by the intermittency and fluctuation of wind speed, which varies significantly over time. This has an impact on the frequency and voltage of the system. It should, however, be noted that by using wind energy storage facilities, these deficiencies can be eliminated.

The conducted study assesses the energy efficiency of wind power plants operating at significant time-varying wind speeds, highlighting solutions based on energy storage systems to ensure the desired outcome: high energy efficiency—performing at the MPP and ensuring electrical grid stability.

The study was carried out at time-varying wind speeds and based on the measurement of three fundamental quantities: wind speed, v, power, P_{EG}, and MAS, ω, at the electric generator.

In order for the wind turbine to operate in the MPP at variable wind speeds over time, it is necessary to change the load on the electric generator, depending on the wind speed value [11].

Capturing maximum wind energy at time-varying wind speeds [28] in harsh weather conditions that can cause damage to the system [29,30], and in the absence of a storage system, significantly disrupts the operation of the power system [12], creating instability in the system.

Considering the problem of storing wind energy at time-varying wind speeds, it becomes necessary to operate the wind turbine at the point of maximum power and at the optimal mechanical angular velocity for the following reasons:

- The operation of the electrical system is not affected;
- Compensation for the fluctuations in wind energy is attained at a local or regional level;
- The discrepancy between the need for grid power and the generation of wind power has been resolved;
- At the local level, collaborative coordination allows for the planning of electricity distribution and expansion in isolated systems;
- It is possible to create a flexible electricity source by harnessing short-term kinetic energy and long-term hydroelectric potential energy.

The authors incorporated the mathematical model of the WT into their simulations in Scientific WorkPlace 4.0, being confident that it accurately reflects the power output of the WT, known as the $P_{WT}(\omega,v)$ function [31]. Different forms of the WT power characteristics $P_{WT}(\omega,v)$ [32,33] have been documented in several sources.

The most advantageous case is determined by three parameters, specifically a, b, and c, in the following format [34]:

$$P_{WT}(\omega,v) = \rho\cdot\pi\cdot R_p{}^2\cdot C_p(\lambda)\cdot v^3 = a(V/\omega - b)\cdot e^{-c(v/\omega)}\cdot v^3 \quad (3)$$

where: ρ—air density in the WT operating location, R_p—rotor blade radius, $C_p(\lambda)$—power conversion coefficient, $\lambda = \omega\cdot R_p/v$.

The determination of parameters a, b, and c is accomplished through the measurement of several factors, including v, P_{EG}, and MAS at EG n/ω. By carefully measuring and analyzing these variables, the specific values for a, b, and c can be established. It is imperative to have a thorough grasp of the mathematical model of the wind turbine (MM-WT) [35], which is also dependent on its constructive characteristics [36,37], in order to effectively regulate the system and maintain the turbine's performance at its maximum power capacity [38].

Considering the experimental data obtained from the GEWE-B2.5 series, a 2.5 MW wind turbine (WT) with a rotor diameter of 100 m is used to establish the mathematical model of the WT (MM-WT). This model takes into account an equivalent inertia moment (turbine + gearbox + generator) of $J = 5372.5$ kgm^2 and the nominal power obtained at the generator level, which operates at a rotational speed of 1500 rpm. The WT is currently operational on the Romanian Black Sea coast [9].

The simulation uses a mathematical representation of the WT, shown in the following format:

$$P_{TV}(\omega,v) = 6.5086\cdot 10^5\ (v/\omega - 1.7488\cdot 10^{-2})\cdot e^{-41.495(v/\omega)}\cdot v^3\ [\mathrm{W}] \quad (4)$$

The $P_{WT}(\omega,v)$ is contingent upon the v and the ω. The maximum power output at v is achieved at ω_{OPTIM}:

$$\omega_{OPTIM}(v) = 24.046\cdot v\ [\mathrm{rad/s}] \quad (5)$$

value obtained by canceling the derivative of the power WT:

$$dP_{WT}(\omega,v)/d\omega = 0 \quad (6)$$

The wind speed [12] undergoes fluctuations as time progresses, with the wind system consistently operating within a transient state. These states are assessed through the equation governing kinetic momentum [15]:

$$J \cdot d\omega/dt = M_{WT} - M_{EG} \quad (7)$$

where: the mechanical angular speed is measured at the shaft of the electric generator, EG; the time derivative of the mechanical angular velocity, $d\omega/dt$; the moment related to the shaft of the electric generator, M_{WT}, is determined by the wind turbine; the electromagnetic moment at the EG shaft, M_{EG}.

The power equation is obtained by multiplying the kinetic moment equation with the velocity, ω:

$$J \cdot (d\omega/dt) \cdot \omega = P_{WT} - P_{EG} \quad (8)$$

or with the inertial power:

$$P_{INERTIAL} = P_{WT} - P_{EG} \quad (9)$$

where P_{WT} refers to the useful power provided by the wind turbine, relative to the electric generator shaft, P_{EG} represents the electromagnetic power of the electric generator at the shaft.

By solving the equation for kinetic moment, one can observe the progression of the process [39,40] through dynamic visualization at various rotational speeds. In order to optimize wind energy capture at a specific wind speed, the wind turbine needs to operate at the ω_{OPTIM} [15].

Considering the relationship (1) between the optimal velocity and the wind speed, Equation (1), follows:

$$J \cdot (d\omega_{OPTIM}/dt) \cdot \omega_{OPTIM} = P_{TW\text{-}MAX} - P_{EG\text{-}OPTIM} \quad (10)$$

Thus, the optimal power of the electric generator is obtained:

$$P_{EG\text{-}OPTIM} = P_{WT\text{-}MAX} - J \cdot (k_V)^2 \cdot dv/dt \cdot v = k_P \cdot v^3 - J \cdot (k_V)^2 \cdot dv/dt \cdot v \quad (11)$$

The power fluctuations $P_{EG\text{-}OPTIM}$ are generated by the power of the WT via the term $k_P \cdot v^3$, inertial power $J \cdot (k_V)^2 \cdot dv/dt \cdot v$ by the wind speed v, and its derivative dv/dt.

At high values of J and significant fluctuations in wind speed, particularly in an upward trend where the derivative of wind speed, dv, is also high, the $P_{EG\text{-}OPTIM}$ value diminishes and may even turn negative. These power fluctuations, in the absence of storage facilities, are transmitted in the grid and create instability in the energy system.

At $J = 511.92$ kgm^2, the optimum of electromagnetic power is reached:

$$P_{EG\text{-}OPTIM} = P_{WT\text{-}MAX} - 24.046^2 \cdot 511.92 \cdot dv/dt \cdot v \quad (12)$$

and with:

$$P_{WT\text{-}MAX} - 2792.8 \cdot v^3 \quad (13)$$

result:

$$P_{EG\text{-}OPTIM} = 2792.8 \cdot v^3 - 2.96 \cdot 10^5 \cdot dv/dt \cdot v \quad (14)$$

The real fluctuation of wind velocity is depicted in Figure 1 [8]. Upon closer examination, the wind speed variation during the time intervals A-B and B-C, as given in Figure 1, is analyzed. This analysis aims to determine the optimal power, denoted as $P_{EG\text{-}OPTIM}$, at the electric generator. The objective is to ensure that the wind turbine operates at the point of maximum power.

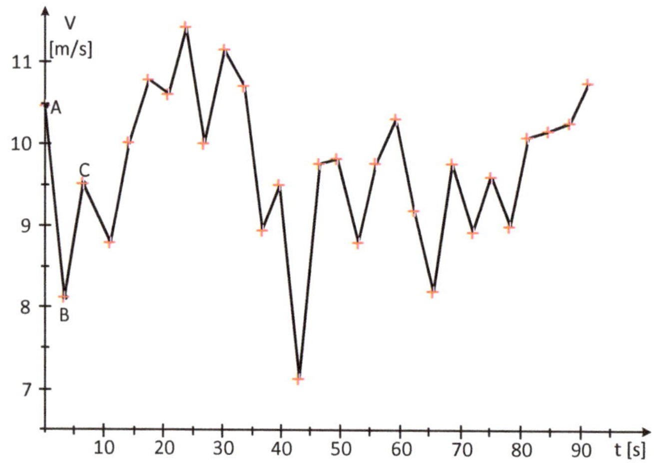

Figure 1. Real variation of wind speed.

The values of wind speed at points A, B, and C are as follows (Table 1):

Table 1. Wind speed at operating points.

Time t [s]	Speed v [m/s]	Point
0	10.47	A
3.433	8.13	B
6.631	9.52	C

A linear change in wind speed between point A and point B yields the following outcome:

$$v(t) = 10.47 + [(8.13 - 10.47)/3.433]\cdot t = 10.47 - 0.732\cdot t \quad (15)$$

The power given by the WT is:

$$P_{TV\text{-}A\text{-}B}(\omega,t) = 650860\cdot[(10.47 - 0.732\cdot t)/\omega - 1.7488\cdot 10^{-2}]\cdot e^{-41.495\cdot[(10.47 - 0.732\cdot t)/\omega]}\cdot[(10.47 - 0.732\cdot t)^3 \quad (16)$$

It is considered that the WT, at point A, does not operate at the point of maximum power, with the value of MAS being:

$$\omega_A = 122\ [\text{rad/s}] < \omega_{opt\text{-}A} = 24.046\cdot 10.47 = 251.76\ [\text{rad/s}] \quad (17)$$

2.1. Maximizing the Power Output of a Wind Turbine by Aligning It with the Maximum Power Point

Achieving the turbine's maximum power point involves accelerating to the optimal MAS, which can be done most efficiently through two methods:

- Disconnecting the EG from the grid (a slower approach);
- Switching to motor operation of the electric generator (a quicker approach).

The EG is considered to be disconnected from the grid and has the MAS of value $\omega(0) = 122$ [rad/s]. From the equation of powers in the form:

$$J\ (d\omega/dt)\cdot\omega = P_{WT}(\omega, t) \quad (18)$$

it is obtained:

$$\begin{cases} 511.92\cdot\frac{d\omega}{dt}\cdot\omega = 650860\cdot\left(\frac{10.47-0.73171\cdot t}{\omega}\right) - 1.7488\cdot 10^{-2} \\ e^{-41.495\left(\frac{10.47-0.73171\cdot t}{\omega}\right)}\cdot(10.47-0.73171\cdot t)^3 \\ \omega(0) = 122 \end{cases} \tag{19}$$

By resolving it, the outcome is the temporal fluctuation of the mechanical angular speed, ω. At time t*, as illustrated in Figure 2, ω—the MAS is measured to have a value of:

$$\omega(3.3) = 197.6\ [\text{rad/s}] \tag{20}$$

which is close to the optimal MAS:

$$\begin{gathered}\omega_{OPTIM}(v) = 24.046\cdot(10.47 - 0.732\cdot t) = \\ 24.046\cdot(10.47 - 0.732\cdot 3.3) = 197.67\ [\text{rad/s}]\end{gathered} \tag{21}$$

At time t = 3.3 [s] when the optimal mechanical angular speed, ω_{OPTIM} and real ω are equal, and the system operates in the turbines maximum power point.

At time t*, the wind speed is measured:

$$v(3.3) = 10.47 + [(8.13 - 10.47)/3.433]\cdot 3.3 = 8.2207\ [\text{m./s.}] \tag{22}$$

and, from the WT power characteristic, $P_{WT}(v, \omega)$ at $\omega(3.3)$:

$$\begin{cases} P = 650860\left(\frac{8.2207}{\omega} - 1.7488\cdot 10^{-2}\right)\cdot e^{-41.495\left(\frac{8.2207}{\omega}\right)}\cdot 8.2207^3 \\ \omega = 197.67 \end{cases} \tag{23}$$

results in power developed by the WT of:

$$P_{WT}(8.2207) = 1.5516\cdot 10^6\ [\text{W}] \tag{24}$$

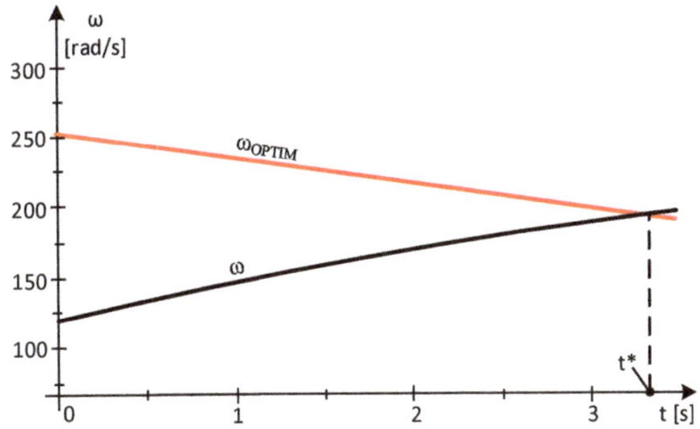

Figure 2. Variations of optimal and actual mechanical angular speeds.

2.2. *Maintaining the Wind System in the Turbine Maximum Power Point*

The P_{EG} at the shaft of the electric generator is derived from the power equation when the generator is operating at its MPP:

$$J(d\omega/dt)\cdot\omega = P_{WT} - P_{EG} \tag{25}$$

The ω_{OPTIM}, during MPP operation, is determined by the wind speed value as:

$$\omega_{OPTIM}(v) = 24.046\cdot v \tag{26}$$

In order to achieve optimal performance, it is necessary for the current mechanical angular velocity to be equal to the optimal mechanical angular velocity. This can be expressed as $\omega = \omega_{OPTIM}$, resulting in the power equation:

$$J \cdot (d\omega_{OPTIM}/dt) \cdot \omega_{OPTIM} = P_{WT\text{-}MAX} - P_{EG\text{-}OPTIM} \tag{27}$$

or:

$$24.046 \cdot J \cdot (d_V/dt) \cdot 24.046 \cdot v = P_{WT\text{-}MAX} - P_{EG\text{-}OPTIM} \tag{28}$$

Thus, based on the knowledge of the dependence of the optimal rotational speed towards wind speed, the optimal power of the electric generator is obtained:

$$P_{EG\text{-}OPTIM} = P_{WT\text{-}MAX} - 24.046^2 \cdot J \cdot (dv/dt) \cdot v \tag{29}$$

At the equivalent moment of inertia of value J = 511.92 [kg·m^2], the optimal electromagnetic power is obtained:

$$\begin{aligned} P_{EG\text{-}OPTIM} &= P_{WT\text{-}MAX} - 24.046^2 \cdot 511.92 \cdot (dv/dt) \cdot v \\ &= 2792.8 \cdot v^3 - 2.96 \cdot 10^5 \cdot (dv/dt) \cdot v \end{aligned} \tag{30}$$

At time t*, once the WT has been brought in MPP, the EG is connected to the grid and charges to the power:

$$P^*_{EG} = P^*_{WT} - J \cdot (d_{\omega\text{-}OPTIM}/dt) \cdot \omega_{OPTIM} = P^*_{WT} - 2.96 \cdot 10^5 \cdot (dv/dt) \cdot v \tag{31}$$

By employing simulation techniques, an examination is conducted within the MPP region to identify the specific instances in time when power gaps emerge.

2.3. *Time Intervals When Wind Speed Increases and Power Gaps Occur*

The examination of the current wind speed fluctuations, as illustrated in Figure 1, reveals distinct time periods during which the wind speed increases and falls. When operating within the MPP range, the electric generator's optimal power output, $P_{EG\text{-}OPTIM}$, will experience fluctuations due to wind conditions, potentially resulting in power gaps at certain times.

Case Study 1—The Appearance of Power Gaps

The analysis focuses on the functioning of the system once it has reached the maximum power point of the turbine. It is assumed that at time t*, the WT operates at point B, which represents the maximum power point as depicted in Figure 1 (Table 1).

At t* = 0 the system operates in state B, at the point of the maximum power of the turbine, when the ω_{OPTIM} is equal to the actual angular velocity ω:

$$\omega_{OPTIM} = \omega = 24.046 \cdot 8.13 = 195.49 \text{ [rad/s]} \tag{32}$$

With the v^* = 8.13 [m/s], the EG load corresponds to the power P^*_{EG}:

$$P^*_{EG}(0) = P^*_{WT\text{-}MAX} = k_p \cdot v^3 = 2792.8 \cdot 8.13^3 = 1.5008 \cdot 10^6 \tag{33}$$

Considering the wind speed variation on the B-C interval as linear, it follows:

$$v_{B\text{-}C}(t) = 8.13 + [(9.52 - 8.13)/6.631 - 3.433)] \cdot t \tag{34}$$

The wind energy that has been harnessed, denoted as E_{wind}, is determined through the integration of the power generated by the wind turbine:

$$E_{wind} = \int P_{WT} \cdot dt \tag{35}$$

or:

$$E_{wind}/dt = P_{WT} \tag{36}$$

The B-C interval is divided into four subintervals:

- Subinterval 1—at a sampling rate of 1s, at t* = 1 s, the wind speed value results in:

$$v^*_1 = 8.13 + [(9.52 - 8.13)/6.631 - 3.433)]\cdot 1 = 8.5646\ [m/s] \tag{37}$$

The wind speed varies linearly from 8.13 m/s to 8. 5646 m/s:

$$v_{B\text{-}C\text{-}1}(t) = 8.13 + (8.5646 - 8.13)\cdot t = 0.4346\cdot t + 8.13 \tag{38}$$

and the ω_{OPTIM} is:

$$\omega_{OPTIM\text{-}1} = \omega = 24.046\cdot(0.4346\cdot t + 8.13) \tag{39}$$

The equations of motion and energy are obtained as:

$$\begin{cases} 511.92\cdot\frac{d\omega}{dt}\cdot\omega = 650860\left[(0.4346\cdot t + 8.13)/\omega - 1.7488\cdot 10^{-2}\right] \\ e^{-41.495[(0.4346\cdot t+8.13)/\omega]}(0.4346\cdot t + 8.13)^3 - 1.5008\cdot 10^6 \\ \frac{dE}{dt} = 650860\left[(0.4346\cdot t + 8.13)/\omega - 1.7488\cdot 10^{-2}\right]\cdot e^{-41.495[(0.4346\cdot t+8.13)/\omega]}(0.4346\cdot t + 8.13)^3 \\ \omega(0) = 195.49 \\ E(0) = 0 \end{cases} \tag{40}$$

In a single second, a wind energy of magnitude $E_{wind}(1) = E(1) = 1.6236\cdot 10^6$ [J] is obtained. If the system is functioning at its peak power output, the captured wind energy within that time frame is:

$$E_{wind-MAX}(1) = \int_0^1 P_{WT-MAX}\cdot dt = \int_0^1 \left[2792.8\cdot(0.4346\cdot t + 8.13)^3\right]\cdot dt = 1.6254\cdot 10^6\ [J] \tag{41}$$

In the same time interval, the debited electricity is:

$$E_{electrical}(1) = \int_0^1 P_{EG}\cdot dt = 1.5008\cdot 10^6\ [J] \tag{42}$$

At t* = 1 [s], as shown in Figure 3, the mechanical angular velocity ω has the value $\omega(1) = 196.71$ [rad/s] compared to the optimal mechanical angular speed:

$$\omega_{OPTIM\text{-}1} = \omega = 24.046\cdot 8.5646 = 205.94\ [rad/s] \tag{43}$$

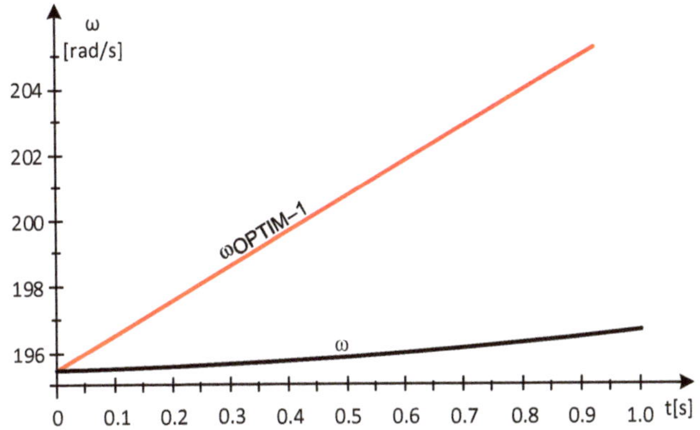

Figure 3. Optimal and real MAS variations on subinterval 1.

Since $\omega_{OPTIM-1} > \omega(1)$, at t = 1 [s], the maximum power point of the turbine has not been reached. When the value of inertial power is positive, $P^*_{INERTIAL} > 0$:

$$P^*_{INERTIAL-1} = J\cdot[(\omega^*_{OPTIM})^2 - (\omega^*)^2]/2\Delta t$$
$$= 511.92\cdot[(205.94)^2 - (196.71)^2]/2 = 9.5126\cdot 10^5\ [W] \quad (44)$$

on subinterval 2, due to the positive value of the inertial power, the load of the electric generator is decreased to the power level:

$$P^{**}_{EG-2} = P^*_{EG} - J\cdot[(\omega^*_{OPTIM})^2 - (\omega^*)^2]/2\Delta t$$
$$= 1.5008\cdot 10^6 - 9.5126\cdot 10^5 = 5.4954\cdot 10^5\ [W] \quad (45)$$

- Subinterval 2, at t = 2 [s], the value of the wind speed is:

$$v^*_2 = 8.13 + [(9.52 - 8.13)/6.631 - 3.433)]\cdot 2 = 8.9993\ [m/s] \quad (46)$$

During subinterval 2, the wind speed experiences a linear change from 8. 5646 m/s to 8. 9993 m/s

$$v_{B\text{-}C\text{-}2}(t) = 8.5646 + (8.9993 - 8.5646)\cdot t = 0.4347\cdot t + 8.5646 \quad (47)$$

while the optimal mechanical angular speed, ω_{OPTIM}, is:

$$\omega_{OPTIM-1} = 24.046\cdot(0.4347\cdot t + 8.5646) \quad (48)$$

In subinterval 2, compared to subinterval 1, the time origin has been shifted by 1 s, and this is repeated thereafter at each analyzed subinterval.

We obtain the equations of motion and energy in the form:

$$\begin{cases} 511.92\cdot\frac{d\omega}{dt}\cdot\omega = 650860\left[(0.4347\cdot t + 8.5646)/\omega - 1.7488\cdot 10^{-2}\right] \\ e^{-41.495[(0.4347\cdot t+8.5646)/\omega]}(0.4347\cdot t + 8.5646)^3 - 5.4954\cdot 10^5 \\ \frac{dE}{dt} = 650860\left[(0.4346\cdot t + 8.5646)/\omega - 1.7488\cdot 10^{-2}\right]\cdot e^{-41.495[(0.4346\cdot t+8.5646)/\omega]}(0.4346\cdot t + 8.5646)^3 \\ E(0) = 0 \\ \omega(0) = 196.71 \end{cases} \quad (49)$$

The wind energy captured in one second is $E_{wind}(1) = E(1) = 1.8883\cdot 10^6$ [J]. When operating in MPP, in the same time interval, the captured wind energy is:

$$E_{wind-MAX}(1) = \int_0^1 P_{WT-MAX}\cdot dt = \int_0^1 \left[2792.8\cdot(0.4346\cdot t + 8.5646)^3\right]\cdot dt = 1.8927\cdot 10^6\ [J] \quad (50)$$

The produced electricity is:

$$E_{electrical}(1) = \int_0^1 P_{EG}\cdot dt = 5.4954\cdot 10^5\ [J] \quad (51)$$

At t = 1 [s], the mechanical angular speed ω has the value $\omega(1) = 209.58$ [rad/s], being lower compared to the optimal mechanical angular speed, as shown in Figure 4.

$$\omega_{OPTIM-2} = 24.046\cdot 8.9993 = 216.4\ [rad/s] \quad (52)$$

Because $\omega_{OPTIM-2} \geq \omega(1)$ at t = 1 [s], the maximum power point of the turbine has not been reached. With the value of inertial power being positive, the EG load should decrease, below subinterval 3, to the value:

$$P^*_{INERTIAL-3} = J\cdot[(\omega^*_{OPTIM})^2 - (\omega^*)^2]/2\Delta t$$
$$= 511.92\cdot[(216.4)^2 - (209.58)^2]/2 = 7.4361\cdot 10^5\ [W] \quad (53)$$

$$P^{**}_{EG\text{-}3} = P^{*}_{EG} - J\cdot[(\omega^{*}_{OPTIM})^2 - (\omega^{*})^2]/2\Delta t = 5.4954\cdot 10^5 - 7.4361\cdot 10^5 = -1.9407\cdot 10^5\ [W] < 0 \quad (54)$$

With the value of the power at EG being negative, it requires its transition into engine mode.

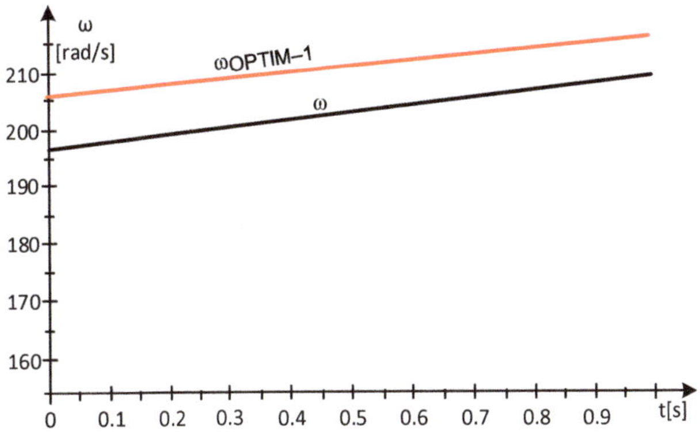

Figure 4. MAS optimal and real variations on subinterval 2.

In order for the value of the actual mechanical angular speed to reach, on subinterval 3, the value of the ω_{OPTIM}, it is necessary to switch EG to motor mode. Practically, this is a difficult operating mode, and, for this reason, the generator is disconnected from the network.

- Subinterval 3, with the power value at the EG being negative on subinterval 3, it discharges. So, $P^{**}_{EG\text{-}3} = 0$.

  At t = 3 [s], the value of the wind speed is:

$$v^{*}_3 = 8.13 + [(9.52 - 8.13)/6.631 - 3.433)]\cdot 3 = 9.4339\ [m/s] \quad (55)$$

On subinterval 3, the wind speed has a linear variation from 8.9993 [m/s] to 9.4339 [m/s].

$$v_{B\text{-}C\text{-}3}(t) = 8.9993 + (9.4339 - 8.9993)\cdot t = 0.4346\cdot t + 8.9993 \quad (56)$$

and the optimal mechanical angular velocity, ω_{OPTIM}, is:

$$\omega_{OPTIM\text{-}3} = 24.046\cdot(0.4346\cdot t + 8.9993)\ [rad/s] \quad (57)$$

So, the system of differential equations governing the motion and energy is obtained in the following form:

$$\begin{cases} 511.92\cdot\frac{d\omega}{dt}\cdot\omega = 650860\left[(0.4346\cdot t + 8.9993)/\omega - 1.7488\cdot 10^{-2}\right] \\ \quad e^{-41.495[(0.4346\cdot t+8.9993)/\omega]}(0.4346\cdot t + 8.9993)^3 \\ \frac{dE}{dt} = 650860\left[(0.4346\cdot t + 8.993)/\omega - 1.7488\cdot 10^{-2}\right]\cdot e^{-41.495[(0.4346\cdot t+8.9993)/\omega]}(0.4346\cdot t + 8.9993)^3 \\ E(0) = 0 \\ \omega(0) = 209.58 \end{cases} \quad (58)$$

In one second, the captured wind energy is $E_{wind}(1) = E(1) = 2.1869\cdot 10^6$ [J].

When operating at the MPP, the captured wind energy, in the same time interval, has the value:

$$E_{wind-MAX}(1) = \int_0^1 P_{WT-MAX}\cdot dt = \int_0^1 \left[2792.8\cdot(0.4346\cdot t + 8.9993)^3\right]\cdot dt = 2.1877\cdot 10^6\ [J] \quad (59)$$

The debited electricity is zero $E_{electrical}(1) = 0$ [J].

At t* = 1 [s], as shown in Figure 5, the mechanical angular speed ω has the value ω(1) = 229.06 [rad/s] and has exceeded the optimal mechanical angular velocity:

$$\omega_{OPTIM\text{-}3} = 24.046 \cdot 9.4339 = 226.85 \text{ [rad/s]} \quad (60)$$

At the time moment t_3, as shown in Figure 5, the ω has reached the optimal value. In this point, the wind turbine operates at the maximum power point at the optimum mechanical angular speed. The value of the captured wind energy is found in the electricity injected into the grid and in the kinetic energy of the rotating masses.

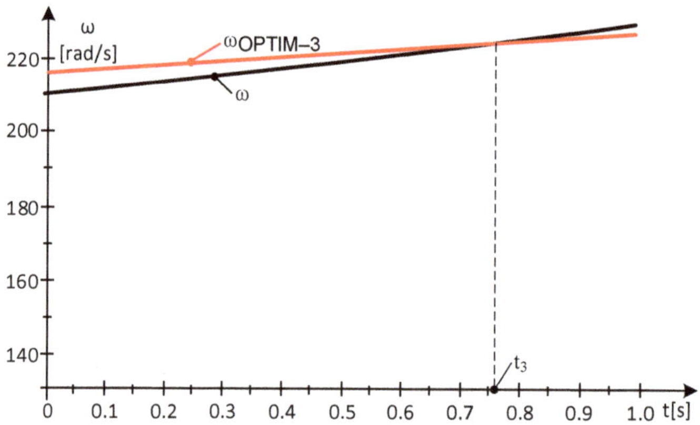

Figure 5. Optimal and real MAS variations on subinterval 3.

On the three subintervals, the kinetic energies are:

$$\Delta E_{kinetic-1} = J\frac{[\omega(1)]^2 - \left[\omega(0)^2\right]}{2} = 511.92\frac{(196.71)^2 - (195.49)^2}{2} = 1.2247 \cdot 10^5 \text{ [J]} \quad (61)$$

$$\Delta E_{kinetic-2} = J\frac{[\omega(2)]^2 - \left[\omega(1)^2\right]}{2} = 511.92\frac{(209.58)^2 - (196.71)^2}{2} = 1.3384 \cdot 10^6 \text{ [J]} \quad (62)$$

$$\Delta E_{kinetic-3} = J\frac{[\omega(3)]^2 - \left[\omega(2)^2\right]}{2} = 511.92\frac{(229.06)^2 - (209.58)^2}{2} = 2.1871 \cdot 10^6 \text{ [J]} \quad (63)$$

In conclusion, on the three subintervals, the debited energies and powers are (Table 2):

Table 2. Energies and powers on the three subintervals.

Subinterval	E_{wind} [J]	$E_{wind\text{-}MAX}$ [J]	P_{EG} [W]	$\Delta E_{kinetic}$ [J]
1	$1.6236 \cdot 10^6$	$1.6254 \cdot 10^6$	$1.5008 \cdot 10^6$	$1.2247 \cdot 10^5$
2	$1.8883 \cdot 10^6$	$1.8927 \cdot 10^6$	$5.4954 \cdot 10^6$	$1.3384 \cdot 10^6$
3	$1.1869 \cdot 10^6$	$1.1877 \cdot 10^6$	0	$2.1871 \cdot 10^6$

On an interval of 3 s, in which the wind speed increases 8.13 [m/s] < v(t) < 9.4339 [m/s], the energy evolutions are reflected in Figure 6.

Through the energy balance, it can be confirmed that the captured wind energy, E_{wind}, is found in the injected electricity, $E_{electric}$, and the kinetic energy of the masses in rotational motion, $\Delta E_{kinetic}$.

Energy balance on subinterval 1, $E_{wind} = 1.6236 \cdot 10^6$ [J] is located in:

$$E_{electrical} + \Delta E_{kinetic} = 1.5008 \cdot 10^6 + 1.2247 \cdot 10^5 = 1.6233 \cdot 10^6 \text{ [J]} \quad (64)$$

Energy balance on subinterval 2, $E_{wind} = 1.8883 \cdot 10^6$ [J] is located in:

$$E_{electrical} + \Delta E_{kinetic} = 5.4954 \cdot 10^5 + 1.3384 \cdot 10^6 = 1.8879 \cdot 10^6 \text{ [J]} \quad (65)$$

Energy balance on subinterval 3, $E_{wind} = 2.1869 \cdot 10^6$ [J] is located in:

$$E_{electrical} + \Delta E_{kinetic} = 0 + 2.1871 \cdot 10^6 = 2.1871 \cdot 10^6 \text{ [J]} \quad (66)$$

The general energy balance on interval $1 \to 3$ is:

$$E_{wind} = 1.6236 \cdot 10^6 + 1.8883 \cdot 10^6 + 2.1871 \cdot 10^6 = 5.6988 \cdot 10^6 \text{ [J]} \quad (67)$$

is found in the injected electricity, $E_{electrical}$, and the kinetic energy of the masses in rotational motion, $\Delta E_{kinetic}$:

$$E_{electrical} + \Delta E_{kinetic} = 1.6233 \cdot 10^6 + 1.8879 \cdot 10^6 + 2.1871 \cdot 10^6 = 5.6983 \cdot 10^6 \text{ [J]} \quad (68)$$

At t_3, maximum wind energy is captured, even if the EG is discharged, ($P^{**}_{EG\text{-}3} = 0$). The turbine operates at the MPP and at the optimal mechanical angular speed. This energy is not injected into the grid ($P^{**}_{EG\text{-}3}$ is zero) but is found in the kinetic energy of the masses in rotational motion. Apparently, the operation of the turbine is, from an energy point of view, not in the optimal zone. This time moment, presenting a power gap, involves disadvantages for network stability.

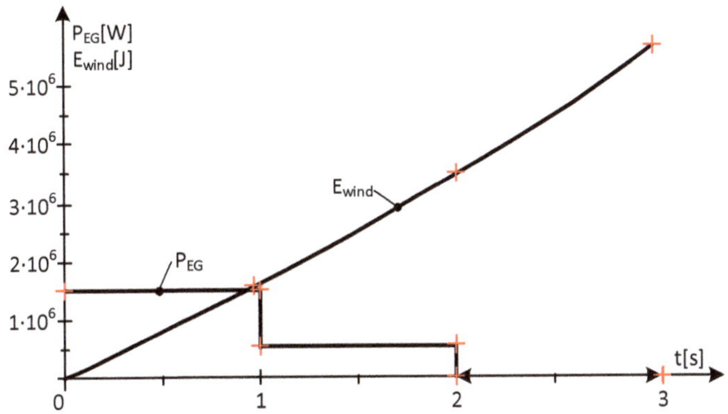

Figure 6. The variations of captured wind energy and electric power.

Based on the variations of the energy as highlighted in Figure 6, we are able to conclude that:

1. The values of the injected electricity, $E_{lectrical}$, decrease (power at the generator decreases);
2. The differences in the values of the kinetic energies $\Delta E_{kinetic}$ increase and compensate for the decrease in the injected electricity;

In conclusion, during increased wind speed, power gaps can occur if the values of the derivative of the wind speed, dv, and the equivalent moment of inertia, J, exceed certain values.

2.4. Time Intervals in Which the Wind Speed Decreases

Case Study 2—MPP Area at Decreasing Wind Speed

From Figure 7, we select the time interval D-E (Table 3), in which the wind speed decreases.

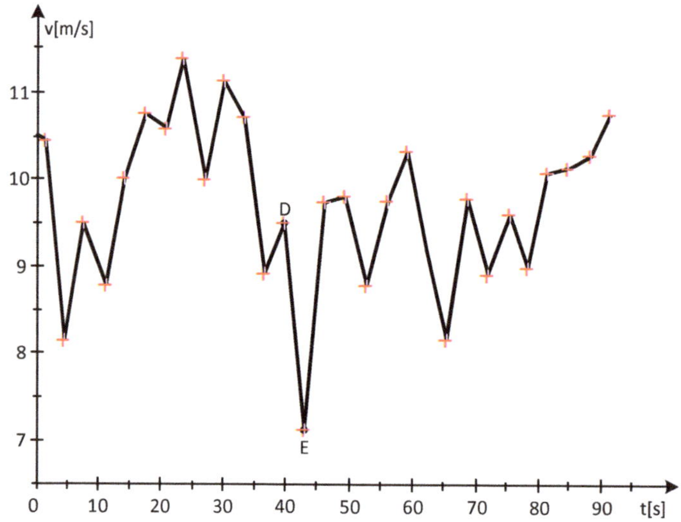

Figure 7. D-E time interval with decreasing wind speed.

Table 3. Speed at operating points.

Time t [s]	Speed v [m/s]	Point
1.198	10.47	A
39.858	9.52	D
43.073	7.145	E

It is considered that at point D, as shown in Figure 7, the WT operates at the MPP and at the same values of the optimal mechanical angular speeds and real MAS:

$$\omega_{OPTIM\text{-}D} = 24.046 \cdot 9.52 = 228.92\ [\text{rad/s}] \quad (69)$$

For the value of the wind speed $v_D = 9.52$ [m/s], the load of the EG is at a $P_{EG\text{-}D}$ power of:

$$P_{EG\text{-}D} = P^*_{WT\text{-}MAX} = k_P \cdot v^3 = 2792.8 \cdot 9.52^3 = 2.4096 \cdot 10^6\ [\text{W}] \quad (70)$$

On the D-E interval, considering the linear wind speed variation results in:

$$v_{D\text{-}E}(t) = 9.52 + [(7.1245 - 9.52)/43.073\text{-}39.858)] \cdot t = 9.52 - 0.73872 \cdot t \quad (71)$$

The D-E interval is divided into three subintervals, similar to the previous case:

- For subinterval 1, with the sampling being 1s, at $t^* = 1$ [s], the wind speed value results in:

$$v_1{}^* = 9.52 + [(7.1245 - 9.52)/43.073\text{-}39.858)] \cdot 1 = 8.7813\ [\text{m/s}] \quad (72)$$

Wind speed has a linear variation from 9.52 m/s to 8. 7813 m/s:

$$v_{D\text{-}E\text{-}1}(t) = 9.52 + (8.7813 - 9.52) \cdot t = 9.52 - 0.7387 \cdot t \quad (73)$$

and the ω_{OPTIM} is:

$$\omega_{OPTIM\text{-}1} = 24.046 \cdot (9.52 - 0.7387 \cdot t) \quad (74)$$

In a second, valuable wind energy is captured: $E_{wind}(1) = E(1) = 2.1382 \cdot 10^6$ [J]

In operation, at the point of maximum power, the wind energy captured over the same time interval has the value $E_{wind\text{-}MAX}(1) = 2.1434 \cdot 10^6$ J, and the electricity flowed has the value $E_{electrical}(1) = 2.4096 \cdot 10^6$ J.

At t* = 1 s, as shown in Figure 8, the mechanical angular velocity has the value $\omega(1) = 226.59$ [rad/s] compared to the optimal mechanical angular velocity:

$$\omega_{OPTIM\text{-}1} = 24.046 \cdot 8.7813 = 211.16 \text{ [rad/s]} \tag{75}$$

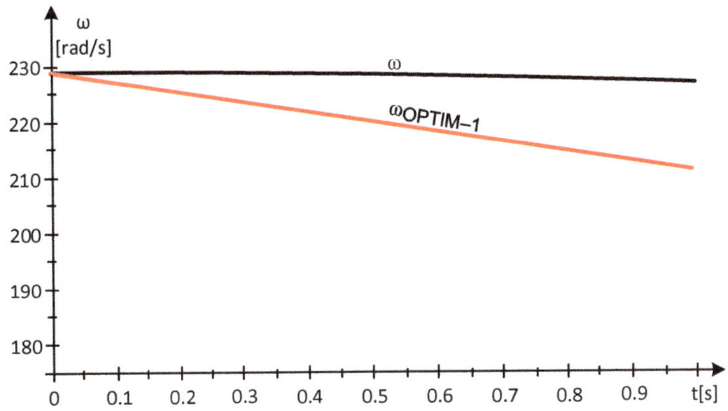

Figure 8. Optimal and actual VUM variations on subinterval 1.

Since $\omega_{OPTIM\text{-}1} < \omega(1)$, the value of the inertial power being negative, $P^*_{INERTIAL} < 0$:

$$P*_{INERTIAL-1} = J\frac{(\omega*_{OPTIM})^2 - (\omega*)^2}{2\Delta t} = 511.92\frac{(211.16)^2 - (226.59)^2}{2} = -1.7289 \cdot 10^6 \text{ [W]} \tag{76}$$

the load of GE to increase power, on subinterval 2, to:

$$P**_{EG-1} = P*_{EG} - J\frac{(\omega*_{OPTIM})^2 - (\omega*)^2}{2\Delta t} = 2.4096 \cdot 10^6 + 1.7289 \cdot 10^6 = 4.1385 \cdot 10^6 \text{ [W]} \tag{77}$$

- Subrange 2 to t = 2 [s], the wind speed value is:

$$v_2^* = 9.52 + (8.7813 - 9.52) \cdot 2 = 8.0436 \text{ [m/s]} \tag{78}$$

The wind speed has a linear variation from 8.7813 m/s to 8.0426 m/s:

$$v_{D\text{-}E\text{-}2}(t) = 8.7813 + (8.0426 - 8.7813) \cdot t = 8.7813 - 0.7387 \cdot t \tag{79}$$

for which the optimal mechanical angular velocity, ω_{OPTIM}, corresponds to:

$$\omega_{OPTIM\text{-}2} = 24.046 \cdot (8.7813 - 0.7387 \cdot t) \tag{80}$$

Proceeding as in the previous case, it follows that in one second, the value of the captured wind energy is $E_{wind}(1) = E(1) = 1.6549 \cdot 10^6$ [J]. The wind energy captured, in the same time interval, at an operation at the maximum power point is $E_{wind\text{-}MAX}(1) = 1.6656 \cdot 10^6$ J, and the produced electricity has the value $E_{electrical}(1) = 4.1385 \cdot 10^6$ J.

At t* = 1 s, the mechanical angular velocity has the value $\omega(1) = 204.06$ [rad/s], compared to the optimal mechanical angular velocity:

$$\omega_{OPTIM\text{-}2} = 24.046 \cdot 8.0426 = 193.39 \text{ [rad/s]} \tag{81}$$

Because $\omega_{OPTIM\text{-}2} < \omega(1)$, and with the value of the inertial power being negative, $P^*_{INERTIAL} < 0$:

$$P{*}_{INERTIAL-2} = J\frac{(\omega{*}_{OPTIM})^2 - (\omega{*})^2}{2\Delta t} = 511.92\frac{(193.39)^2 - (204.06)^2}{2} = -1.0855{\cdot}10^6\ [\text{W}] \quad (82)$$

increase, on subinterval 3, the load of GE to power:

$$P{*}{*}_{EG-2} = P{*}_{EG} - J\frac{(\omega{*}_{OPTIM})^2 - (\omega{*})^2}{2\Delta t} = 4.1385{\cdot}10^6 + 1.0855{\cdot}10^6 = 5.224{\cdot}10^6\ [\text{W}] \quad (83)$$

- Subinterval 3 to t = 3 [s], the wind speed value is:

$$v_3{*} = 9.52 + (8.7813 - 9.52){\cdot}3 = 7.3039\ [\text{m/s}] \quad (84)$$

Wind speed drops from 8.0426 m/s to 7.3039 m/s:

$$v_{D\text{-}E\text{-}3}(t) = 8.0426 + (7.3039 - 8.0426){\cdot}t = 8.0426 - 0.7387{\cdot}t \quad (85)$$

and the optimal mechanical angular velocity, ω_{OPTIM}, is:

$$\omega_{OPTIM\text{-}3} = 24.046{\cdot}(8.0426 - 0.7387{\cdot}t) \quad (86)$$

In one second, the captured wind energy is $E_{wind}(1) = E(1) = 1.2619{\cdot}10^6$ [J]. The wind energy captured, in the same time interval, at an operation at the maximum power point is $E_{wind\text{-}MAX}(1) = 1.2647{\cdot}10^6$ J, and the produced electricity has the value $E_{electrical}(1) = 5.224{\cdot}10^6$ J.

At t* = 1 s, the mechanical angular velocity ω has the value $\omega(1) = 161.74$ [rad/s], compared to the optimal mechanical angular velocity:

$$\omega_{OPTIM\text{-}3} = 24.046{\cdot}7.3039 = 175.63\ [\text{rad/s}] \quad (87)$$

At time t_3, as shown in Figure 9, the optimal mechanical angular velocity, $\omega_{OPTIM\text{-}3}$, equals the mechanical angular velocity, reaching the maximum power point, and thus the tuning algorithm is validated.

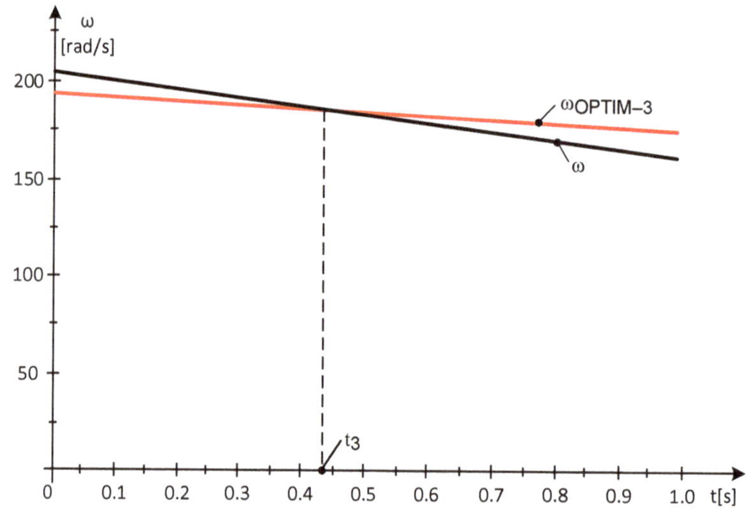

Figure 9. Optimal and actual MAS variations on subinterval 3.

Because $\omega_{OPTIM\text{-}3} > \omega(1)$, the value of the inertial power is positive, $P{*}_{INERTIAL} > 0$:

$$P{*}_{INERTIAL-3} = J\frac{(\omega{*}_{OPTIM})^2 - (\omega{*})^2}{2\Delta t} = 511.92\frac{(175.63)^2 - (161.74)^2}{2} = 1.1994{\cdot}10^6\ [\text{W}] \quad (88)$$

should, over the next sub-time frame, decrease the GE load to:

$$P**_{EG-3} = P*_{EG} - J\frac{(\omega*_{OPTIM})^2 - (\omega*)^2}{2\Delta t} = 5.224\cdot10^6 - 1.1994\cdot10^6 = 4.026\cdot10^6 \text{ [W]} \tag{89}$$

The value of the captured wind energy is found in the electricity flowed into the system and in the kinetic energy of the rotating masses.

On the three subintervals, the variations of kinetic energies are:

$$\Delta E_{kinetic-1} = J\frac{(\omega(1))^2 - (\omega(0))^2}{2} = 511.92\frac{(226.59)^2 - (228.92)^2}{2} = -2.7166\cdot10^5 \text{ [J]} \tag{90}$$

$$\Delta E_{kinetic-2} = J\frac{(\omega(2))^2 - (\omega(1))^2}{2} = 511.92\frac{(204.06)^2 - (226.59)^2}{2} = -2.4835\cdot10^6 \text{ [J]} \tag{91}$$

$$\Delta E_{kinetic-3} = J\frac{(\omega(3))^2 - (\omega(2))^2}{2} = 511.92\frac{(161.74)^2 - (204.06)^2}{2} = -3.9624\cdot10^6 \text{ [J]} \tag{92}$$

In conclusion, on the three subintervals, the powers and energies are (Table 4):

Table 4. Power and energy values on the three subintervals.

Subinterval	E_{wind} [J]	$E_{wind\text{-}MAX}$ [J]	P_{EG} [W]	$\Delta E_{kinetic}$ [J]
1	$2.1382\cdot10^6$	$2.1434\cdot10^6$	$2.4096\cdot10^6$	$-2.7166\cdot10^5$
2	$1.6549\cdot10^6$	$1.6656\cdot10^6$	$4.1385\cdot10^6$	$-2.4835\cdot10^6$
3	$1.2619\cdot10^6$	$1.2647\cdot10^6$	$5.224\cdot10^6$	$-3.9624\cdot10^6$

Over an interval of 3 s, during which the wind speed decreases, 7.145 m/s < v(t) < 9.52 m/s, developments in captured wind energy and power output are given in Figure 10.

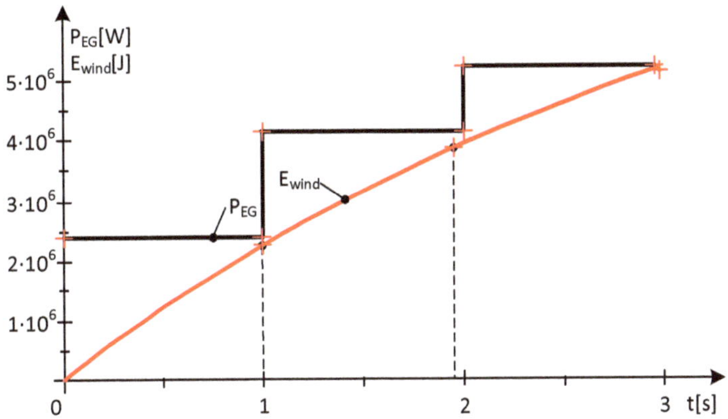

Figure 10. Variations in captured wind energy and electrical power.

The energy balance in this case is:

$$E_{wind} = E_{electric} + \Delta E_{kinetic} \tag{93}$$

and on the three subintervals, this results in:

$$E_{wind\text{-}1} = 2.1382\cdot10^6 \approx E_{electric} + \Delta E_{kinetic} = 2.4096\cdot10^6 - 2.7166\cdot10^5 = 2.1379\cdot10^6 \text{ [J]} \tag{94}$$

$$E_{wind\text{-}2} = 1.6549\cdot10^6 \approx E_{electric} + \Delta E_{kinetic} = 4.1385\cdot10^6 - 2.4835\cdot10^6 = 1.655\cdot10^6 \text{ [J]} \tag{95}$$

$$E_{wind\text{-}3} = 1.2619\cdot10^6 \approx E_{electric} + \Delta E_{kinetic} = 5.224\cdot10^6 - 3.9624\cdot10^6 = 1.2616\cdot10^6 \text{ [J]} \tag{96}$$

In time intervals, when wind speed decreases, no power gaps occur, although mechanical angular velocity tends towards the optimal mechanical angular velocity. The turbine does not operate at the MPP, and yet the power output into the system is greater than the power given by the WT, the difference being from the kinetic energy of the rotating masses. In terms of the stability of the energy system, this is an advantage.

Observations resulting from variations in energy values given in Figure 10:

1. The values of the electrical power flow, P_{EG}, increase, although the wind speed decreases;
2. The differences in the kinetic energy values $\Delta E_{kinetic}$, in absolute value, increase and compensate for the decrease in the values of the captured wind energies;
3. The values of captured wind energies decrease in direct proportion to the decrease in wind speed.

In conclusion, in time intervals, when wind speed decreases, no power gaps occur and, on the contrary, there is an increase in electrical power, whatever the values of the velocity derivative wind, dv/dt, and the equivalent moment of inertia J. This extra electrical power comes, as mentioned, from the kinetic energy of rotating masses. The surplus power must be stored, and only the electricity forecasted in advance given to the national energy system.

3. Results and Discussion

3.1. Fundamental Aspects

With the results obtained, the following relevant aspects can be highlighted:

1. It has been shown that during time intervals when wind speed increases, the power output in the system decreases if the turbine is operating at the point of maximum power;
2. It has been shown that, in the time intervals when the speed decreases, the power debited in the system is higher than the power given by the turbine, which is an advantage for the stability of the energy system;
3. It has been demonstrated that the turbine can operate at its maximum power point at any wind speed value, regardless of restrictions, if a wind energy storage system is available;
4. Bringing the system to the point of maximum turbine power is achieved in the shortest time by decoupling EG from the grid or switching EG to engine mode;
5. By estimating the difference between the optimal mechanical angular velocity and the current mechanical angular velocity, one can prescribe the power value at the generator;
6. The operation of the WT at the MPP is ensured by the correlation of the EG power with the wind speed and its variation;
7. During periods when wind speed increases, power gaps occur (disconnect the generator from the mains) if operation in the energy-optimal area is required;
8. During periods when wind speed decreases, no power gaps occur, although operation in the energy optimal zone is required.

3.2. Discussion

1. The method presented lays the foundation for a control that takes into account wind speed and mechanical angular speed;
2. The system shall be brought to the point of maximum power by calculating the values of the variations in the kinetic energies of the rotating masses, values obtained by measurements;
3. At increasing wind speeds over time, power gaps occur under conditions where operation at the maximum power point is required;
4. At decreasing wind speeds over time, no power gaps occur, although operation is required at the maximum power point and the electrical power debited into the system is greater than the power given by the turbine, the surplus power coming from the various kinetic energies of the rotating masses.

4. Conclusions

In this paper, it was demonstrated that wind energy storage is necessary if the wind turbine is operating at its maximum power point. The power generated by the electric generator comes from the power developed by the wind turbine and from the variation in the kinetic energies of the rotating masses and the energy stored in them during the period of time in which a significant decrease in wind speed occurs (an interesting statement that may seem contradictory). It has been shown that operating in the energy-optimal zone at the peak power point of the turbine requires a system to store captured wind energy. The behavior of the wind turbine at wind speeds variable in time was analyzed, and it was observed that, at significant variations in wind speed, the value prescribed to the generator may be negative, in which case a power gap is created in the national electricity system. The optimal area from an energy point of view was visualized. The simulations presented were based on real variations of wind speed at turbine hub height in the Romanian Black Sea coast area.

Author Contributions: Conceptualization, C.P.C. and G.-O.T.; methodology, C.P.C. and E.S.; software, C.P.C.; validation, C.P.C. and E.S.; formal analysis, E.S.; investigation, C.P.C. and G.-O.T.; resources, C.P.C. and E.S.; data curation, G.-O.T.; writing—original draft preparation, C.P.C.; writing—review and editing, E.S.; visualization, C.P.C. and E.S.; supervision, C.P.C.; project administration, C.P.C. and E.S.; funding acquisition, C.P.C. and E.S. All authors have read and agreed to the published version of the manuscript.

Funding: The publication of this article was supported by the 2024 Development Fund of the UBB.

Data Availability Statement: The data presented in this study are available on request from the first author.

Conflicts of Interest: The authors declare no conflict of interest. The funders had no role in the design of the study; in the collection, analyses, or interpretation of data; in the writing of the manuscript, or in the decision to publish the results.

Abbreviations

Symbol/Abbreviation	Meaning
WT	wind turbine
MPP	maximum power point
MAS	mechanical angular speed
EG	electric generator
P_{EG}	power at the electric generator
P_{WT}	power of wind turbine
v	wind speed
J	inertia moment
n	rotational speed
ω	mechanical angular speed
ω_{OPTIM}	optimum mechanical angular speed
n_{OPTIM}	optimum rotational speed
k_v	proportionality factor
$P_{INERTIAL}$	inertial power
ρ	air density in the wind turbine
R_p	rotor blade radius
$C_p(\lambda)$	power conversion coefficient
M_{WT}	moment related to the shaft of the electric generator
M_{EG}	electromagnetic torque at the electric generator
k_p	the proportionality factor for power
E_{wind}	wind energy
$E_{electrical}$	electrical energy
$\Delta E_{kinetic}$	kinetic energy of the masses in rotational motion
NPS	National Power System

References

1. Adebiyi, A.A.; Moloi, K. Renewable Energy Source Utilization Progress in South Africa: A Review. *Energies* **2024**, *17*, 3487. [CrossRef]
2. Kaczmarczyk, B.; Lis, K.; Bogucka, A. Renewable Energy Management in European Union Member States. *Energies* **2023**, *16*, 5863. [CrossRef]
3. Früh, W.G. Assessing the Performance of Small Wind Energy Systems Using Regional Weather Data. *Energies* **2023**, *16*, 3500. [CrossRef]
4. Guo, D.; Piao, Z.L.; Wang, L.D.; Wang, J.; Lv, Q.Q.; Cao, D.X. The research on the integration of energy storage and generating wind turbine system model. In Proceedings of the China International Conference on Electricity Distribution (CICED), Xi'an, China, 10–13 August 2016.
5. Thapa, S.; Karki, R. Reliability benefit of energy storage in wind integrated power system operation. *IET Gener. Transm. Distrib.* **2016**, *10*, 807–814. [CrossRef]
6. Roche, N.; Courtney, J. Optimal Sizing of Energy Storage with Embedded Wind Power Generation. In Proceedings of the 55th International Universities Power Engineering Conference (UPEC), Turin, Italy, 1–4 September 2020.
7. Zhang, H.L.; Liu, J.; Li, W. Optimized Design of Energy-Storage Capacity for Wind Power Generating. In Proceedings of the Asia-Pacific Power and Energy Engineering Conference (APPEEC), Shanghai, China, 27–29 March 2012.
8. Jamii, J.; Mimouni, F. Model of wind turbine-pumped storage hydro plant. In Proceedings of the 9th International Renewable Energy Congress (IREC), Hammamet, Tunisia, 20–22 March 2018.
9. MONSSON. Available online: https://www.monsson.eu/ (accessed on 3 October 2022).
10. Edrah, M.; Zhao, X.; Hung, W.; Qi, P.; Marshall, B.; Karcanias, A.; Baloch, S. Effects of POD Control on a DFIG Wind Turbine Structural System. *IEEE Trans. Energy Convers.* **2020**, *35*, 765–774. [CrossRef]
11. Hussain, J.; Mishra, M.K. An Efficient Wind Speed Computation Method Using Sliding Mode Observers in Wind Energy Conversion System Control Application. *IEEE Trans. Ind. Appl.* **2020**, *56*, 730–739. [CrossRef]
12. Sorandaru, V.; Musuroi, S.; Frigura-Iliasa, F.M.; Vatau, D.; Dordescu, M. Analysis of the Wind System Operation in the Optimal Energetic Area at Variable Wind Speed over Time. *Sustainability* **2019**, *11*, 1249. [CrossRef]
13. Babescu, M. Sisteme automate de reglare. In *Maşina Sincronă: Modelare-Identificare-Simulare*; Editura Politehnica: Timisoara, Romania, 2003; pp. 154–196.
14. Papakonstantinou, A.G.; Konstanteas, A.I.; Papathanassiou, S.A. Solutions to Enhance Frequency Regulation in an Island System with Pumped-Hydro Storage Under 100% Renewable Energy Penetration. *IEEE Access* **2023**, *11*, 76675–76690. [CrossRef]
15. Sorandaru, C.; Ancuti, M.-C.; Musuroi, S.; Svoboda, M.; Muller, V.; Erdodi, M.G.; Edu, E.R. Fundamental Issue for Wind Power Systems Operating at Variable Wind Speeds: The Dependence of the Optimal Angular Speed on the Wind Speed. In Proceedings of the 2022 International Conference and Exposition on Electrical and Power Engineering (EPE), Iasi, Romania, 20–22 October 2022.
16. Rebello, E.; Watson, D.; Rodgers, M. Performance Analysis of a 10 MW Wind Farm in Providing Secondary Frequency Regulation: Experimental Aspects. *IEEE Trans. Power Syst.* **2019**, *34*, 3090–3097. [CrossRef]
17. Dobrin, E.; Ancuti, M.-C.; Musuroi, S.; Sorandaru, C.; Ancuti, R.; Lazar, M.A. Dynamics of the Wind Power Plants at Small Wind Speeds. In Proceedings of the IEEE 14th International Symposium on Applied Computational Intelligence and Informatics (SACI), Timisoara, Romania, 21–23 May 2020.
18. Najih, Y.; Adar, M.; Aboulouard, A.; Khaouch, Z.; Bengourram, M.; Zekraoui, M.; Mabrouki, M. Mechatronic Control Model of a Novel Variable Speed Wind Turbine Concept with Power Splitting Drive Train: A Bond Graph Approach. In Proceedings of the 2020 IEEE 6th International Conference on Optimization and Applications (ICOA), Beni Mellal, Morocco, 20–21 April 2020.
19. Vadi, S.; Gürbüz, F.B.; Bayindir, R.; Hossain, E. Design and Simulation of a Grid Connected Wind Turbine with Permanent Magnet Synchronous Generator. In Proceedings of the 8th International Conference on Smart Grid (icSmartGrid), Paris, France, 17–19 June 2020.
20. Yingyu, A.; Li, Y.; Zhang, J.; Wang, T.; Liu, C. Enhanced Frequency Regulation Strategy for Wind Turbines Based on Over-speed De-loading Control. In Proceedings of the 5th Asia Conference on Power and Electrical Engineering (ACPEE), Chengdu, China, 4–7 June 2020.
21. Chioncel, C.P.; Erdodi, G.M.; Petrescu, D.I.; Elisabeta, S.; Gillich, N. Control of Wind Power Systems Imposing the Current in the Intermediate Circuit of the Converter at Variable Wind Speed. In Proceedings of the 9th International Symposium on Advanced Topics in Electrical Engineering (ATEE), Bucharest, Romania, 7–9 May 2015.
22. Sorandaru, C.; Musuroi, S.; Ancuti, M.C.; Erdodi, G.M.; Petrescu, D.I. The Control of the Wind Power Systems by Imposing the DC Current. In Proceedings of the 10th Jubilee IEEE International Symposium on Applied Computational Intelligence and Informatics, Timisoara, Romania, 21–23 May 2015.
23. Ye, H.; Ding, Y.; Xiao, L.; Li, Q.; Zheng, C.; Wang, Y. Hydro-Tower Pumped Storage-based Wind Power Generator: Short-term Dynamics and Controls. In Proceedings of the 2nd Asian Conference on Frontiers of Power and Energy (ACFPE), Chengdu, China, 20–22 October 2023.
24. Huynh, P.; Tungare, S.; Banerjee, A. Maximum Power Point Tracking for Wind Turbine Using Integrated Generator–Rectifier Systems. *IEEE Trans. Power Electron.* **2020**, *36*, 504–512. [CrossRef]

25. Carstea, C.; Butaru, F.; Ancuti, M.-C.; Musuroi, S.; Deacu, A.; Babaita, M.; Stanciu, A.M. Wind Power Plants Operation at Variable Wind Speeds. In Proceedings of the 14th International Symposium on Applied Computational Intelligence and Informatics (SACI), Timisoara, Romania, 21–23 May 2020.
26. Deng, X.; Yang, J.; Sun, Y.; Dongran, S.; Yang, Y.; Joo, Y.H. An Effective Wind Speed Estimation Based Extended Optimal Torque Control for Maximum Wind Energy Capture. *IEEE Access* **2020**, *8*, 65959–65969. [CrossRef]
27. Karthik, R.; Sri Hari, A.; Pavan Kumar, Y.V.; John Pradeep, D. Modelling and Control Design for Variable Speed Wind Turbine Energy System. In Proceedings of the 2020 International Conference on Artificial Intelligence and Signal Processing (AISP), Amaravati, India, 10–12 January 2020.
28. Fan, X.; Crisostomi, E.; Zhang, B.; Thomopulos, D. Rotor Speed Fluctuation Analysis for Rapid De-Loading of Variable Speed Wind Turbines. In Proceedings of the 2020 IEEE 20th Mediterranean Electrotechnical Conference (MELECON), Palermo, Italy, 16–18 June 2020.
29. Ma, Y.; Zhang, L.; Yan, J.D.; Guo, Z.; Li, Q.; Fang, Z.; Siew, W.H. Inception mechanism of lightning upward leader from the wind turbine blade and a proposed critical length criterion. *Proc. Chin. Soc. Electr. Eng.* **2016**, *36*, 5975–5982.
30. Yokoyama, S.; Yasuda, Y.; Masayuki, M.; Sekioka, S.; Yamamoto Member, K.; Honjo, N.; Sato, T. Clarification of the mechanism of wind turbine blade damage taking lightning characteristics into consideration and relevant research project. In Proceedings of the 2012 International Conference on Lightning Protection (ICLP), Vienna, Austria, 2–7 September 2012.
31. Xingjia, Y.; Yingming, L.; Jieqiu, B.; Zuoxia, X. Research and simulation of direct drive wind turbine VSCF characteristic. In Proceedings of the 2008 IEEE International Conference on Automation and Logistics, Qingdao, China, 1–3 September 2008.
32. Srinivasa Sudharsan, G.; Vishnupriyan, J.; Vijay Anand, K. Active flow control in Horizontal Axis Wind Turbine using PI-R controllers. In Proceedings of the 6th International Conference on Advanced Computing and Communication Systems (ICACCS), Coimbatore, India, 6–7 March 2020.
33. Chioncel, C.P.; Spunei, E.; Tirian, G.O. Visualizing the Maximum Energy Zone of Wind Turbines Operating at Time-Varying Wind Speeds. *Sustainability* **2024**, *16*, 2659. [CrossRef]
34. Chioncel, C.P.; Dordescu, M.; Lazar, M.A.; Tirian, O.G. Wind turbine power and optimum energy, at variable wind speeds. In Proceedings of the 24th IEEE International Conference on Intelligent Engineering Systems (INES), Reykjavík, Iceland, 8–10 July 2020; pp. 197–202.
35. Ragheb, M.; Ragheb, A.M. Wind Turbines Theory—The Betz Equation and Optimal Rotor Tip Speed Ratio. In *Fundamental and Advanced Topics in Wind Power*; Carriveau, R., Ed.; IntechOpen: London, UK; University of Windsor: Windsor, ON, Canada, 2011.
36. Singh, M.; Santoso, S. *Dynamic Models for Wind Turbines and Wind Power Plants*; University of Texas at Austin: Austin, TX, USA, 2021.
37. Vestas. Available online: https://www.vestas.com (accessed on 1 September 2024).
38. Wei, D.X.; Wang, J.Z.; Li, Z.W.; Wang, R. Wind Power Curve Modeling with Hybrid Copula and Grey Wolf Optimization. *IEEE Trans. Sustain. Energy* **2022**, *13*, 265–276. [CrossRef]
39. Goyal, S.; Maurya, S.K.; Kumar, S.; Agarwal, D. A Consolidation Review of Major Wind Turbine Models in Global Market. In Proceedings of the 2020 International Conference on Power Electronics & IoT Applications in Renewable Energy and Its Control (PARC), Mathura, India, 28–29 February 2020.
40. Baala, Y.; Bri, S. DFIG-Based Wind Turbine Control Using High-Gain Observer. In Proceedings of the 1st International Conference on Innovative Research in Applied Science, Engineering and Technology (IRASET), Meknes, Morocco, 16–19 April 2020.

energies

MDPI

Article

Designing a Photovoltaic–Wind Energy Mix with Energy Storage for Low-Emission Hydrogen Production

Arkadiusz Małek [1,*], Agnieszka Dudziak [2], Andrzej Marciniak [1] and Tomasz Słowik [2,*]

1 Department of Transportation and Informatics, WSEI University, 20-209 Lublin, Poland; andrzej.marciniak@wsei.lublin.pl
2 Department of Power Engineering and Transportation, Faculty of Production Engineering, University of Life Sciences in Lublin, 20-612 Lublin, Poland; agnieszka.dudziak@up.lublin.pl
* Correspondence: arkadiusz.malek@wsei.lublin.pl (A.M.); tomasz.slowik@up.lublin.pl (T.S.)

Abstract: In the introduction to this article, a brief overview of the generated energy and the power produced by the photovoltaic systems with a peak power of 3 MWp and different tilt and orientation of the photovoltaic panels is given. The characteristics of the latest systems generating energy by wind turbines with a capacity of 3.45 MW are also presented. In the subsequent stages of the research, the necessity of balancing the energy in power networks powered by a mix of renewable energy sources is demonstrated. Then, a calculation algorithm is presented in the area of balancing the energy system powered by a photovoltaic–wind energy mix and feeding the low-emission hydrogen production process. It is analytically and graphically demonstrated that the process of balancing the entire system can be influenced by structural changes in the installation of the photovoltaic panels. It is proven that the tilt angle and orientation of the panels have a significant impact on the level of power generated by the photovoltaic system and, thus, on the energy mix in individual hourly intervals. Research has demonstrated that the implementation of planned design changes in the assembly of panels in a photovoltaic system allows for a reduction in the size of the energy storage system by more than 2 MWh. The authors apply actual measurement data from a specific geographical context, i.e., from the Lublin region in Poland. The calculations use both traditional statistical methods and probabilistic analysis. Balancing the generated power and the energy produced for the entire month considered in hourly intervals throughout the day is the essence of the calculations made by the authors.

Keywords: renewable energy source; photovoltaic systems; wind turbines; energy mix; energy grid balancing; low-emission hydrogen production

1. Introduction

In an era of growing energy demand and the need to reduce greenhouse gas emissions, renewable energy sources (RESs) are becoming a key element of the energy transition [1]. In the third decade of the 21st century, humans can draw energy from many renewable sources available on land and at sea [2]. Photovoltaics (PV) [3] and wind turbines [4] are currently among the most popular RES technologies. Despite their obvious advantages, balancing energy in the power system, the stability of which is crucial for the reliable operation of the grid, remains a challenge [5,6]. Solar and wind energy are considered as rather unstable sources which are heavily dependent on the weather conditions [7]. Photovoltaic systems work best on sunny days, reaching peak production at noon, but their efficiency decreases during cloud cover, and they cease work at night [8]. Wind turbines generate energy depending on the wind speed, which means that production is variable both on an hourly and daily scale. This variability causes difficulties in ensuring the continuity of energy supply and requires the use of advanced power balancing technologies [9]. Energy balancing methods in a mix based on PV and wind are as follows:

1. Energy storage. Energy storage systems, such as lithium-ion batteries, allow for the storage of surplus generated energy during periods of high production and its use during periods of energy shortage [10]. Energy storage technologies in the form of compressed air (CAES) or hydrogen (Power-to-Gas) are also being developed [11].
2. Integration with flexible energy sources [12]. Mix balancing requires combining unstable RESs with more predictable energy sources, such as gas or hydroelectric power plants [13]. These can act as power reserves in the event of a drop in PV and wind generation.
3. Digital solutions and smart grids (Smart Grid). Advanced network management systems enable dynamic control of energy flows [14]. By using predictive algorithms, it is possible to optimize the operation of the network depending on the forecasts of the energy production from RES and the demand of the recipients [15,16].
4. Diversification of location and scale [17,18]. Distributed photovoltaic systems and wind farms located in different parts of the country can partially compensate for local differences in the weather conditions [19].

Balancing the energy mix based on PV and wind turbines is technically feasible but requires an integrated approach [20]. Investments in energy storage, intelligent grid management systems, and flexible backup sources are key [21]. In the long term, the development of such technologies will contribute to the creation of a stable, sustainable, and ecological energy system. Photovoltaic systems and wind turbines play an essential role in the decarbonization of the energy sector [22]. Although balancing the mix dominated by these sources is a challenge, the development of storage technologies, smart grids, and forecasting gives hope for its full integration [23]. An example of investment in renewable energy sources in the form of photovoltaic systems and wind turbines is shown in Figure 1. Energy transformation is an ongoing process; however, the direction of the change is unmistakably positive.

Figure 1. Components of the photovoltaic and wind energy mix located in the Lublin region in Poland.

Poland is the fifth largest producer of hydrogen in the world and the third largest in Europe, but it is mainly gray hydrogen. A significant part of this hydrogen is produced in Zakłady Azotowe in Puławy for the production of ammonia and artificial fertilizers, which in itself is an excellent base for the development of innovative technologies in the field of green hydrogen [24]. The Lublin region is also characterized by human capital in the form of highly qualified specialists and graduates of technical and natural science universities, has industrial traditions, and, crucial for green hydrogen, is characterized by excellent conditions for the development of renewable energy sources.

Despite its great potential, the technologies for producing green hydrogen are not yet mature enough to give this energy carrier an economic justification, i.e., to allow it to compete with fuels based on hydrocarbons or gray hydrogen in industry [25]. In connection with this, active actions are being taken in the Lublin region in Poland to create a supply of green hydrogen at the place of its future use, and preparations are being made for the implementation of the following:

- Infrastructure for the distribution of power from renewable energy sources located in the region to future locations for the production of green hydrogen [26].
- Creation of infrastructure for the production of green hydrogen on an industrial and municipal scale, importantly located near future hydrogen recipients [27].

The infrastructure for the distribution of power, referred to above, has been internally called the half-ring of the Lublin Hydrogen Valley and, according to the currently implemented project, has a length of up to 100 km [28]. According to the assumptions of the main investor, the half-ring of the Lublin Hydrogen Valley will be open to connecting the renewable energy sources and energy storage facilities and to power the electrolyzers belonging to external entities [29]. The idea of the half-ring as an open integrator of green hydrogen technology is to create economies of scale through joint effort; to create a strong substantive, technical, and production base for hydrogen technologies in the region; and to disperse the risk of innovation among all partners. The business goal of the half-ring is to optimize the entire value chain aimed at reducing the price of produced hydrogen [30]. Due to sharing the same connection infrastructure by many RES installations, energy storage facilities, and many electrolyzer installations connected to the half-ring, their CAPEX savings will be able to translate into a reduction in the cost of producing a unit of electricity, which, in the case of green hydrogen, is its main price-setting factor [31]. In addition, the half-ring will enable, regardless of the location of RES sources, the production of hydrogen directly at the recipient, i.e., without losses on compression, expansion, and transport [32] and without the need to invest in the construction and operation of expensive infrastructure for this purpose [33,34]. Savings related to balancing and transmitting electricity within the half-ring will be very beneficial for the price of hydrogen and for all participants in the Lublin Hydrogen Valley half-ring [35].

The availability of renewable energy around the world means it already plays a leading role in the decarbonization of the energy sector [36]. A thorough review of the scientific literature and the state of the art in the market shows that hybrid solar–wind power systems have been the main solutions to the challenges centered around reliable power supply, sustainability, and energy costs for several years [37]. However, there are still various challenges in the renewable energy industry, especially with regard to limited peak periods. According to the authors, the main challenges for hybrid solar–wind systems at present are overproduction, enabling policies, and storage of electricity.

In balancing networks powered by RES, it is very important to have an appropriate strategy for managing the energy produced and consumed. Nordström et al. [38], conducting case studies from several countries around the world, prove that there is no "one size fits all" approach to continuous balancing with high shares of Variable Renewable Energy. The main challenges depend on the system capacity and its connection possibilities, available technologies, and basic balancing principles. According to the authors, in the case of energy systems performing energy balancing with fine time granularity close to real time, the main challenge is frequency balancing at low levels of inertia. In the case of other systems considered, the main emphasis is on operating the system in the most economically viable way. However, the most important conclusion from the presented analyses should be the pursuit of engaging more technologies in order to contribute to continuous balancing.

Shavolkin et al. [39] proposed to improve the operation of a hybrid solar–wind system by equipping it with an energy storage with control of the power drawn from the network. The aim of the research was to increase the degree of use of energy from own renewable sources for self-consumption purposes while limiting the degree of battery discharge, taking into account deviations in the load schedule and generation of energy sources in relation to the calculated (forecasted) values. A mathematical 24 h model of energy processes was developed, taking into account the error in estimating the state of charge of the energy storage. The results of modeling using archive data on generation from renewable sources confirm the effectiveness of the proposed solutions.

The challenges of power balancing processes in power grids depend largely on the geographical location of the system and the prevailing conditions for renewable energy production. Fang et al. [40] assessed the potential and temporal complementarity of wind and solar energy in the northwestern provinces of China. Using the ERA5 reanalysis data on wind speed and solar irradiance, an assessment was conducted to determine the potential and spatial distribution of wind and solar energy in these provinces.

Nefabas et al. presented an hourly dispatch model to analyze the challenges of balancing the system and curtailing wind power in the future Ethiopian power grid system [41]. The model presented by the authors was validated using historical data and used to analyze the grid system in 2030 under different scenarios. The influence of hydropower generation in network balancing was considered.

Boubii et al. noticed a gap in existing renewable energy systems, especially in the area of stability and efficiency under variable environmental conditions [42]. This helped to develop a new hybrid system combining photovoltaic (PV) and wind energy. The novelty of this study lies in the adopted methodological approach, integrating dynamic modeling with a sophisticated control mechanism. The mechanism presented in this paper is a combination of model predictive control (MPC) and particle swarm optimization (PSO). It is designed to take into account the fluctuations inherent to PV and wind energy sources.

In addition, a new trend has emerged that is very helpful in balancing energy systems. This is the virtual power plant (VPP), which can play a very important role in energy system management, offering dynamic solutions to the challenges of renewable energy integration, grid stability, and demand management [43]. Today, PPVs are sophisticated devices that combine a variety of energy assets, including solar panels, wind turbines, battery storage systems, and demand response units.

Singh et al. [44] proposed the use of advanced machine learning algorithms, in particular support vector regression. This resulted in significant improvement in the efficiency and reliability of power systems. The SVR algorithm utilizes extensive archival data in the form of energy production, detailed weather patterns, and dynamic network load conditions. The aim of this advanced modeling was to accurately forecast the energy production from renewable energy sources. The authors' approach allowed for increased grid stability by better balancing supply and demand, mitigating the variability and irregularity of the generated power from renewable energy sources. Many authors promote more balanced integration of renewable energy into the microgrid, contributing to cleaner, more resilient, and efficient energy infrastructure. The results of such studies provide valuable insights into the development of smart energy systems capable of adapting to changing conditions, paving the way for future innovations in energy management.

Many publications indicate the need to coordinate the use of hydrogen technologies and electricity in order to change the direction and shape of energy use in the power grid [45]. Electrolytic systems for hydrogen production are quite demanding receivers of electricity, while hydrogen fuel cells are very good generators of electricity. Therefore, hydrogen technologies can be used to balance power grids in the area of energy supply and reception. Moreover, hydrogen is a good medium for storing energy for longer periods ranging from several weeks to several months. The above literature analysis shows that the research presented in the article is consistent with global trends in balancing power grids powered by RESs using hydrogen technologies and energy storage systems. The computational algorithm presented in the article fills the gap in the research results consisting of the accurate calculation of the size of the energy storage system using statistical and probabilistic methods. The authors follow current trends in science and in the commercialization of research results in the form of new technologies appearing on the market. Participation in the Intersolar Europe 2024 trade fair in Munich allowed us to notice that the offer of companies selling energy storage for RES systems is already extensive. However, the price of both small home energy storage (up to 20 kWh capacity) and large industrial ESS (over 1 MWh) remains relatively high. Therefore, the size of the energy storage must be calculated precisely, and its cost must be included in the business plan of the entire investment. RES developers must develop and present such calculations to the banks that are financing such multi-million investments. This article presents an algorithm that allows to balance the energy produced in the photovoltaic–wind mix and calculate the size of the energy storage system needed. Moreover, the methods of influencing the performance of each RES (photovoltaic system and wind turbine) are presented, which confirms that it is possible to construct the performance of the photovoltaic–wind mix. The measurement data from the real systems were employed for calculations and analyses. The power generated by a wind turbine with a maximum power of 3.45 MW was supplemented by the power generated by a photovoltaic system with a peak power of 3 MWp. Two variants of panel assembly in a photovoltaic system were presented and compared. The first one includes the traditional method of assembly towards the south (because the system under study is located in Poland on the northern hemisphere of the Earth), which allows to maximize the amount of energy produced. The second variant consists of the assembly of panels in a direction close to east–west, which is used to balance the power networks powered by RESs. This article presents precise statistical and probabilistic calculations of the power generated by individual energy sources and their mix. A major achievement is the analytical presentation of the process of power balancing and determining the size of the energy storage system. The authors argue that it is possible to have a structural influence on the construction and performance of photovoltaic systems, which may have a significant impact on the power balancing of the entire system and contribute to reducing the capacity of the energy storage system.

2. Materials and Methods

The authors used the Anaconda3 2024.10-1 platform in their research. It is a comprehensive open-source platform designed for Python users working in the fields of data analysis, machine learning, and broadly defined computational sciences. Due to its comprehensiveness and user-friendliness, Anaconda has become one of the most popular tools among data scientists, software engineers, and researchers. The Anaconda platform provides everything one needs to start working with the data, both the Python interpreter itself and a wide set of libraries and tools that are necessary for processing, analyzing, and visualizing the data. Any user can commence work immediately, avoiding the need to manually install libraries such as NumPy, pandas, Matplotlib, or Scikit-learn. Anaconda includes over 1500 built-in libraries, which allow for starting projects quickly without worrying about compatibility issues or installing additional packages. The authors used the Anaconda platform for Python 3.7 programming because of their familiarity with this programming environment. The platform uses the Conda package manager, which allows for easy installation, updating, and management of packages and their dependencies. It works independently of the system package manager, which reduces the risk of conflicts. The authors value Anaconda for its ability to quickly launch Jupyter Notebook v7 for working with interactive code, data analysis, and visualizations. Anaconda has a large user community and rich documentation, which makes it easy to troubleshoot and learn.

Figure 2 graphically depicts the data processing algorithm used in this article. The calculations were performed for a 3.45 MW wind turbine and for two design variants of the photovoltaic systems.

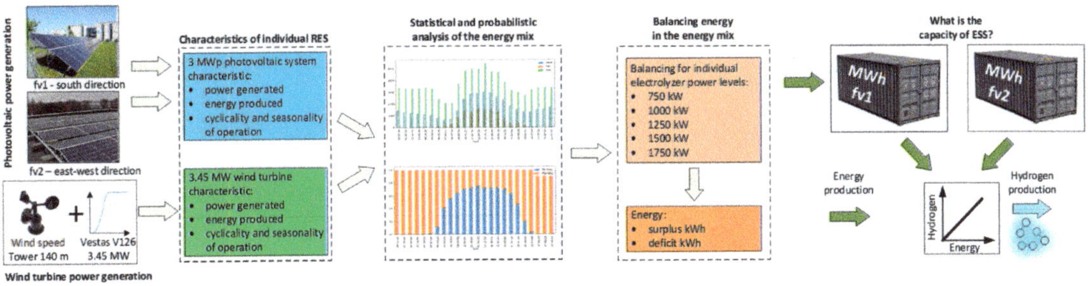

Figure 2. Data flow diagram.

The variant termed fv1 (in Figure 2) concerns the installation of panels in the south direction, while the variant fv2 concerns the installation of panels in the direction close to east–west. In the first phase, the characteristics of individual RES sources were made, taking into account the generated power, produced energy, and cyclicality and seasonality in the operation of the systems. Then, statistical and probabilistic analyses related to the generation of power and energy production for the created energy mixes (for two variants of the photovoltaic panel arrangement fv1 and fv2) were presented. Probabilistic calculations are of particular importance, since they allow to determine both the generated power and the amount of energy produced with accuracy to the probability distribution. The next phase of the research included the process of balancing the energy production system from the photovoltaic–wind mix, taking into account the energy receivers in the form of hydrogen generation systems. The process of balancing the energy production and reception system was presented for several power levels of hydrogen electrolyzers. In this way, the amount of excess energy and its deficit in individual hourly intervals of the day were calculated. Energy surplus and deficit modeling allowed us to determine the size of the energy storage system required. The final discussion includes determining the impact of individual PV system design variants on the amount of hydrogen produced from the PV–wind mix.

3. Component Characteristics

In this section, the authors characterize all the components of the low-emission hydrogen production system using a mix of renewable energy sources and an energy storage. Particular attention is paid to the influence of the geographical and construction context on the performance of the entire system. The geographical context includes the exact selection of the location for the wind turbine and the photovoltaic system. The engineering context includes the selection of the power of individual sources, the selection of components for their construction, and aspects related to the tilt angle and azimuth of the photovoltaic panels. The authors attempt to prove in an analytical and graphical way that it is possible to design a system for generating energy from RESs, which affects the size of the energy storage and the amount of low-emission hydrogen produced.

3.1. Power Generation by Photovoltaic Systems

The location of the photovoltaic system has a huge impact on its performance in the form of instantaneous generated power and the amount of energy produced. The decision to build a photovoltaic system in a given country in the world is of particular importance. The longitude and latitude of the photovoltaic system location (called the geographic context) determine the amount of electricity produced in individual months and has been the subject of many scientific works. The geographic context primarily affects the seasonality of the overlap of seasons and the height of the sun above the horizon, which affects the length of the day and, thus, the amount of energy produced. However, even within a small country, the geographic context can have a dozen or so percent impact on the performance of a photovoltaic system. This is due to the different local climatic conditions related to sunlight, cloudiness, and wind conditions. Photovoltaic systems located in Poland near Lublin are characterized by several percent higher efficiency than those located in the vicinity of nearby Warsaw (located 170 km away).

The Lublin region has the best solar conditions in Poland, and this is confirmed by the data on the amount of energy produced from the peak power of installed photovoltaic systems. In our region, 1200 kWh of energy can be obtained from 1 kWp of installed power, as shown in Figure 3.

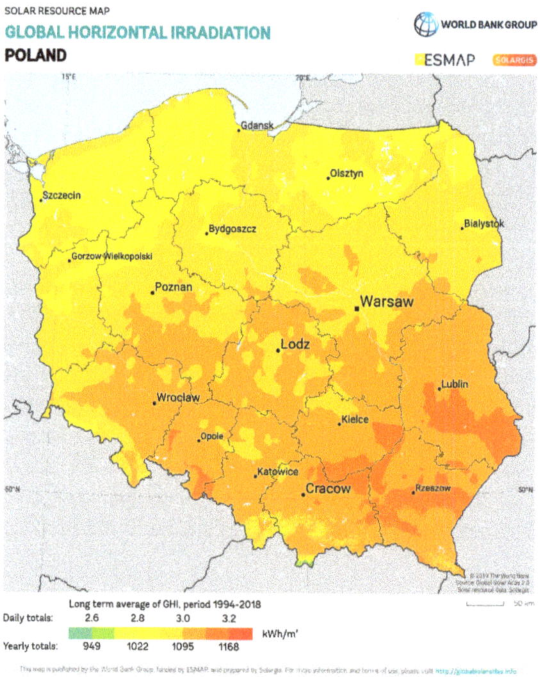

Figure 3. Solar exposure conditions in Poland and their impact on the performance of photovoltaic systems [46].

Figure 4 depicts two time series describing the generation of instantaneous power by two photovoltaic systems that differ in the engineering context. The first one (fv1) is a classic approach to ground mounting of the photovoltaic systems facing exactly south (Lublin in Poland is located in the northern hemisphere) with an optimal tilt angle of 40°. The peak power of the fv1 photovoltaic system is 3 kWp. Its performance has been scaled to a peak power of 3 MWp to achieve an appropriate level to the power of the wind turbine. The second engineering context (fv2) includes the installation of photovoltaic systems on flat roofs of buildings. The panels are mounted towards the east and west at a small angle of 15°, azimuth 295°. The physical appearance of both variants of the development is shown in Figure 5. It was decided to present such a comparative analysis in order to confirm the possibility of shaping the generated power by photovoltaic systems during the day.

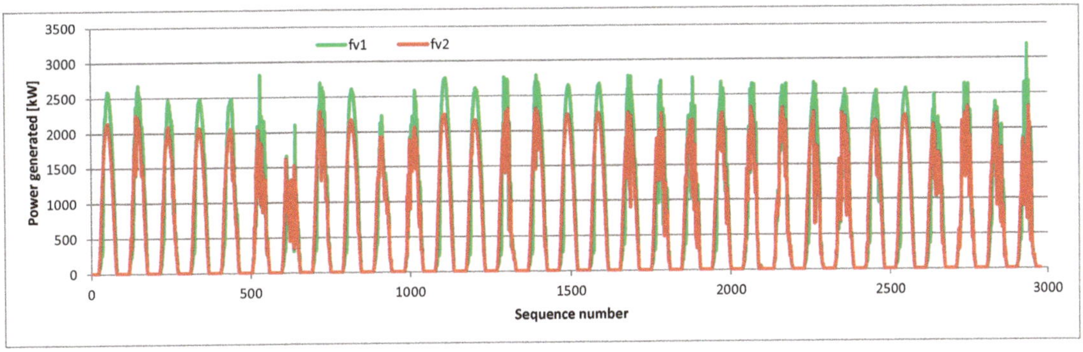

Figure 4. Time series of power generated every 15 min by the photovoltaic systems with different panel mounting orientations during the entire month of May 2024.

(**a**) (**b**)

Figure 5. Various engineering contexts of panel assembly in photovoltaic systems: (**a**) photovoltaic panels mounted towards the south; (**b**) photovoltaic panels mounted in the east–west direction.

The differences in power generation during the day by both systems are shown in Figure 6. The system mounted on the roof starts generating power significantly earlier than the ground system. However, it achieves significantly lower maximum power at high noon. The authors' task is to compare and describe quantitatively the differences in the operation of both systems and the impact of these differences on energy balancing in the low-emission hydrogen production system and on the size of the energy storage system needed for this.

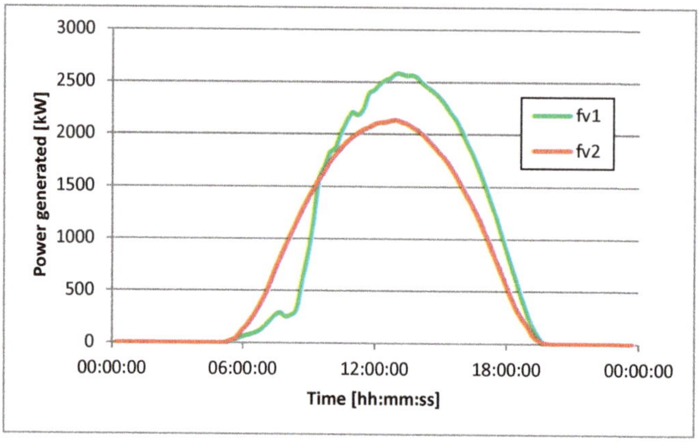

Figure 6. Daily course of generated power by photovoltaic systems with different panel mounting directions on 1 May 2024.

The first analytical method used to describe the similarity (or lack thereof) of measurement data is their mutual correlation. The correlation is presented in Figure 7.

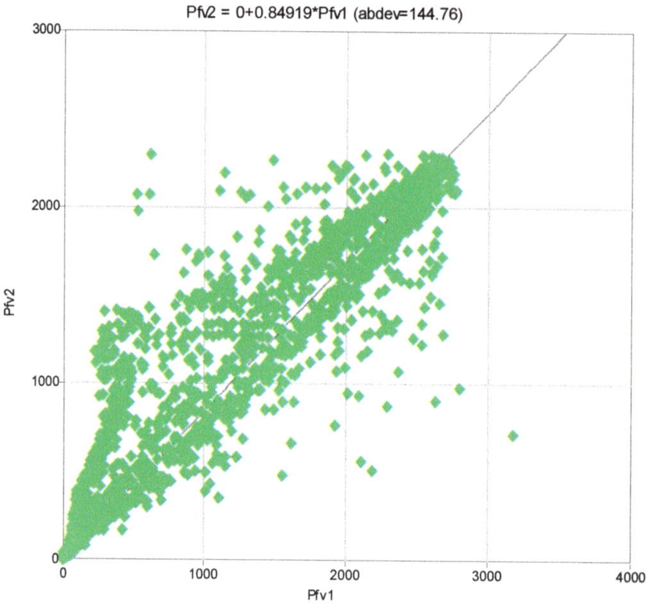

Figure 7. Correlation in the generated power by photovoltaic systems with different orientation of photovoltaic panels.

The difference in energy production between south-facing and east- and west-facing photovoltaic systems is due to the characteristics of solar radiation and the angle of incidence of sunlight during the day. South-facing panels achieve the highest efficiency in the northern hemisphere regions because of the following:

- They maximize the total amount of energy reaching the panels during the day.
- Production peaks at noon, when the sun is at the highest point in the sky and the angle of incidence of sunlight is most optimal.

- The typical annual production for a well-optimized system in Poland is around 1000–1200 kWh per kWp.

East- and west-facing panels generate energy at different times of the day:

- East: energy production is maximum in the morning and decreases in the afternoon.
- West: production increases in the afternoon and decreases in the morning.
- These systems spread production better throughout the day but do not achieve as high total production as south-facing systems.

The typical annual production for such systems in Poland is around 900–1000 kWh per kWp, which is 10–15% lower than for south-facing panels. The differences in production during the day for both solutions are as follows:

- For south-facing PV panels: they generate peak power at noon, which often coincides with periods of high energy demand, but production before noon and in the afternoon is much lower.
- For east- and west-facing PV panels: production is more evenly distributed throughout the day, which can better match consumer demand in the morning (east) and afternoon (west).

Although the energy production is lower, the east–west orientation has some advantages:

- Optimal use of space: on flat roofs, east–west systems can accommodate more panels because there is no need to maintain large spacing between rows (less shading problems).
- Better match to demand: in some cases, the production spread better meets the needs of households and businesses.

The optimal orientation depends on your energy consumption profile and local conditions. If your energy needs are concentrated at certain times of the day, the orientation can be adjusted to maximize efficiency during these times of the day.

The course of the mean hourly power generated by the fv2 photovoltaic system during the entire month of May 2024 is shown in Figure 8.

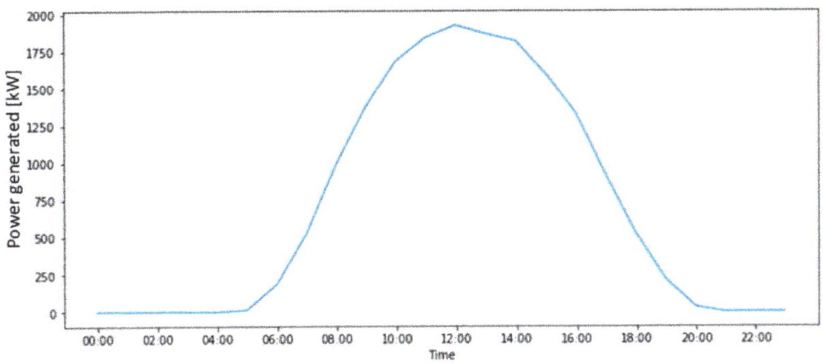

Figure 8. The course of the mean hourly power generated by the fv2 photovoltaic system during the entire month of May 2024.

Further analyses will be conducted for an energy mix consisting of a 3.45 MW wind turbine and a 3 MWp peak photovoltaic system with different panel orientations (south—fv1 and east–west—fv2).

3.2. Generating Power with Wind Turbines

In the case of wind turbines, the key decision is whether they are to be installed on land or offshore. Offshore and onshore wind turbines differ in many ways, mainly due to their location and environmental conditions [47]. Offshore wind turbines are characterized by the following features [48,49]:

- They are usually located far from the coast, where winds are stronger and more stable.
- Extreme environmental conditions, such as salt water, moisture, and strong waves, require additional protection against corrosion.
- The infrastructure must be resistant to the effects of sea currents, waves, and potential ice.

Onshore wind turbines are characterized as follow:

- They are located on hills, on open plains, or in mountainous regions where winds are sufficiently strong.
- Less extreme environmental conditions, which reduces construction and maintenance costs.

Wind turbines installed at sea are usually larger than onshore ones, with rotors with diameters exceeding 200 m [50]. The power of a single turbine can reach up to 15–20 MW, which is necessary to use expensive infrastructure. Smaller turbines are usually installed on land, with rotor diameters of 100–150 m. Their typical power is 1.5–4 MW [51].

Investments in turbine systems installed at sea are characterized by higher construction and installation costs, mainly due to the need to use ships, deep-sea foundations, and underwater infrastructure [52]. Maintenance and service are also more expensive, requiring specialist equipment, e.g., service vessels [53]. Land-based wind turbines are characterized by lower construction and service costs, with easier access to the turbines and lower technological requirements. Land-based investment is cheaper to implement, but it is more difficult to find suitable locations with good wind conditions [54].

Land-based and offshore wind turbines also differ significantly in terms of efficiency. Offshore turbines are characterized by a higher capacity utilization factor (40–60%) as a result of more stable winds [55]. They also have a smaller impact on the local community, for example, in the form of noise [56,57]. Wind turbines installed on land have a lower capacity factor (20–40%) due to more variable winds [58]. They also have a greater impact on the landscape and local communities (e.g., noise, impact on animals). Offshore wind turbines are more expensive but have a greater potential for energy production, while onshore turbines are cheaper and easier to maintain, but they are limited by local wind and social conditions [59]. The choice of technology depends on the energy and budget goals of the project [60].

Similarly to the case of insolation, the Lublin region has good wind conditions [61]. In almost all general studies, wind conditions are indicated as rather favorable, as shown in Figure 9. However, in the local context, they are favorable and allow for obtaining an annual power factor of 30 to 35%.

Figure 10 presents a time series of power generated by the wind turbine during the month of May. The data were obtained as a result of processing wind speed measurements on a 140 m high measuring tower. Wind speeds were mean values measured in 10 min time intervals. As a result of sampling measurements in such time intervals, almost 4500 records were obtained during the entire month of May. The measured wind speed was used to calculate the power generated by the Vestas 3.45 MW turbine (Vestas Wind Systems A/S, Aarhus, Denmark) using its characteristics. A wind turbine with the assumed power is currently operating at the location of the measuring tower. The course of generated power clearly shows that the measured wind speeds repeatedly allow the turbine to operate at maximum power. Calculations show that during a month, the wind turbine is able to produce 729.663 kWh of energy at a mean wind speed of 6.4 m/s. The monthly power factor is therefore 28%. The authors' experience shows that May is a rather average month for wind energy production. May is not a representative month characterizing the production of a given turbine in a given location throughout the year. It was chosen at random to present the possibilities of balancing the energy produced in the photovoltaic–wind mix.

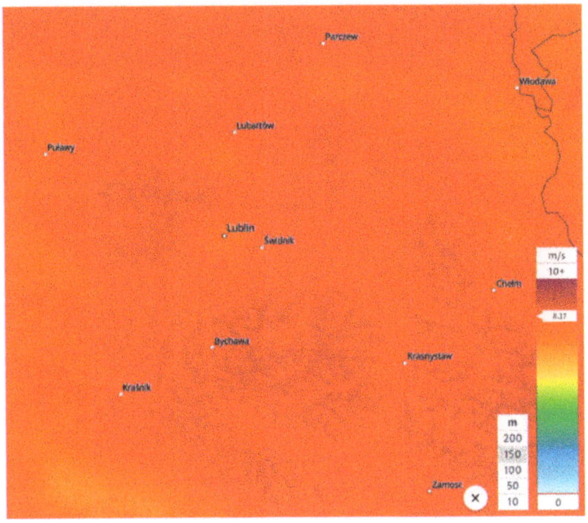

Figure 9. Wind conditions in the Lublin region in Poland [62].

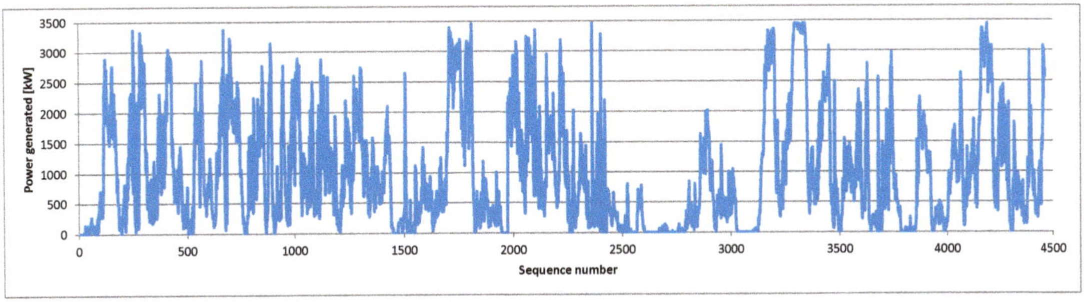

Figure 10. Time series of power generated by a wind turbine during the month of May [9].

The course of the mean hourly power generated by the 3.45 MW wind turbine during the entire sample month of May is shown in Figure 11. The graph below covers a daily time window.

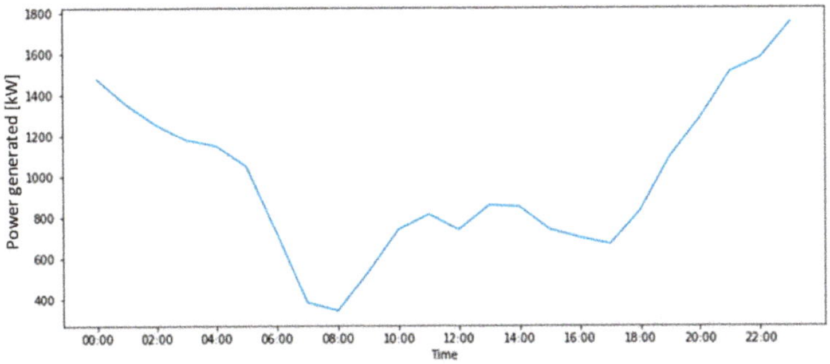

Figure 11. The course of the mean hourly power generated by the 3.45 MW wind turbine during the entire sample month of May [9].

4. Results

Photovoltaic and wind system operators and managers typically analyze the performance of their systems over a one-month period. This is usually the period of settlement with the energy supplier for the energy taken by the customer and delivered to the power grid. A period of one month is also long enough that the daily performance of photovoltaic and wind systems can be treated as invariant within the scope of seasonality [63]. Figure 12 depicts the hourly mean values of power generated by a 3 MWp peak photovoltaic system oriented east–west and a 3.45 MW wind turbine and their mix.

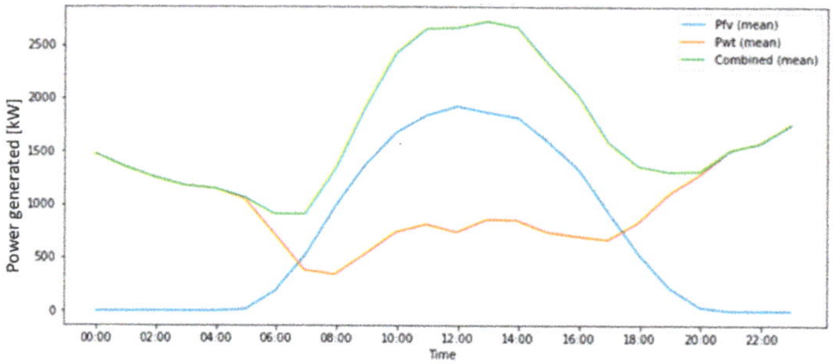

Figure 12. Mean hourly power generated by a 3 MWp peak east–west oriented photovoltaic system and a 3.45 MW wind turbine and their mix in May 2024.

Data on the generation of these two RESs and their mix were subjected to traditional statistical analysis, and quantile calculations were made [63]. Quantile analysis carries a lot of information for people involved in power balancing in the power grid because its results were calculated with accuracy to the probability distribution. The data presented in Table 1 show that in an hourly perspective covering the entire month of operation, the wind turbine is a much more stable source of energy than the photovoltaic system. The probability that the power generated by the photovoltaic system will be equal to or less than 372.26 kW is 50%. With almost 30% probability, the power produced will be equal to 0 kW. It should be noted that the mean hourly values of generated power during the entire month of May are still being considered.

Table 1. Basic statistics and quantiles for individual renewable energy sources and their combinations.

	Country	Mean	Std	Min	10%	20%	30%	40%	50%	60%	70%	80%	90%	95%	99%	Max
Pfv	24.00	701.58	767.69	0.00	0.00	0.00	0.02	62.59	371.26	840.46	1335.10	1628.42	1834.47	1862.38	1910.59	1923.91
Pwt	24.00	981.98	381.12	340.18	571.07	713.29	738.33	815.31	853.63	1082.67	1184.00	1308.77	1499.95	1567.79	1713.21	1753.61
Combined	24.00	1683.56	594.34	906.09	1086.54	1221.41	1312.26	1352.76	1493.21	1583.53	1921.28	2367.06	2659.17	2666.06	2710.45	2723.47

The hourly power values generated by the mix of a 3 MWp peak east–west-oriented PV system and a 3.45 MW wind turbine are shown in a box plot in Figure 13. The box plot shows the distribution of the data using several key statistical measures. The box plot contains Minima and Maxima (non-outlier values), Lower Quartile (Q1), Median, and Upper Quartile (Q3). This tool is useful for quick analysis of the distribution of the data, detecting asymmetries, and identifying outliers.

Figure 14 represents the share in the energy mix of the mean hourly power values generated by a 3 MWp peak photovoltaic system oriented to the south (fv1) and a 3.45 MW wind turbine. Figure 15 shows the share in the energy mix of the mean hourly power values generated by a 3 MWp peak photovoltaic system oriented to the east–west (fv2) and a 3.45 MW wind turbine.

A comparison of the data presented in Figures 14 and 15 clearly indicates that there were increases in the share of photovoltaic energy in the mix at 6:00 a.m. (from 11 to 20%), at 7:00 a.m. (from 40 to 58%), at 8:00 a.m. (from 50 to 74%), and at 9:00 a.m. (from 63 to 72%). Then, we observe the decline in the share of photovoltaic energy in the mix at 10:00 a.m. (from 70 to 69%), at 11:00 a.m. (from 72 to 69%), at 12:00 p.m. (from 75 to 72%), etc. The largest percentage of the decrease occurred at 6:00 p.m. (from 46 to 39%). These data clearly indicate that, as a result of the east–west orientation of the photovoltaic system, the share of energy from it in the energy mix increased significantly in the morning hours. The share of energy coming from this photovoltaic system simultaneously decreased by several percent from 10:00.

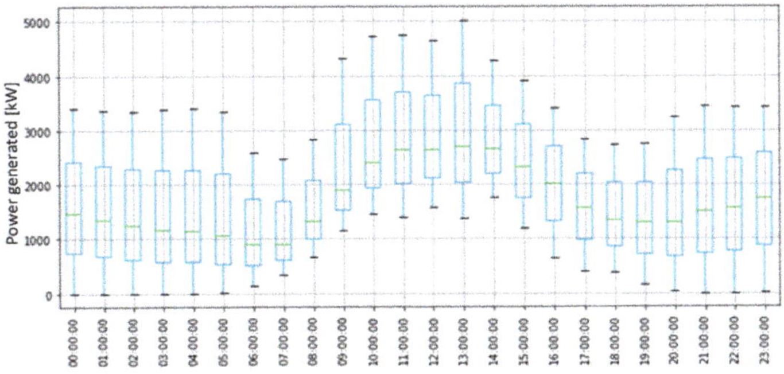

Figure 13. Hourly values of the power generated by the mix of a 3 MWp peak east–west-oriented PV system and a 3.45 MW wind turbine in box plot form.

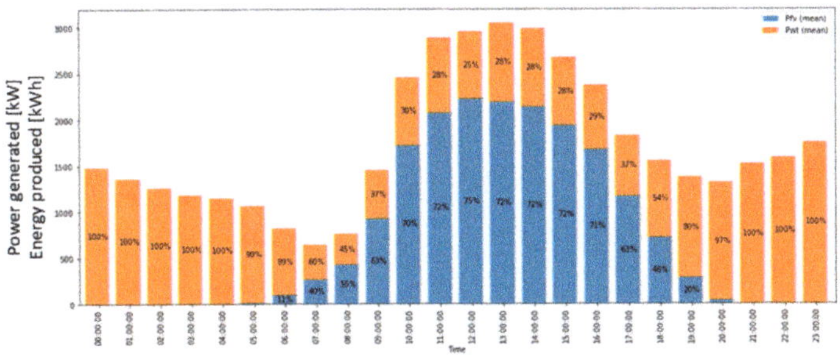

Figure 14. Share in the energy mix of mean hourly values of power generated by a 3 MWp peak photovoltaic system oriented towards the south (fv1) and a 3.45 MW wind turbine [9].

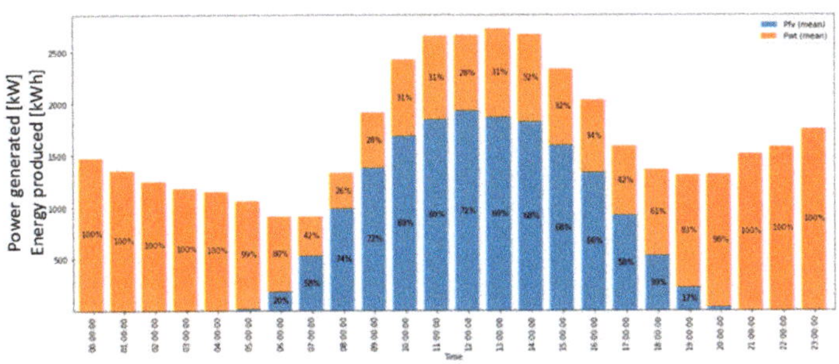

Figure 15. Share in the energy mix of mean hourly values of power generated by a 3 MWp peak photovoltaic system oriented east–west (fv2) and a 3.45 MW wind turbine.

The share of the mean hourly power values generated by a 3 MWp peak photovoltaic system oriented east–west (fv2) and a 3.45 MW wind turbine in the energy mix in a standardized form in the range 0–1 is presented in Figure 16.

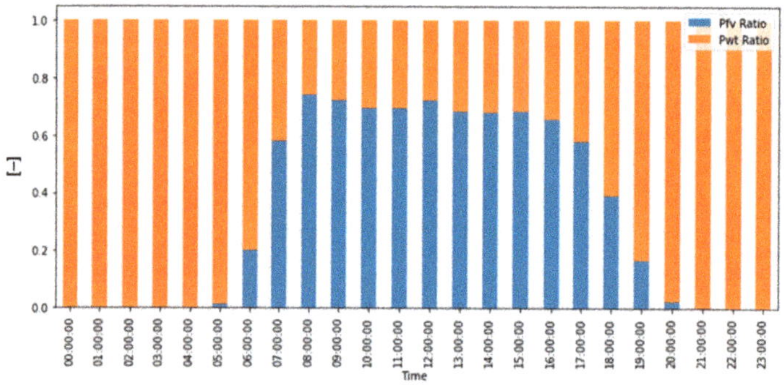

Figure 16. Share in the energy mix of mean hourly values of power generated by a 3 MWp peak photovoltaic system oriented east–west (fv2) and a 3.45 MW wind turbine—standardized presentation in the range 0–1.

The balancing of energy in the power grid is not influenced only by the amount of energy produced from the photovoltaic–wind mix. In this complicated process, the load on the network, which results from the type of electrical receivers used, is of great importance. In this article, the only receiver will be the low-emission hydrogen generation system. It usually takes the form of an ISO container-mounted water electrolyzer in alkaline, PEM, or, increasingly, AEM technology [64]. This section of the article will present the simulations related to the use of energy obtained from the previously presented photovoltaic–wind energy mix to power low-emission hydrogen production systems with a capacity of 750 to 1750 kW [65]. Figure 17 shows the probability of powering the hydrogen generation system with energy from the photovoltaic–wind mix. Table 2 presents the same data but in tabular form. Their analysis shows that the energy produced in the mix is sufficient to power the selected load levels needed to power water electrolyzers at specific hours of the day. For example, in the hourly interval marked as 14:00, the energy from the mix is able to supply with probability 1 electrolyzers operating at all possible power levels from 750 to 1750 kW. However, it is already able to supply with probability 1 electrolyzer powers of 750, 1000, and 1250 kW between 10:00 and 14:00. For some analyses, it is more convenient to present the calculation results in tabular form. In the case of a probability of less than 1, it means that the energy from the mix is not sufficient to meet the energy demand of the green hydrogen generation systems. The missing energy must be drawn from the power grid. Significant amounts of energy are clearly missing at night, in the morning, and in the evening. From this shape of the graph, we immediately conclude that the excess energy produced from the mix should be stored and returned to the hydrogen generation process when required.

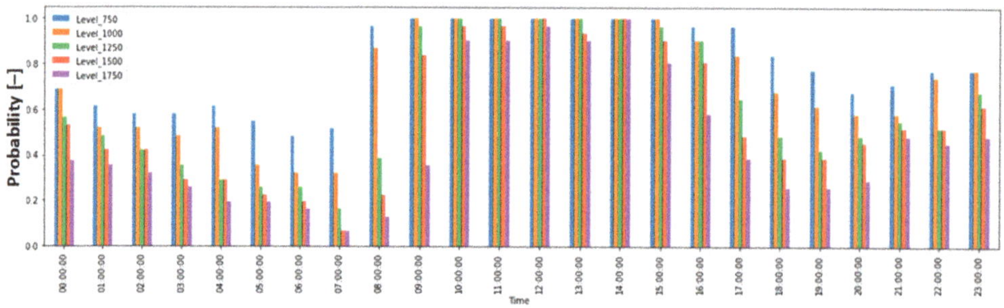

Figure 17. Probability of generating individual power levels from a solar–wind mix.

Table 2. Probability of generating individual power levels from a solar–wind mix.

Time	00:00	01:00	02:00	03:00	04:00	05:00	06:00	07:00	08:00	09:00	10:00	11:00	12:00	13:00	14:00	15:00	16:00	17:00	18:00	19:00	20:00	21:00	22:00	23:00
Level_750	0.69	0.61	0.58	0.58	0.61	0.55	0.48	0.52	0.97	1.00	1.00	1.00	1.00	1.00	1.00	1.00	0.97	0.97	0.84	0.77	0.68	0.71	0.77	0.77
Level_1000	0.69	0.52	0.52	0.48	0.52	0.35	0.32	0.32	0.87	1.00	1.00	1.00	1.00	1.00	1.00	1.00	0.90	0.84	0.68	0.61	0.58	0.58	0.74	0.77
Level_1250	0.56	0.48	0.42	0.35	0.29	0.26	0.26	0.16	0.39	0.97	1.00	1.00	1.00	1.00	1.00	0.97	0.90	0.65	0.48	0.42	0.48	0.55	0.52	0.68
Level_1500	0.53	0.42	0.42	0.29	0.29	0.23	0.19	0.06	0.23	0.84	0.97	0.97	1.00	0.94	1.00	0.90	0.81	0.48	0.39	0.39	0.45	0.52	0.52	0.61
Level_1775	0.38	0.35	0.32	0.26	0.19	0.19	0.16	0.06	0.13	0.35	0.90	0.90	0.97	0.90	1.00	0.81	0.58	0.39	0.26	0.26	0.29	0.48	0.45	0.48

5. Discussion

The Python programming language tool can generate very useful graphs to visualize the energy balancing strategies in power systems. For example, Figure 18 presents the excess and the deficit of energy production from the mix on an hourly basis for different levels of the electrical load.

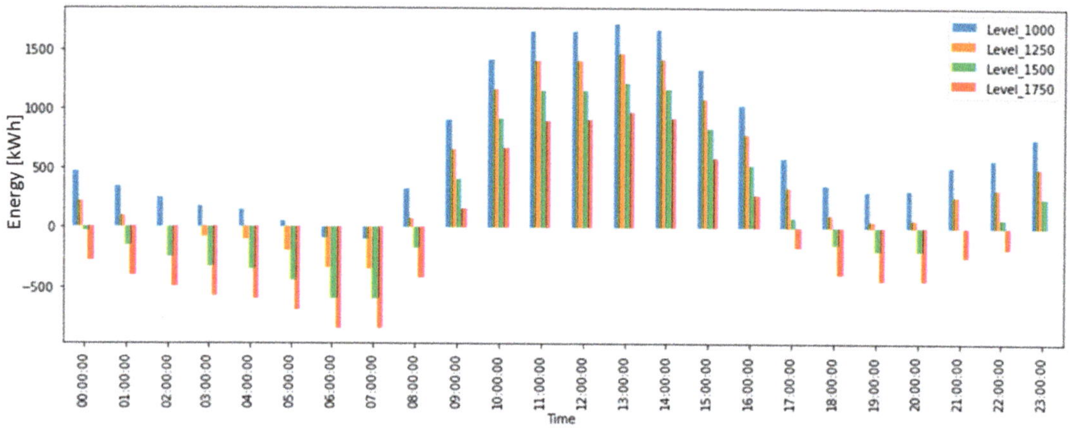

Figure 18. Excess and deficit energy production from the mix on an hourly basis for different electrical load levels.

Table 3 summarizes both the overproduction and the deficit in energy production in the photovoltaic–wind mix over the daily period. Table 3 also includes the results of their modeling.

Table 3. Energy balance for individual levels of generated power from the solar–wind mix.

Level	Energy Surplus [kWh]	Energy Deficit [kWh]	Balance [kWh]	Energy Surplus (Model) [kWh]	Energy Surplus (Model) [kWh]
Level_1000	16,265.17	140.37	16,273	132.9	16,265.17
Level_1250	11,380.81	−975.27	11,133.25	−727.35	11,380.81
Level_1500	7271.36	−2865.82	7331	−2925.1	7271.36
Level_1750	5074.42	−6668.88	4866.25	−6460.35	5074.42
Level_1683.58				5391	−5391

The balancing characteristic of the energy system powered by the photovoltaic–wind mix for different external load levels is presented in Figure 19. Both graphical visualization and mathematical calculations allow for the precise calculation of the point of intersection of the characteristic with the x-axis. Perfect balancing occurs for the receiver power of 1724.38 kW for (fv1) and 1683.58 kW for (fv2). The receiver power needed for ideal energy balance for variant (fv2) decreased as a result of the photovoltaic system (fv2) producing smaller amounts of energy during the day. This situation will affect the smaller amount of green hydrogen produced during the day. However, this is only less than 2.5% of production.

Figure 20 shows the characteristics of energy surplus and deficit for individual power levels of hydrogen electrolyzers. From the graphical presentation and mathematical calculations, it is evident that for the ideal balancing point of 1724.38 kW, the surplus energy and the deficit energy are equal and have values of 7409 kWh. The simplest and most accurate interpretation of the obtained results is that they constitute an ideal calculation of the energy capacity of the required energy storage. For variant (fv2), the same calculations were made, and the energy storage capacity was 5391 kWh. This means a fairly large decrease compared to the energy storage system (ESS) capacity for variant (fv1).

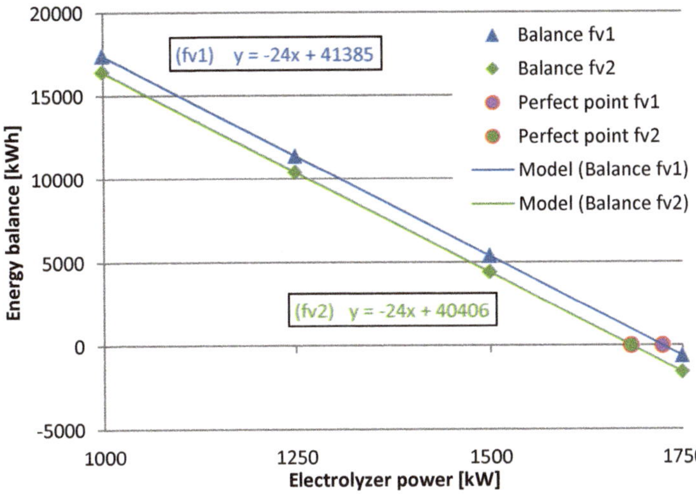

Figure 19. Balancing characteristics of an energy system powered by a photovoltaic–wind mix for different external load levels.

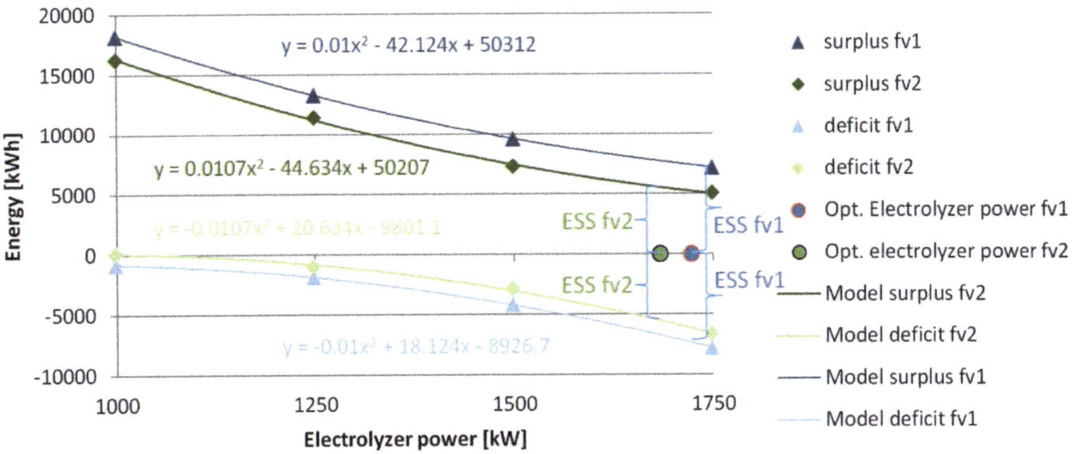

Figure 20. Characteristics of energy surplus and deficit for individual power levels of hydrogen electrolyzers.

6. Conclusions

This article was introduced with a brief description of the generated energy and power produced by the photovoltaic systems with a peak power of 3 MWp and different tilt and orientation of photovoltaic panels. The characteristics of the latest systems for generating energy by wind turbines were also presented. In the subsequent stages, the necessity of balancing energy in power networks powered by a mix of renewable energy sources was demonstrated. Then, a calculation algorithm was presented in the area of balancing the energy system powered by a photovoltaic–wind energy mix and powering the low-emission hydrogen production process. It was analytically and graphically demonstrated that the process of balancing the entire system can be influenced by structural changes in the installation of the photovoltaic panels. It was proven that the tilt angle and orientation of the panels have a significant impact on the level of power generated by the photovoltaic system and thus by the energy mix in individual hourly intervals. It is worth emphasizing that the authors used the actual measurement data from a specific geographical context, i.e., from the Lublin region in Poland. Both traditional statistical methods and probabilistic analysis were employed in the calculations. This means that the results obtained were accurate to the probability distribution. Balancing the generated power and the energy produced for the entire month considered in hourly intervals throughout the day is the essence of the calculations made by the authors. The results obtained can be the basis for conceptual assumptions and economic analyses related to investments in the design and creation of new renewable energy generation capacities for the production of low-emission hydrogen. Based on the calculations and analyses carried out, the following conclusions can be drawn:

1. The orientation of photovoltaic panels in the east–west direction significantly affects the amount of energy produced in individual hourly intervals compared to the orientation of the panels to the south. As a result of the orientation of the panels in the east–west direction, a significant increase in the share of energy from the photovoltaic system in the energy mix was observed in the hourly interval from 6:00 to 9:00 and a slight decrease in the remaining hours. However, it is worth noting that the hourly interval of the increase in production falls on the morning peak hours, when the cost of energy drawn from the power grid is very high.
2. As a result of the orientation of the panels in the east–west direction, the power of the powered electrolyzers needed for perfect system balancing was reduced by less than 2.5%. This will reduce the production of green hydrogen by less than 1 kg per day. However, as a result of the orientation of the panels in the east–west direction, better energy balancing in the system was achieved, and the required energy storage system was reduced by over 27% from 7409 kWh to 5391 kWh. In this way, the thesis put forward at the beginning of this article that a structural influence on the construction and performance of photovoltaic systems is possible, having a very large impact on the power balancing of the entire system and contributing to the reduction in the capacity of the energy storage system, was positively verified. This means that it is possible to design the performance of the photovoltaic–wind mix for the production of low-emission hydrogen.

According to the authors, in the future, the presented analyses should be extended to include economic calculations. By assigning the cost of energy drawn from the network and energy returned to the network by the energy mix to individual time intervals, specific amounts related to earning on effective balancing of the energy network can be obtained. According to the applied calculation algorithm, it is worth performing the analyses for all months of the year, because the article made calculations for only one month of the year.

Author Contributions: Conceptualization, A.M. (Arkadiusz Małek) and A.M. (Andrzej Marciniak); methodology, A.M. (Andrzej Marciniak); software, A.M. (Andrzej Marciniak); validation, A.M. (Arkadiusz Małek), A.M. (Andrzej Marciniak) and A.D.; formal analysis, A.M. (Andrzej Marciniak); investigation, A.M. (Arkadiusz Małek); resources, A.M. (Arkadiusz Małek), T.S.; data curation, A.D.; writing—original draft preparation, A.M. (Arkadiusz Małek) and A.D., T.S.; writing—review and editing, A.M. (Arkadiusz Małek), A.D. and T.S.; visualization, A.M. (Andrzej Marciniak); supervision, A.M. (Arkadiusz Małek), A.D. and T.S.; project administration, A.M. (Arkadiusz Małek) and A.D.; funding acquisition, A.D. All authors have read and agreed to the published version of the manuscript.

Funding: This research received no external funding.

Data Availability Statement: The original contributions presented in this study are included in the article; further inquiries can be directed to the corresponding author.

Conflicts of Interest: The authors declare no conflicts of interest.

References

1. Hoody, P.; Chiasson, A.; Brecha, R.J. The Transition to a Renewable Energy Electric Grid in the Caribbean Island Nation of Antigua and Barbuda. *Energies* **2023**, *16*, 6206. [CrossRef]
2. Lian, J.; Fu, Q.; Cui, L.; Liu, R.; Guo, B. Offshore Renewable Energy. *J. Mar. Sci. Eng.* **2024**, *12*, 749. [CrossRef]
3. Qiu, Z.; Tian, Y.; Luo, Y.; Gu, T.; Liu, H. Wind and Photovoltaic Power Generation Forecasting for Virtual Power Plants Based on the Fusion of Improved K-Means Cluster Analysis and Deep Learning. *Sustainability* **2024**, *16*, 10740. [CrossRef]
4. Seward, W.; Chi, L.; Qadrdan, M.; Allahham, A.; Alawasa, K. Sizing, economic and reliability analysis of photovoltaics and energy storage for an off-grid power system in Jordan. *IET Energy Syst. Integr.* **2023**, *5*, 393–404. [CrossRef]
5. Gryniewicz-Jaworska, M. Analysis of the Selection of Compensation Devices Determining the Reactive Power Balance of Wind Farms. *Adv. Sci. Technol. Res. J.* **2020**, *14*, 30–36. [CrossRef]
6. Gryniewicz-Jaworska, M. Application of Heuristic Optimization for the Selection of Parameters of the Generation System and Reactive Power Consumption of Wind Farms. *Adv. Sci. Technol. Res. J.* **2020**, *14*, 9–14. [CrossRef]
7. Dolge, K.; Blumberga, D. Transitioning to Clean Energy: A Comprehensive Analysis of Renewable Electricity Generation in the EU-27. *Energies* **2023**, *16*, 6415. [CrossRef]
8. Małek, A.; Dudziak, A.; Caban, J.; Matijošius, J. Probabilistic Analysis of Low-Emission Hydro-gen Production from a Photovoltaic Carport. *Appl. Sci.* **2024**, *14*, 9531. [CrossRef]
9. Małek, A.; Marciniak, A.; Bednarczyk, T. Probabilistic Analysis of Electricity Production from a Photovoltaic–Wind Energy Mix for Sustainable Transport Needs. *Sustainability* **2024**, *16*, 10164. [CrossRef]
10. Li, N.; Lukszo, Z.; Schmitz, J. An approach for sizing a PV–battery–electrolyzer–fuel cell energy system: A case study at a field lab. *Renew. Sustain. Energy Rev.* **2023**, *181*, 113308. [CrossRef]
11. Burian, O.; Dančová, P. Compressed Air Energy Storage (CAES) and Liquid Air Energy Storage (LAES) Technologies—A Comparison Review of Technology Possibilities. *Processes* **2023**, *11*, 3061. [CrossRef]
12. Loza, B.; Minchala, L.I.; Ochoa-Correa, D.; Martinez, S. Grid-Friendly Integration of Wind Energy: A Review of Power Forecasting and Frequency Control Techniques. *Sustainability* **2024**, *16*, 9535. [CrossRef]
13. Clifton, A.; Kilcher, L.; Lundquist, J.K.; Fleming, P. Using machine learning to predict wind tur-bine power output. *Environ. Res. Lett.* **2013**, *8*, 024009. [CrossRef]
14. Güven, A.F.; Yücel, E. Sustainable energy integration and optimization in microgrids: Enhancing efficiency with electric vehicle charging solutions. *Electr. Eng.* **2024**. [CrossRef]
15. Tsai, W.-C.; Hong, C.-M.; Tu, C.-S.; Lin, W.-M.; Chen, C.-H. A Review of Modern Wind Power Generation Forecasting Technologies. *Sustainability* **2023**, *15*, 10757. [CrossRef]
16. Jency, W.G.; Judith, J.E. Homogenized Point Mutual Information and Deep Quantum Reinforced Wind Power Prediction. *Int. Trans. Electr. Energy Syst.* **2022**, *2022*, 3686786. [CrossRef]
17. Gil-García, I.C.; García-Cascales, M.S.; Fernández-Guillamón, A.; Molina-García, A. Categorization and Analysis of Relevant Factors for Optimal Locations in Onshore and Offshore Wind Power Plants: A Taxonomic Review. *J. Mar. Sci. Eng.* **2019**, *7*, 391. [CrossRef]
18. Möller, M.C.; Krauter, S. Investigation of Different Load Characteristics, Component Dimensioning, and System Scaling for the Optimized Design of a Hybrid Hydrogen-Based PV Energy System. *Hydrogen* **2023**, *4*, 408–433. [CrossRef]
19. Olczak, P.; Matuszewska, D. Energy Storage Potential Needed at the National Grid Scale (Poland) in Order to Stabilize Daily Electricity Production from Fossil Fuels and Nuclear Power. *Energies* **2023**, *16*, 6054. [CrossRef]
20. Lee, Y.H.; Jeong, I.W.; Sung, T.H. An Assessment of the Optimal Capacity and an Economic Evaluation of a Sustainable Photovoltaic Energy System in Korea. *Sustainability* **2021**, *13*, 12264. [CrossRef]
21. Aldosari, O.; Ali, Z.M.; Abdel Aleem, S.H.E.; Mostafa, M.H. Optimizing microgrid performance: Strategic integration of electric vehicle charging with renewable energy and storage systems for total operation cost and emissions minimization. *PLoS ONE* **2024**, *19*, e0307810. [CrossRef] [PubMed]
22. Li, Y.; Dong, Z. Coordinated optimization of source-grid-load-storage for wind power grid-connected and mobile energy storage characteristics of electric vehicles. *IET Gener. Transm. Distrib.* **2024**, *18*, 1528–1547. [CrossRef]
23. Li, D.; Shao, Z.; Huang, W.; Zhang, B.; He, J.; Liu, X. Optimal power–heat–carbon scheduling strategy for interconnected heterogeneous multi-microgrid considering hydrogen fuel cell vehicles. *PLoS ONE* **2024**, *19*, e0301333. [CrossRef] [PubMed]
24. Available online: https://pulawy.grupaazoty.com/ (accessed on 20 December 2024).

25. Di Micco, S.; Romano, F.; Jannelli, E.; Perna, A.; Minutillo, M. Techno-economic analysis of a multienergy system for the co-production of green hydrogen, renewable electricity and heat. *Int. J. Hydrogen Energy.* **2023**, *48*, 31457–31467. [CrossRef]
26. Perna, A.; Jannelli, E.; Di Micco, S.; Romano, F.; Minutillo, M. Designing, sizing and economic feasibility of a green hydrogen supply chain for maritime transportation. *Energy Convers. Management.* **2023**, *278*, 116702. [CrossRef]
27. Duan, L.; Taylor, G.; Lai, C.S. Solar–Hydrogen-Storage Integrated Electric Vehicle Charging Stations with Demand-Side Management and Social Welfare Maximization. *World Electr. Veh. J.* **2024**, *15*, 337. [CrossRef]
28. Available online: https://lpnt.pl/biuletyn/biuletyn-lpnt-4-2023/ (accessed on 20 December 2024).
29. Afanasiev, A.; Kandinskaia, O.; Eminidou, S.; Sousanis, A. Hydrogen Valley in Cyprus: Insights and Strategies for Citizen Engagement. *Sustainability* **2025**, *17*, 804. [CrossRef]
30. Lu, Z.; Li, Y.; Zhuo, G.; Xu, C. Configuration Optimization of Hydrogen-Based Multi-Microgrid Systems under Electricity Market Trading and Different Hydrogen Production Strategies. *Sustainability* **2023**, *15*, 6753. [CrossRef]
31. Janocha, M.J.; Ong, M.C.; Lee, C.F.; Chen, K.; Ye, N. Reference Power Cable Models for Floating Offshore Wind Applications. *Sustainability* **2024**, *16*, 2899. [CrossRef]
32. Bartolucci, L.; Cennamo, E.; Cordiner, S.; Mulone, V.; Pasqualini, F.; Tatangelo, C. Hydrogen Fuel Cell Hybrid Electric Vehicles: The Impact of Commercial Vehicle Fleets on the Penetration of Renewable Energy Sources. *J. Phys. Conf. Ser.* **2022**, *2385*, 012079. [CrossRef]
33. Shahabuddin, M.; Rhamdhani, M.A.; Brooks, G.A. Technoeconomic Analysis for Green Hydrogen in Terms of Production, Com-pression, Transportation and Storage Considering the Australian Perspective. *Processes* **2023**, *11*, 2196. [CrossRef]
34. Konrad, J.; Bernt, A.O.; Hofmann, P. Life cycle assessment of MHP (Mobile Hydrogen Powersupply), an off-grid system to charge battery electric vehicles. *Int. J. Life Cycle Assess.* **2023**, *28*, 304–319. [CrossRef]
35. Villena-Ruiz, R.; Honrubia-Escribano, A.; Gómez-Lázaro, E. Solar PV and Wind Power as the Core of the Energy Transition: Joint Integration and Hybridization with Energy Storage Systems. *Energies* **2023**, *16*, 2917. [CrossRef]
36. Babaremu, K.; Olumba, N.; Chris-Okoro, I.; Chuckwuma, K.; Jen, T.-C.; Oladijo, O.; Akinlabi, E. Overview of Solar–Wind Hy-brid Products: Prominent Challenges and Possible Solutions. *Energies* **2022**, *15*, 6014. [CrossRef]
37. Nordström, H.; Söder, L.; Flynn, D.; Matevosyan, J.; Kiviluoma, J.; Holttinen, H.; Vrana, T.K.; van der Welle, A.; Mo-rales-España, G.; Pudjianto, D.; et al. Strategies for Continuous Balancing in Future Power Systems with High Wind and Solar Shares. *Energies* **2023**, *16*, 5249. [CrossRef]
38. Shavolkin, O.; Shvedchykova, I.; Kolcun, M.; Medved', D. Improvement of a Hybrid Solar-Wind System for Self-Consumption of a Local Object with Control of the Power Consumed from the Grid. *Energies* **2023**, *16*, 5851. [CrossRef]
39. Fang, W.; Yang, C.; Liu, D.; Huang, Q.; Ming, B.; Cheng, L.; Wang, L.; Feng, G.; Shang, J. Assessment of Wind and Solar Power Potential and Their Temporal Complementarity in China's Northwestern Provinces: Insights from ERA5 Reanalysis. *Energies* **2023**, *16*, 7109. [CrossRef]
40. Nefabas, K.L.; Mamo, M.; Söder, L. Analysis of System Balancing and Wind Power Curtailment Challenges in the Ethiopian Power System under Different Scenarios. *Sustainability* **2023**, *15*, 11400. [CrossRef]
41. Boubii, C.; Kafazi, I.E.; Bannari, R.; El Bhiri, B.; Bossoufi, B.; Kotb, H.; AboRas, K.M.; Emara, A.; Nasiri, B. Synergizing Wind and Solar Power: An Advanced Control System for Grid Stability. *Sustainability* **2024**, *16*, 815. [CrossRef]
42. Abdelkader, S.; Amissah, J.; Abdel-Rahim, O. Virtual power plants: An in-depth analysis of their advancements and im-portance as crucial players in modern power systems. *Energy Sustain. Soc.* **2024**, *14*, 52. [CrossRef]
43. Singh, A.R.; Kumar, R.S.; Bajaj, M.; Khadse, C.B. Machine learning-based energy management and power forecasting in grid-connected microgrids with multiple distributed energy sources. *Sci. Rep.* **2024**, *14*, 19207. [CrossRef] [PubMed]
44. Dai, S.; Shen, P.; Deng, W.; Yu, Q. Hydrogen Energy in Electrical Power Systems: A Review and Future Outlook. *Electronics* **2024**, *13*, 3370. [CrossRef]
45. Cai, X.; Wang, Y.; Xu, B.; Feng, J. Performance and Effect of Load Mitigation of a Trailing-Edge Flap in a Large-Scale Offshore Wind Turbine. *J. Mar. Sci. Eng.* **2020**, *8*, 72. [CrossRef]
46. Available online: https://solargis.com/resources/free-maps-and-gis-data?locality=poland.Solarresourcemap%C2%A92021 Solargis (accessed on 7 February 2025).
47. Tumse, S.; Bilgili, M.; Yildirim, A.; Sahin, B. Comparative Analysis of Global Onshore and Offshore Wind Energy Characteristics and Potentials. *Sustainability* **2024**, *16*, 6614. [CrossRef]
48. Das, P.; Amin, S.M.M.; Lipu, M.S.H.; Urooj, S.; Ashique, R.H.; Al Mansur, A.; Islam, M.T. Assessment of Barriers to Wind Energy Development Using Analytic Hierarchy Process. *Sustainability* **2023**, *15*, 15774. [CrossRef]
49. Pender, K.; Bacharoudis, K.; Romoli, F.; Greaves, P.; Fuller, J. Feasibility of Natural Fibre Usage for Wind Turbine Blade Com-ponents: A Structural and Environmental Assessment. *Sustainability* **2024**, *16*, 5533. [CrossRef]
50. Bošnjaković, M.; Katinić, M.; Santa, R.; Marić, D. Wind Turbine Technology Trends. *Appl. Sci.* **2022**, *12*, 8653. [CrossRef]
51. Liang, X.; Li, Z.; Han, X.; Fu, S.; Zhu, W.; Pu, T.; Sun, Z.; Yang, H.; Shen, W. Study on Aerodynamic Performance and Wake Characteristics of a Floating Offshore Wind Turbine in Wind–Wave Coupling Field. *Sustainability* **2024**, *16*, 5324. [CrossRef]

52. Wang, Y.; Xu, C.; Yu, M.; Huang, Z. Lightweight Design of Vibration Control Devices for Offshore Substations Based on Inerters. *Sustainability* **2024**, *16*, 3385. [CrossRef]
53. Olczak, P.; Surma, T. Energy Productivity Potential of Offshore Wind in Poland and Cooperation with Onshore Wind Farm. *Appl. Sci.* **2023**, *13*, 4258. [CrossRef]
54. Backhus, J.; Rao, A.R.; Venkatraman, C.; Padmanabhan, A.; Kumar, A.V.; Gupta, C. Equipment Health Assessment: Time Series Analysis for Wind Turbine Performance. *Appl. Sci.* **2024**, *14*, 3270. [CrossRef]
55. Bošnjaković, M.; Hrkać, F.; Stoić, M.; Hradovi, I. Environmental Impact of Wind Farms. *Environments* **2024**, *11*, 257. [CrossRef]
56. Narayanamoorthy, S.; Ramya, L.; Kang, D.; Baleanu, D.; Kureethara, J.V.; Annapoorani, V. A new extension of hesitant fuzzy set: An application to an offshore wind turbine technology selection process. *IET Renew. Power Gener.* **2021**, *15*, 2340–2355. [CrossRef]
57. Gkeka-Serpetsidaki, P.; Skiniti, G.; Tournaki, S.; Tsoutsos, T. A Review of the Sustainable Siting of Offshore Wind Farms. *Sustainability* **2024**, *16*, 6036. [CrossRef]
58. Kiyoki, S.; Yoshida, S.; Rushdi, M.A. Estimation of Hub Center Loads for Individual Pitch Control for Wind Turbines Based on Tower Loads and Machine Learning. *Electronics* **2024**, *13*, 3648. [CrossRef]
59. Talarek, K.; Knitter-Piątkowska, A.; Garbowski, T. Wind Parks in Poland—New Challenges and Perspectives. *Energies* **2022**, *15*, 7004. [CrossRef]
60. Zamanidou, A.; Giannakopoulos, D.; Pappa, D.; Pitsilis, V.; Makropoulos, C.; Manolitsis, K. Day-ahead scheduling of a hybrid renewable energy system based on generation forecasting using a deep-learning approach. *Energy Sci. Eng.* **2023**, *11*, 1688–1704. [CrossRef]
61. Fovell, R.G.; Capps, S.B. Sustained Wind Forecasts from the High-Resolution Rapid Refresh Model: Skill Assessment and Bias Mitigation. *Atmosphere* **2025**, *16*, 16. [CrossRef]
62. Davis, N.N.; Badger, J.; Hahmann, A.N.; Hansen, B.O.; Mortensen, N.G.; Kelly, M.; Larsén, X.G.; Olsen, B.T.; Floors, R.; Lizcano, G.; et al. The Global Wind Atlas: A high-resolution dataset of climatologies and associated web-based application. *Bull. Am. Meteorol. Soc.* **2023**, *104*, E1507–E1525. [CrossRef]
63. Krakauer, N.Y.; Cohan, D.S. Interannual Variability and Seasonal Predictability of Wind and Solar Resources. *Resources* **2017**, *6*, 29. [CrossRef]
64. Ma, B.; Zheng, J.; Xian, Z.; Wang, B.; Ma, H. Optimal Operation Strategy for Wind–Photovoltaic Power-Based Hydrogen Production Systems Considering Electrolyzer Start-Up Characteristics. *Processes* **2024**, *12*, 1756. [CrossRef]
65. Lyu, S.; Miu, H.; Yuan, X.; Wang, M.; Zeng, F.; Pan, Y. Electric Vehicle Charging Guidance Strategy with Dual-Incentive Mechanisms for Charging and Discharging. *Electronics* **2024**, *13*, 4676. [CrossRef]

energies

MDPI

Article

Disturbance Rejection and Uncertainty Analysis in Wind Turbines Using Model Predictive Control

Alok Kumar and Atul Kelkar *

Mechanical Engineering Department, Clemson University, Clemson, SC 29631, USA; aalok3740@gmail.com
* Correspondence: atul@clemson.edu

Abstract: For effective wind turbine operations, it is essential to maintain the power limit and reduce the stress on the drive train in the presence of disturbance and uncertain conditions. In our work, we propose a Model Predictive Control (MPC) framework with quadratic cost functions, incorporating control input and state constraints to mitigate the challenge of disturbance rejection and uncertainty analysis for the wind turbine operation. We have tailored the algorithm to the practical parameters of the National Renewable Energy Laboratory's (NREL) Controls Advanced Research Turbine (CART) model. We illustrate the impact of wind disturbances on achieving the optimal control law and evaluate the performance of integral MPC in disturbance rejection for the wind turbine operation, comparing it with the constrained optimal control law outcomes. The simulation results also show the efficacy of integral MPC for the uncertainty in the initial conditions of the wind turbines. This is shown by the propagation of the first two moments, i.e., mean and variance, for the states of the wind turbine. Further, we obtained the control law and mean–variance propagation for the variation in disturbance intensity. The overall results prove the efficacy of using the MPC framework for uncertainty analysis and disturbance rejection to obtain optimal operation in wind turbines.

Keywords: disturbance rejection; model predictive control; uncertainty analysis; renewable energy

1. Introduction

Wind power is a highly promising sustainable renewable energy source, poised to make a substantial contribution to global energy needs [1]. According to the International Energy Agency (IEA), in 2022, wind power generation surged by 265 TWh (14% increase), surpassing 2100 TWh [2]. However, this growth, second only to solar PV, must still catch up to the trajectory needed to achieve the goal of Net Zero Emissions by 2050. The 2022 Inflation Reduction Act in the United States of America allocated significant funding for wind power [3] to various places. As a result, the East Coast of the United States of America launched its first large-scale offshore wind project, which is expected to have 62 operational turbines in the near future. The variable power output of wind turbines is heavily influenced by environmental disturbances such as wind speed, wind direction, and turbulence [4]. Aerodynamic disturbances [5] present a major challenge in different forms, such as turbulent wind inflow, wind shear, and gusts. These challenges have the potential to lead to unsteady aerodynamic loading on the blades of the wind turbine. These fluctuations reduce the energy conversion efficiency of the wind turbine system and induce fatigue loads that can compromise the structural health of the wind turbine components over time. Another challenge is structural disturbances [6] such as tower shadow effects, blade tower interactions, and dynamic coupling between different wind turbine components, further complicating the behavior of the whole system. Addressing disturbances is crucial to enhancing the efficiency and longevity of wind turbines. The challenge lies in minimizing the fatigue caused by these disturbances while maximizing the power output of the whole system [7,8]. Larger wind turbines, designed to produce more power, are particularly susceptible to significant fatigue, which can shorten their operational lifespan. Therefore, effective disturbance rejection not only improves the wind turbine efficiency but also extends the lifetime of components, leading to better cost benefits of operating the wind turbine. This focus on mitigating environmental disturbances encourages researchers to develop methods that optimize the power output of wind turbines without compromising their durability.

Wind turbines typically operate in four distinct regions, each with specific control objectives determined by the wind speed and power, as illustrated in Figure 1. We focus on the operation of wind turbine in Region 3, the full load region, where the turbine generates rated power at wind speeds exceeding the rated speed but below the cut-out speed. In this region, the primary control objective is maintaining power output at the rated level while ensuring the turbine's safe operation.

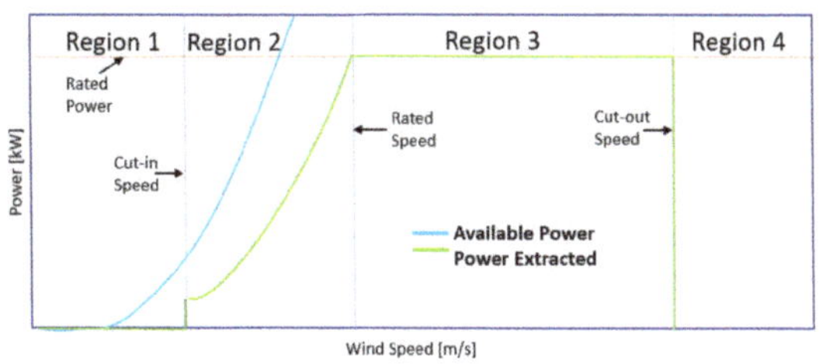

Figure 1. Working region profile for the wind turbine.

In a wind turbine, the aerodynamic torque is governed by the pitch angle and blade position, which are regulated to achieve the desired power output. To enhance wind turbine performance and mitigate structural fatigue due to disturbance, pitch angle control methods have been explored in [8,9]. Traditionally, Proportional Integral Derivative (PID) controllers have been used for wind turbine control [10], with controller gain adjusted for different requirements such as power output and drive train twist. However, the effectiveness of these controllers can vary depending on the operational regions, with some being more suited to specific regions and others transitioning between regions [11]. Various other methods have been explored to tackle this challenge, including adaptive controllers [12], feedback linearization techniques [13], robust control [14], a nonlinear controller [15], and methods based on Maximum Power Point Tracking (MPPT) [16] to enhance the turbine efficiency.

Since the late 2000s, methods employing optimization techniques in controlling wind turbines have been explored [17]. As a result, various methodologies utilize MPC formulations to enhance the wind turbine operation [18–20]. Using various practical constraints on parameters of the system, these methods achieve the control objectives like enhancing power generation or mitigating tower loads [21] in the wind turbine operation. MPC proves highly beneficial for wind turbine control because it manages multivariable constraints on both system inputs and states. It also provides predictive functionalities, allowing for optimized control inputs by considering future predictions for disturbance rejection [22,23]. However, the MPC approach must be appropriately explored for effective disturbance rejection problems in wind turbine operations, which is the main purpose of our work.

In this work, we have formulated an optimal control framework using a linear MPC-based approach for wind turbine operation in Region 3, focusing on designing a pitch angle controller for disturbance rejection and uncertain initial conditions. We used practical parameters from the NREL's Controls Advanced Research Turbine (CART) model and defined the optimal control problem with a quadratic cost function involving state variables and control inputs. We have considered the constraints on the state and input of the wind turbine system and compared the results with those of unconstrained MPC to analyze the effectiveness of the MPC framework.

Recognizing that the operation of wind turbines is continuously exposed to disturbances, we previously discussed the MPC framework for disturbance rejection in [24]. Our findings showed that MPC performs satisfactorily even in the presence of disturbances. In this work, we have utilized integral MPC for disturbance rejection to address the losses due to steady-state error. Although integral MPC may slow down the system, it significantly reduces the steady-state error, enhancing overall performance in managing disturbances of the system. We have also shown the integral MPC results for the wind turbine's varying initial conditions and the change in disturbance intensity to analyze the effectiveness of MPC for uncertainty analysis. In our work, we have utilized a deterministic approach for computationally efficient and systematic exploration of how disturbance inputs affect a wind turbine's dynamic response. This can be helpful in isolating specific uncertainty sources and their impact on system behavior, enhancing reproducibility and clarity in assessing stability. The mean and variance propagation plot provides the uncertainty analysis for the initial conditions by analyzing the expected value and the corresponding dispersion of states of wind turbines about the mean at future times. The results strengthen the efficacy of the proposed MPC controller design in the case of uncertainty in initial conditions and change in disturbance intensity. It facilitates the scope to evaluate and improve our control action strategy, providing wind turbines with more efficient and reliable operation.

The structure of the paper is outlined as follows. Section 2 presents the formulation of wind turbine dynamics. In Section 3, we detail the optimal control formulation for disturbance rejection and provide the preliminaries for the statistical analysis. We provide the MPC problem formulation for the wind turbine in Section 4. Section 5 showcases the simulation results. Finally, Section 6 summarizes the entire paper and highlights potential directions for future research.

2. Dynamics of the Wind Turbine

In this section, we will discuss the dynamics of the wind turbine model used in our work. We will first discuss the details of the nonlinear equations of the system, which are further linearized around the operating point of the wind turbine system.

The rotor of a wind turbine includes blades and a hub, which convert wind energy into mechanical energy. In high wind speed conditions, the turbine's control system adjusts the pitch angle of the blades to maintain a consistent power output from the system. The drive train of the wind turbine transmission can be simplified using a two-mass model, as illustrated in Figure 2. It is assumed that the gearbox operates with perfect mechanical efficiency.

The wind turbine model's nonlinear generalized dynamics are described by the following mathematical equation:

$$\dot{\mathbf{z}} = \mathbf{F}(\mathbf{z}) + \mathbf{B}u. \tag{1}$$

Here , $\mathbf{F}(\mathbf{z})$ is given as below and whose variables are discussed ahead:

$$\mathbf{F}(\mathbf{z}) = \begin{bmatrix} f_1 \\ f_2 \\ f_3 \\ f_4 \end{bmatrix} = \begin{bmatrix} \frac{P_r(z_1,z_4,V)}{z_1 J_r} - \frac{z_1 D_s}{J_r} + \frac{z_2 D_s}{J_r} - \frac{z_3 K_s}{J_r} \\ \frac{z_1 D_s}{N_g J_g} - \frac{z_2 D_s}{N_g^2 J_g} + \frac{z_3 K_s}{N_g J_g} - \frac{T_g}{J_g} \\ z_1 - \frac{z_2}{N_g} \\ -\frac{1}{\tau_\beta} z_4 \end{bmatrix}. \quad (2)$$

The control input matrix for the wind turbine is denoted by $\mathbf{B} = \begin{bmatrix} 0 & 0 & 0 & \frac{1}{\tau_\beta} \end{bmatrix}^T$, with the state vector $\mathbf{z} = \begin{bmatrix} \omega_r & \omega_g & \delta & \beta \end{bmatrix}^T$ and control input $u = \beta_r$. Using these variables, we can write Equation (1) as:

$$\dot{\mathbf{z}} = \begin{bmatrix} \frac{P_r(z_1,z_4,V)}{z_1 J_r} - \frac{z_1 D_s}{J_r} + \frac{z_2 D_s}{J_r} - \frac{z_3 K_s}{J_r} \\ \frac{z_1 D_s}{N_g J_g} - \frac{z_2 D_s}{N_g^2 J_g} + \frac{z_3 K_s}{N_g J_g} - \frac{T_g}{J_g} \\ z_1 - \frac{z_2}{N_g} \\ -\frac{1}{\tau_\beta} z_4 \end{bmatrix} + \begin{bmatrix} 0 \\ 0 \\ 0 \\ \frac{1}{\tau_\beta} \end{bmatrix} \beta_r. \quad (3)$$

The mechanical power (P_r) captured by the wind turbine is given by $P_r = \frac{1}{2}\pi\rho R^2 V^3 C_p(z_1, z_4, V)$. The nonlinear power coefficient (C_p), which depends on the pitch angle (β) and tip-speed ratio (λ), is shown in Figure 3, and the expression is given as:

$$C_p = 0.22(116\lambda_t - 0.4z_4 - 5)e^{-12.5\lambda_t}, \quad (4)$$

$$\lambda_t = \frac{1}{\lambda + 0.08\beta} - \frac{0.035}{\beta^3 + 1}, \quad \lambda = \frac{\omega_r R}{V}. \quad (5)$$

Other parameters of the wind turbine dynamical equation are given in Table 1.

Table 1. Parameters of the model.

Parameter	Definition
ω_r, ω_g	Rotor speed , Generator speed
δ, β	Twist angle , Pitch angle
τ_β	Time constant of pitch actuator
β_r, T_g	Pitch angle control, Generator torque
J_r, J_g	Rotor inertia, Generator inertia
N_g, λ	Gear ratio, Tip-speed ratio
D_s, K_s	Drive-train damping, Spring constant
R, ρ, V	Rotor radius, Air density, Wind speed
C_p	Power conversion coefficient

We linearize the nonlinear dynamics of the wind turbine model given in (3) around an operating point $\tilde{\mathbf{z}} = \begin{bmatrix} 2.14 & 188.4 & 0 & 25 \end{bmatrix}^T$ to obtain the system matrix $\mathbf{A}$ for the linearized model of wind turbine. The elements of $\mathbf{A}$ are computed as the partial derivatives of $\mathbf{F}$ with respect to $\mathbf{z}$, evaluated at $\tilde{\mathbf{z}}$ with values as follows:

$$\frac{\partial f_1(\mathbf{z})}{\partial z_1} = -\frac{D_s}{J_r} - \frac{1}{J_r z_1}[\frac{P_r}{z_1} + 0.11\pi\rho R^3 V^2 \frac{178.5 - 1450\lambda_t + 5z_4}{(\lambda + 0.08z_4)^2} e^{-12.5\lambda_t}],$$
$$\frac{\partial f_1(\mathbf{z})}{\partial z_2} = \frac{D_s}{N_g J_r},$$
$$\frac{\partial f_1(\mathbf{z})}{\partial z_3} = -\frac{K_s}{J_r},$$
$$\frac{\partial f_1(\mathbf{z})}{\partial z_4} = \frac{0.11\pi\rho R^2 V^3}{z_1 J_r}\{(178.5 - 1450\lambda_t + 5z_4)[\frac{-0.08}{(\lambda + 0.08z_4)^2} + \frac{0.105z_4^2}{(z_4^3+1)^2}] - 0.4\}e^{-12.5\lambda_t},$$
$$\frac{\partial f_2(\mathbf{z})}{\partial z_1} = \frac{D_s}{N_g J_g},$$
$$\frac{\partial f_2(\mathbf{z})}{\partial z_2} = -\frac{D_s}{N_g^2 J_g},$$
$$\frac{\partial f_2(\mathbf{z})}{\partial z_3} = \frac{K_s}{N_g J_g},$$
$$\frac{\partial f_2(\mathbf{z})}{\partial z_4} = 0,$$
$$\frac{\partial f_3(\mathbf{z})}{\partial z_1} = 1,$$
$$\frac{\partial f_3(\mathbf{z})}{\partial z_2} = -\frac{1}{N_g},$$
$$\frac{\partial f_3(\mathbf{z})}{\partial z_3} = 0,$$
$$\frac{\partial f_3(\mathbf{z})}{\partial z_4} = 0,$$
$$\frac{\partial f_4(\mathbf{z})}{\partial z_1} = \frac{\partial f_4(\mathbf{z})}{\partial z_2} = \frac{\partial f_4(\mathbf{z})}{\partial z_3} = 0, \frac{\partial f_4(\mathbf{z})}{\partial z_4} = -\frac{1}{\tau_\beta}.$$

Our work uses the linearized state-space representation of the wind turbine dynamics as $\dot{\mathbf{z}} = \mathbf{A}\mathbf{z} + \mathbf{B}u$. For the simulation results in MATLAB, we adopt discrete-time dynamics with a zero-order hold on inputs. Next, we will discuss the theory of constructing the MPC formulation aimed at disturbance rejection and a quick overview of uncertainty analysis.

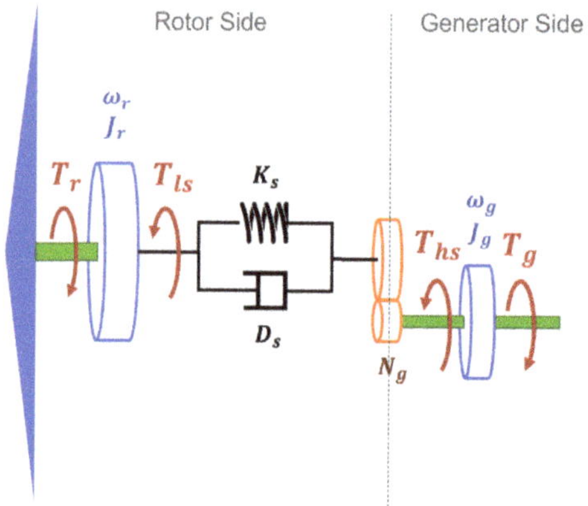

Figure 2. Two-mass model for wind turbine operation.

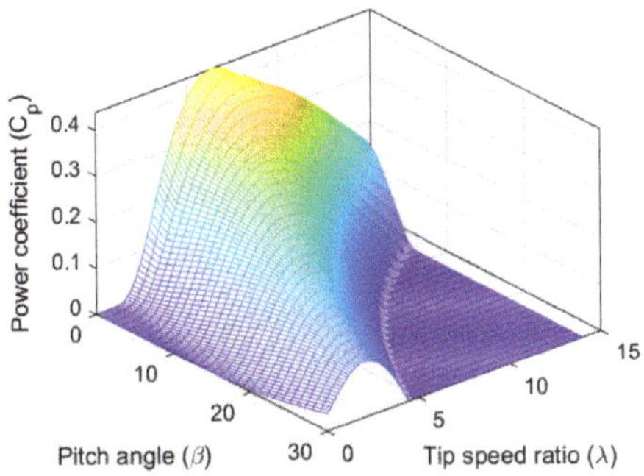

Figure 3. Power coefficient.

3. MPC Problem Formulation for Disturbance Rejection

In our work, we propose an MPC framework for the wind turbine system's disturbance rejection and uncertainty analysis. In this section, we will discuss the various MPC frameworks that are utilized in our work. These are based on the linearized state-space representation of the wind turbine dynamics obtained in the previous section.

3.1. MPC Framework for Constraints

MPC is a robust tool for solving optimal control design problems, providing significant advantages over traditional PID controllers, such as handling constraints and rejecting disturbances. This section presents a formal methodology for applying MPC to determine the optimal control law for disturbance rejection for a discrete-time linear dynamical system. We consider the dynamics of the form:

$$\mathbf{z}(k+1) = \mathbf{A}\mathbf{z}(k) + \mathbf{B}\mathbf{u}(k), \quad \mathbf{y}(k) = \mathbf{C}\mathbf{z}(k), \tag{6}$$

where $\mathbf{z} \in \mathbb{R}^n$ represents the state variables, $\mathbf{u} \in \mathbb{R}^m$ is the control input, and $\mathbf{y}(k)$ denotes the measured outputs.

Assumption 1.

A1. *The pair $(\mathbf{A}, \mathbf{B})$ is stabilizable, and the pair $(\mathbf{A}, \mathbf{C})$ is detectable. The dynamical systems focus on full-state feedback.*

A2. *Consider constant constraints across the prediction horizon. Therefore, at each time step, we have constant matrices given as $\mathbf{M}_i = \mathbf{M}$, $\mathbf{E}_i = \mathbf{E}$, and $\mathbf{b}_i = \mathbf{b}$.*

For now, we assume the dynamics are free from disturbances, delays, model errors, and noise. Our goal is to solve the finite horizon optimal control problem by minimizing the cost function $\mathcal{J}(\mathbf{z}(0), \mathbf{u}) = \mathbf{z}^\top(N)\mathbf{P}\mathbf{z}(N) + \sum_{i=0}^{N-1}\{\mathbf{z}(i)^\top\mathbf{Q}\mathbf{z}(i) + \mathbf{u}(i)^\top\mathbf{R}\mathbf{u}(i)\}$, where, $\mathbf{Q} \succeq \mathbf{0}, \mathbf{P} \succeq \mathbf{0}, \mathbf{R} \succ \mathbf{0}, \mathbf{u} = [\mathbf{u}(0), \ldots, \mathbf{u}(N-1)]$, and N is the prediction horizon. The control horizon is often chosen to be shorter than the prediction horizon. In such cases, the remaining control inputs are either held constant or set to zero. However, in our work, we set the control horizon equal to the prediction horizon. By utilizing the recursive relation of the state equation, we can express the following for Equation (6):

$$\begin{bmatrix} \mathbf{z}(1) \\ \mathbf{z}(2) \\ \vdots \\ \mathbf{z}(N) \end{bmatrix} = \begin{bmatrix} \mathbf{A} \\ \mathbf{A}^2 \\ \vdots \\ \mathbf{A}^N \end{bmatrix} \mathbf{z}(0) + \begin{bmatrix} \mathbf{B} & 0 & \cdots & 0 \\ \mathbf{A}\mathbf{B} & \mathbf{B} & \cdots & 0 \\ \vdots & \vdots & \cdots & \\ \mathbf{A}^{N-1}\mathbf{B} & \mathbf{A}^{N-2}\mathbf{B} & \cdots & \mathbf{B} \end{bmatrix} \begin{bmatrix} \mathbf{u}(0) \\ \mathbf{u}(1) \\ \vdots \\ \mathbf{u}(N-1) \end{bmatrix}. \tag{7}$$

The above dynamic model using the stack variables is given as:

$$\mathbf{Z} = \Phi\mathbf{z} + \Gamma\mathbf{U}. \tag{8}$$

Similarly, the cost function by stacking the variables from above is given as $\mathcal{J}(\mathbf{z}(0), \mathbf{U}) = \mathbf{z}^\top(N)\mathbf{P}\mathbf{z}(N) + \Sigma_{i=0}^{N-1}[\mathbf{z}^\top(i)\mathbf{Q}(i)\mathbf{z}(i) + \mathbf{u}^\top(i)\mathbf{R}(i)\mathbf{u}(i)]$. The matrix form for the cost function is given as:

$$\mathcal{J}(\mathbf{z}(0), \mathbf{U}) = \mathbf{z}^\top(0)\mathbf{Q}(0)\mathbf{z}(0) + \begin{bmatrix} \mathbf{z}(1) \\ \mathbf{z}(2) \\ \vdots \\ \mathbf{z}(N) \end{bmatrix}^\top \begin{bmatrix} \mathbf{Q}(1) & 0 & \cdots & 0 \\ \vdots & \mathbf{Q}(2) & \vdots & 0 \\ & & \ddots & \\ 0 & 0 & \cdots & \mathbf{P} \end{bmatrix} \begin{bmatrix} \mathbf{z}(1) \\ \mathbf{z}(2) \\ \vdots \\ \mathbf{z}(N) \end{bmatrix} +$$

$$\begin{bmatrix} \mathbf{u}(0) \\ \mathbf{u}(1) \\ \vdots \\ \mathbf{u}(N-1) \end{bmatrix}^\top \begin{bmatrix} \mathbf{R}(0) & 0 & \cdots & 0 \\ 0 & \mathbf{R}(2) & \cdots & 0 \\ \vdots & \vdots & \ddots & \vdots \\ 0 & \cdots & \cdots & \mathbf{R}(N-1) \end{bmatrix} \begin{bmatrix} \mathbf{u}(0) \\ \mathbf{u}(1) \\ \vdots \\ \mathbf{u}(N-1) \end{bmatrix}. \tag{9}$$

We can also write the compact form as $\mathcal{J}(\mathbf{z}, \mathbf{U}) = \mathbf{z}^\top\mathbf{Q}\mathbf{z} + \mathbf{Z}^\top\Omega\mathbf{Z} + \mathbf{U}^\top\Psi\mathbf{U}$. Here, $\mathbf{z} = \mathbf{z}(0)$. Following from $\mathbf{Q}(i) \succeq \mathbf{0}$, and $\mathbf{P} \succeq \mathbf{0}$, we obtain $\Omega \succeq \mathbf{0}$. Similarly, from $\mathbf{R}(i) \succ \mathbf{0}$, we obtain $\Psi \succ \mathbf{0}$. Now, we substitute for $\mathbf{Z}$ from (8). On further simplification, the quadratic cost function in $\mathbf{U}$ is given as:

$$\mathcal{J}(\mathbf{z}, \mathbf{U}) = \frac{1}{2}\mathbf{U}^\top\mathbf{G}\mathbf{U} + \mathbf{U}^\top\mathbf{F}\mathbf{z} + \mathbf{z}^\top(\mathbf{Q} + \Phi^\top\Omega\Phi)\mathbf{z}, \tag{10}$$

where $\mathbf{G} := 2(\Psi + \Gamma^\top\Omega\Gamma)$ and $\mathbf{F} := 2\Gamma^\top\Omega\Phi$. Additionally, $\mathbf{G} \in \mathbb{R}^{(m*N)\times(m*N)}$ where m is the number of the control inputs at each stage. The last term of the above expression does not depend on $\mathbf{U}$ and, therefore, is not considered for the optimization. Since $\mathbf{G}$ is positive definite, the optimization problem is convex. The analytical solution for optimal control law is obtained by solving $\nabla_u\mathcal{J} = 0$ and given as $\mathbf{U}^\star(\mathbf{z}) = -\mathbf{G}^{-1}\mathbf{F}\mathbf{z}$. Here, we consider the Receding Horizon (RH) approach, so we only apply the first entry of the above optimal control sequence as:

$$\mathbf{u}^\star_{RH}(\mathbf{z}) = [I_{m\times m} \quad 0_{m\times m}\ldots 0_{m\times m}] - \mathbf{G}^{-1}\mathbf{F}\mathbf{z} = -\mathbf{K}_{RH}\mathbf{z}, \tag{11}$$

where $\mathbf{K}_{RH}$ is the linear state-feedback control law.

The linear inequality constraints for the optimal control problem can be expressed as $\mathbf{M}_i\mathbf{z}(i) + \mathbf{E}_i\mathbf{u}(i) \leq \mathbf{b}_i \quad \forall i = 0,\ldots,N-1$ and $\mathbf{M}_N\mathbf{z}(N) \leq \mathbf{b}_N$. We observe that for $\mathbf{M}_i = 0$, the constraints exclusively involve control inputs. Conversely, for $\mathbf{E}_i = 0$, the constraints solely pertain to states. Furthermore, we can also consider the output constraints in terms of states as $\mathbf{y}(i) = \mathbf{C}\mathbf{z}(i)$. Following Assumption A2. and combining all inequality constraints, we obtain:

$$\begin{bmatrix} \mathbf{M}_0 \\ 0 \\ \vdots \\ \vdots \\ 0 \end{bmatrix} \mathbf{z}(0) + \begin{bmatrix} 0 & 0 & \cdots & 0 \\ \mathbf{M}_1 & 0 & \cdots & 0 \\ & & \ddots & \\ 0 & \mathbf{M}_2 & \ddots & 0 \\ 0 & 0 & \cdots & \mathbf{M}_N \end{bmatrix} \begin{bmatrix} \mathbf{z}(1) \\ \mathbf{z}(2) \\ \vdots \\ \vdots \\ \mathbf{z}(N) \end{bmatrix} + \begin{bmatrix} \mathbf{E}_0 & 0 & \cdots & 0 \\ & \ddots & & 0 \\ 0 & \ddots & \cdots & 0 \\ 0 & 0 & \cdots & \mathbf{E}_{N-1} \\ 0 & 0 & \cdots & 0 \end{bmatrix} \begin{bmatrix} \mathbf{u}(0) \\ \mathbf{u}(1) \\ \vdots \\ \vdots \\ \mathbf{u}(N-1) \end{bmatrix} \leq \begin{bmatrix} \mathbf{b}_0 \\ \mathbf{b}_1 \\ \vdots \\ \vdots \\ \mathbf{b}_N \end{bmatrix}. \tag{12}$$

In its condensed form, the expression from above becomes $\mathcal{D}\mathbf{z} + \mathcal{M}\mathbf{Z} + \Sigma\mathbf{U} \leq \mathcal{C}$. By substituting $\mathbf{Z}$ from (8), and further simplification, we derive $\mathcal{S}\mathbf{U} \leq \mathcal{C} + \mathcal{W}$. Subsequently, we incorporate this constraint equation into the cost function defined in (10). The optimization problem with constraints is then solved using quadratic programming solvers like `quadprog` in MATLAB. The solver iteratively solves the optimization problem at each time step, computes $\mathbf{U}$, implements it into the system, advances the horizon by one step, and repeats the process.

Remark 1. *The constraint optimization infeasibility problems can be solved using slacking [25]. The hard constraint of* $\mathbf{u}(i) \preceq \bar{\mathbf{u}}$ *is replaced with the soften constraint as* $\mathbf{u}(i) \preceq \bar{\mathbf{u}} + \xi_i$. *Here,* ξ_i *denotes the slack variable and is kept as small as possible. As a result, the cost function is given as* $\min_{\mathbf{U},\xi_i} \mathcal{J}_{new} = \mathcal{J} + \kappa_\xi \xi^2$, *where* κ_ξ *denotes the weights assigned some large value.*

We have discussed the constrained MPC framework above. Next, we will use the above MPC algorithm to incorporate disturbance rejection and integral control.

3.2. MPC with Disturbance Rejection and Integral Control

On adding disturbance to the dynamical equations given in (6), we obtain the dynamics with disturbances as follows:

$$\mathbf{z}(k+1) = \mathbf{A}\mathbf{z}(k) + \mathbf{B}\mathbf{u}(k) + \mathbf{B}_d\mathbf{d}(k), \tag{13}$$

$$\mathbf{y}(k) = \mathbf{C}\mathbf{z}(k) + \mathbf{D}_d\mathbf{d}(k). \tag{14}$$

The augmented state-space model is given as:

$$\mathbf{z}_a(k+1) = \mathbf{A}_a\mathbf{z}_a(k) + \mathbf{B}_a\mathbf{u}(k), \quad \mathbf{y}(k) = \mathbf{C}_a\mathbf{z}_a(k). \tag{15}$$

Here, $\mathbf{z}_a(k) = \begin{bmatrix}\mathbf{z}(k) & \mathbf{d}(k)\end{bmatrix}^T$, $\mathbf{A}_a = \begin{bmatrix}\mathbf{A} & \mathbf{B}_d \\ 0 & I\end{bmatrix}$, $\mathbf{B}_a = \begin{bmatrix}\mathbf{B} & 0\end{bmatrix}^T$, and $\mathbf{C}_a = \begin{bmatrix}\mathbf{C} & \mathbf{D}_d\end{bmatrix}$. At each point k, the disturbance $\mathbf{d}(k)$ is measurable and assumed to remain constant over the prediction horizon, meaning $\mathbf{d}(k+1) = \mathbf{d}(k)$.

Remark 2. *The assumption of constant disturbance augments integrating modes in the model, i.e., eigenvalues at 1. It ensures that the constant disturbances are rejected.*

For the above dynamical model, the augmented cost is given as $\mathcal{J}(\mathbf{z}(0), \mathbf{d}(0), \mathbf{U}) = \mathbf{z}^\top(N)\mathbf{P}\mathbf{z}(N) + \sum_{k=0}^{N-1}[\mathbf{z}^\top(k)\mathbf{Q}\mathbf{z}(k) + \mathbf{u}^\top(k)\mathbf{R}\mathbf{u}(k)]$. In terms of the augmented variables, the cost is given as $\mathcal{J}(\mathbf{z}(0), \mathbf{d}(0), \mathbf{U}) = \mathbf{z}_a^\top(N)\begin{bmatrix}\mathbf{P} & 0 \\ 0 & 0\end{bmatrix}\mathbf{z}_a(N) + \sum_{k=0}^{N-1}[\mathbf{z}_a^\top(k)\begin{bmatrix}\mathbf{Q} & 0 \\ 0 & 0\end{bmatrix}\mathbf{z}_a(k) + \mathbf{u}^\top(k)\mathbf{R}\mathbf{u}(k)]$. Now, using the state-space model and the cost function, we can obtain the optimal control law by using the approach from the previous subsection. To understand the impact of control with disturbance rejection, we consider the control law obtained in (11). For the disturbance rejection, we obtain the augmented control law $\mathbf{u}(i) = -\mathbf{K}_{aug}\begin{bmatrix}\mathbf{x}(i) & \mathbf{d}(i)\end{bmatrix}^T$. On further simplification, we obtain $\mathbf{u}(i) = -\mathbf{K}_{aug}\mathbf{x}(i) - \mathbf{K}_{aug}\mathbf{d}(i)$, where the first term corresponds to the feedback control and the second term to feedforward control action. A similar analysis can be extended for the constrained MPC case as well.

Now, to ensure that the system's output can accurately follow a constant desired reference value without any steady-state error in the optimal control problem, we add integral control to the original control problem discussed so far. This augmented control law is represented as $u(t) = -Kz(t) + u_I(t)$. Here, $u_I(t) = -K_I \int_0^t e(\tau)d\tau$ denotes the integral control component. In discrete-time settings, we express the dynamics for this integral control variable as $\mathbf{q}(k+1) = \mathbf{q}(k) + [\mathbf{y}(k) - \mathbf{r}(k)]$. This formulation allows us to derive the augmented state dynamics given as $\mathbf{z}_I(k+1) = \mathbf{A}_I\mathbf{z}_I(k) + \mathbf{B}_I\mathbf{u}(k) + \mathbf{G}_I\mathbf{r}(k)$. Here, $\mathbf{z}_I(k) = \begin{bmatrix}\mathbf{z}(k) & \mathbf{q}(k)\end{bmatrix}^T$, $\mathbf{A}_I = \begin{bmatrix}\mathbf{A} & 0 \\ \mathbf{C} & I\end{bmatrix}$, $\mathbf{B}_I = \begin{bmatrix}\mathbf{B} & 0\end{bmatrix}^T$, and $\mathbf{G}_I = \begin{bmatrix}0 & -I\end{bmatrix}^T$. Using the MPC framework, we can use the augmented dynamic model with constraints to derive the optimal control law.

So far, we have discussed the various MPC framework for the disturbance rejection in the wind turbine system. Next, we will briefly discuss the scope of uncertainty analysis in our work.

3.3. *Uncertainity Propogation*

In this work, our main motivation is to use a deterministic approach for uncertainty analysis to achieve a computationally efficient and systematic exploration of the wind turbine system's response to structured variations in disturbance inputs to the dynamic model. It will allow us to clearly isolate and understand the influence of specific uncertainty sources on the wind turbine's dynamic behavior [5]. These techniques also provide reproducibility and clarity in tracing cause–effect relationships, which is particularly important when analyzing the wind turbine system's stability and control margins [26]. Next, we provide some fundamentals of the uncertainty analysis.

For a discrete random variable Z with outcomes $(z_i, i = 1, \ldots, n)$ and corresponding probabilities $(p_i, i = 1, \ldots, n)$, the expected value of Z is calculated as $E(Z) = \sum_{i=1}^{n} z_i p_i$. It gives the weighted average of all possible values of Z, with the weights being the probabilities of those values. Similarly, the higher k-th moments of Z are computed as $E(Z^k) = \sum_{i=1}^{n} z_i^k p_i$. These moments provide information about the shape of the distribution of Z. For example, the second moment ($k = 2$) can be used to calculate the variance, which measures the spread of the values around the mean.

Similarly, for a continuous random variable $Z : \Omega \to \mathbb{R}$ in the sample space Ω and its Probability Density Function (PDF) $P_Z(z)$, the expected value of Z is defined as $E(Z) = \int_{-\infty}^{\infty} z P_Z(z)\, dz$. It gives the continuous weighted average of all possible values of Z, where the weights are provided by the PDF, $P_Z(z)$. Higher moments of Z for a continuous random variable are given as $E(Z^k) = \int_{-\infty}^{\infty} z^k P_Z(z)\, dz$ and provide insights into the characteristics of the distribution of Z.

For the uncertainty analysis of the wind turbine system, we have analyzed the controller's performance under varying initial conditions and disturbance intensities for the wind turbine operation. We conducted the study by scaling the disturbance amplitude both positively and negatively. Time-domain simulations were performed for each case, and we computed the propagation of the wind turbine system's state (rotor speed, generator speed, twist angle, and pitch angle) mean and variance over the simulation horizon. The analysis provided the statistical characterization, where the mean captured the nominal trajectory and the variance indicated the wind turbine system's sensitivity to disturbances.

So far, we have provided the mathematical framework for the MPC formulation of disturbance rejection and the scope of uncertainty analysis for the wind turbine system. Next, we will discuss the framework to obtain the optimal control law for wind turbine operation.

4. MPC Formulation for Wind Turbine Operation

This section discusses formulating an optimal control problem framework for disturbance rejection and uncertainty analysis in wind turbine operation, employing a linear MPC approach. Our approach involves formulating a quadratic cost function, as discussed in Section 3. The goal of the control regulation problem is to find an optimal control strategy for the wind turbine system that minimizes the cost function $\mathcal{J}$. Specifically, our cost function aggregates the deviations of all wind turbine dynamics system states from a reference value r. This regulatory framework is well-suited for our purposes, given that we have linearized the original nonlinear wind turbine dynamics around an operational point $\tilde{\mathbf{z}} = \begin{bmatrix} 2.14 & 188.4 & 0 & 25 \end{bmatrix}^T$, as detailed in Section 2. To quantify the penalties for states and the control input of the system, we utilize penalty weight matrices denoted as **P**, **Q**, and **R**. It is important to note that these penalty weights remain constant across the control prediction horizon.

We have considered limitations on the control input u and its variation Δu, alongside constraints on the pitch angle z_4 and its deviation Δz_4. These constraints are specified as:

$$-1^\circ \leq z_4, u \leq 90^\circ, \qquad -10 \leq \Delta z_4, \Delta u \leq +10. \tag{16}$$

Precise control over pitch angle is crucial in the operational context of wind turbines in Region 3. Enforcing strict boundaries on both the pitch angle state and control input is essential to ensure optimal wind turbine performance. Consequently, we have applied the constraints exclusively to the pitch angle state out of the four states involved. Next, we will discuss the simulation results obtained based on the abovementioned problem formulation.

5. Simulation Results

For simulation results using our proposed MPC framework, we have considered the Controls Advanced Research Turbine (CART) model [27], an experimental wind turbine developed by the National Renewable Energy Laboratory (NREL). This turbine is a 1.5 MW variable-speed pitch-controlled model. The parameters used in our work for the CART model are outlined in Table 2.

Table 2. Parameters of the model [15].

Parameter	Value	Parameter	Value
ω_r (rad/s)	2.1428	N_g	87.965
P_e (MW)	1.5	D_s (N m s/rad)	1.0×10^7
τ_β (s)	1	K_s (N m/rad)	5.6×10^9
J_r (kg m^2)	2.96×10^6	R (m)	35
J_g (kg m^2)	53	ρ (kg/m^3)	1.225

Based on the theory discussed in the introduction section, the turbine is expected to operate in Region 3, to regulate power output to the rated level by adjusting the pitch angle. This region corresponds to wind speeds ranging from 12 m/s to 24 m/s, and a mean wind speed of 18 m/s is selected for control input calculations in our simulation results. Furthermore, the linear model is discretized using a sampling time of $T_s = 0.1$ s, and the penalty weights are set as $\mathbf{P} = I_{4\times4}$, $\mathbf{Q} = I_{4\times4}$, and $\mathbf{R} = 1$. The simulation period is 300 s, and the figures are displayed for a reduced time span for clarity and better visualization of the results.

Initially, the unconstrained MPC was simulated using two distinct prediction horizons, denoted as $N = 3$ and $N = 10$, to analyze the impact of the prediction horizon. The eigenvalues of the resultant closed-loop system are presented in Table 3. The eigenvalues represent the stability characteristics of the closed-loop wind turbine model under the unconstrained MPC with prediction horizons, $N = 3$ and $N = 10$. Each eigenvalue corresponds to the poles of the wind turbine model's closed-loop transfer function, which are directly linked to the stability of the wind turbine system. Since the placement of the eigenvalues in the complex plane determines whether the system is stable or unstable, we can see from the values of Table 3 that all the eigenvalues lie inside the unit circle in the complex plane (for the discrete-time model). Therefore, the wind turbine CART model considered in our work operates under stable conditions. The stability condition is further verified by all the simulation results discussed in the following section. The eigenvalues have been computed using MATLAB's eig() function by simulating the closed-loop wind turbine dynamics using prediction horizons $N = 3$ and $N = 10$.

Table 3. Eigenvalues for a closed-loop wind turbine system.

Eigenvalues	$N = 3$	$N = 10$
λ_1	0.9251 + 0.0125i	0.8883 + 0.0769i
λ_2	0.9251 − 0.0125i	0.8883 − 0.0769i
λ_3	0.2452 + 0.0435i	0.2452 + 0.0435i
λ_4	0.2452 − 0.0435i	0.2452 − 0.0435i

Figure 4 shows the convergence of the states of the wind turbine dynamics to the operating point. It is evident that, when choosing a higher prediction horizon ($N = 10$), the value converges quickly as the control algorithm decides by looking further into the future than before. The longer-term view allows time to make the adjustments as necessary. However, the increase in the convergence rate by choosing a larger prediction horizon results in higher control input for the system, as evident from Figure 5. It is supported numerically from the value of receding horizon control gain in Table 4. This is justified as the controller tries to compensate for the potential disturbances to the turbine operation over a longer time span and takes more aggressive control action.

Table 4. Receding horizon control gain.

N = 3	−6.8072	−0.0166	16.0869	0.2603
N = 10	−40.1319	−0.0689	15.2870	0.9769

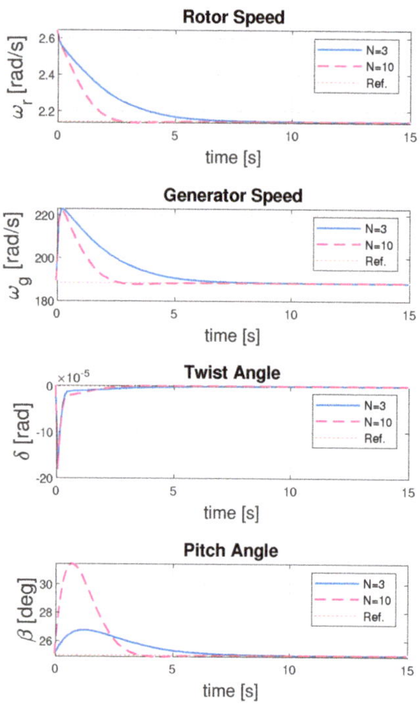

Figure 4. Propagation of states for the unconstrained case.

In real-world operation, the pitch angle actuator places strict limitations on the wind turbine system in terms of both the magnitude of its movement and how quickly it can change position. The MPC problem takes the constraints of the system as outlined in Table 5. The propagation of the states of the system is illustrated in Figure 6. As the prediction horizon extends, the convergence rate also improves, although at a slower pace compared to the unconstrained scenario because of the constraints imposed on the pitch angle rate of change, i.e., how quickly the pitch angle rate of the wind turbine can change. This is considered for the physical limitation and safe operation of the turbine. Similarly, the control input is larger on increasing the prediction horizon, as shown in Figure 7. However, it is less aggressive than the unconstrained case discussed before.

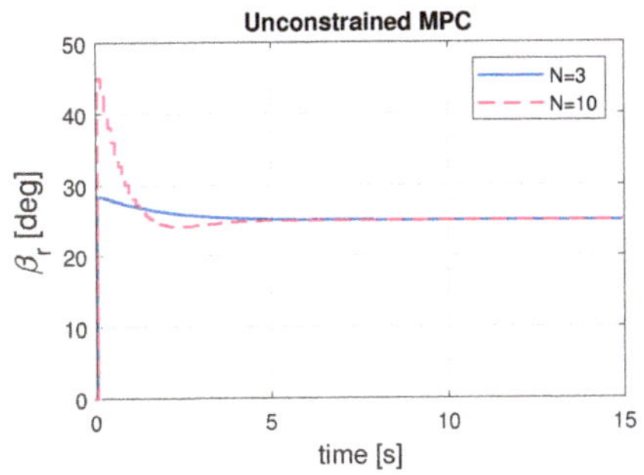

Figure 5. Control action for the unconstrained case.

Table 5. Constraints.

Parameter	Description	Value
β_{lim}	Pitch angle limit	$-1°$ to $90°$
$\dot{\beta}_{lim}$	Pitch rate limit	$\pm 10°/s$

As highlighted in the introduction section, in practice, environmental disturbances impact wind turbines' operation. The variation could be in wind speed, wind direction, turbulence, etc. We have considered the disturbance signal shown in Figure 8 into the system Equation (13) to study and mitigate the effect of the disturbances on the performance of the wind turbine. The simulation is done with a prediction horizon $N = 3$. In Figure 9, the results are shown in solid lines for the convergence of the system states for the constrained MPC with disturbance rejection. It measures the controller's adaptability to changing wind conditions and analyzes how quickly the controller acts to the disturbance. The control input is shown in solid blue lines in Figure 10. The controller is working effectively in response to the disturbance to the wind turbine.

Moreover, to further enhance the control performance of the wind turbine for disturbance rejection, we have implemented integral control to reduce the steady-state error of the system. Firstly, it is helpful in rotor speed regulation to maximize the energy capture while maintaining safe operation. Further, it is useful for optimizing the blade pitch angle to control power production and mitigate loads on turbine control. It also regulates the generator torque to maintain the grid stability and maximize the output power. In Figure 9, it can be seen that the steady-state error has decreased by using the integral MPC. This comes at the expense of a reduced convergence rate of the system because the accumulated error influences the control input gradually in the integral MPC, leading to a slower adjustment process. The integral term, as given in Section 3.2, reacts to past errors, which slows down the immediate response to changes. To further quantify the performance comparison of the controllers for the wind turbine system, the value of Integral Absolute Error (IAE) is computed as $IAE = \int |e| \, dt$. It is calculated for a 300 s simulation time and presented in Table 6. It verifies the objective of reducing the steady-state error in the wind turbine operation by using the integral control MPC.

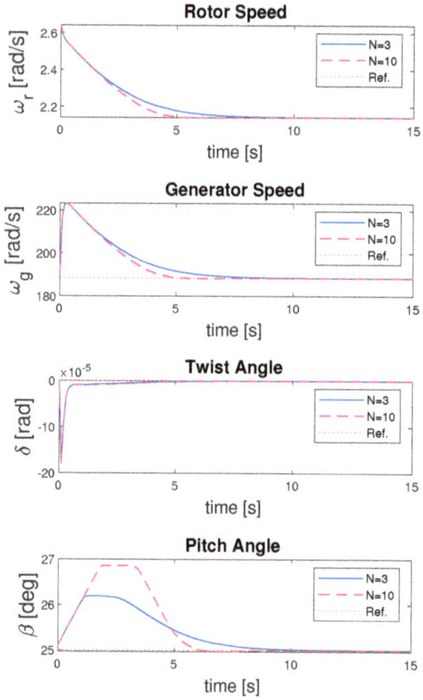

Figure 6. Propagation of states for the constrained case.

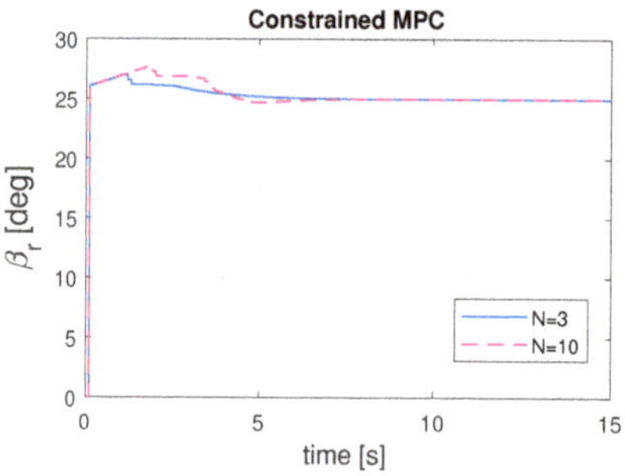

Figure 7. Control action for the constrained case.

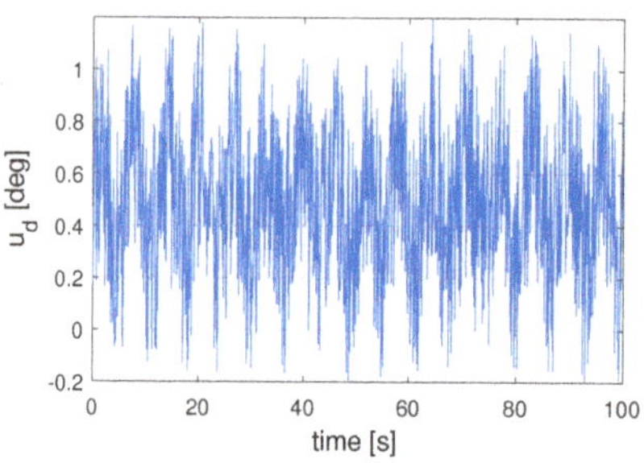

Figure 8. Disturbance signal injected to the system.

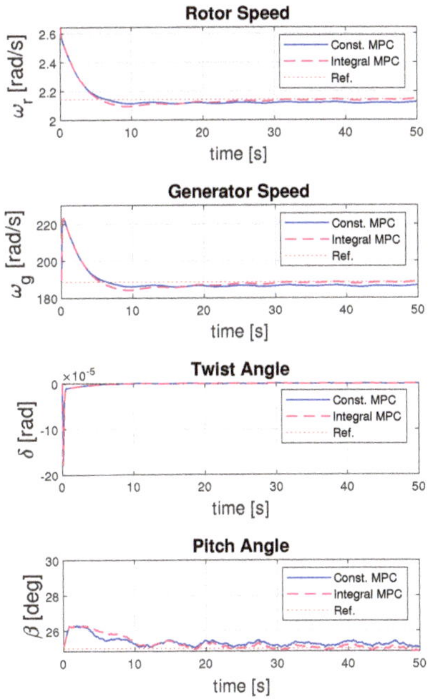

Figure 9. State propagation with MPC disturbance rejection.

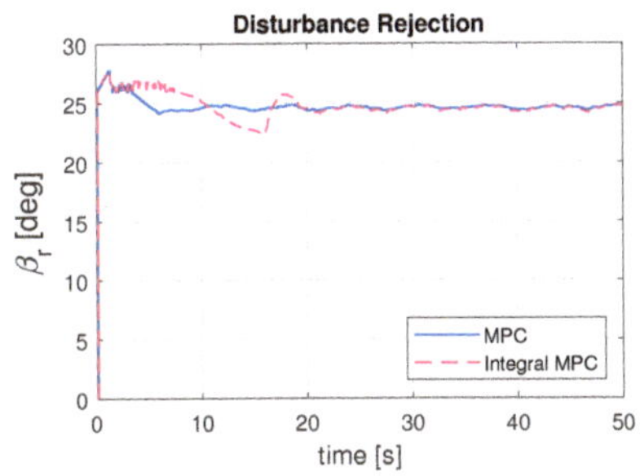

Figure 10. Control input for the disturbance rejection case.

Table 6. Integral Absolute Error of States.

Variable	Constrained MPC	Integral MPC
ω_r	8.0671	2.2428
ω_g	708.2074	195.8734
δ	0.0001	0.0001
β	76.4240	37.3294

To further verify the effectiveness of the controller action for disturbance rejection, we have obtained the statistical analysis for varying initial conditions of the wind turbine operation. Figure 11 shows the first moment, i.e., mean propagation, and it is evident that the mean propagation of all the states of the wind turbine converges to the designated operating point of the wind turbine, i.e., $\tilde{\mathbf{z}} = \begin{bmatrix} 2.14 & 188.4 & 0 & 25 \end{bmatrix}^T$. It shows that the controller can guide the turbine to its desired operational state regardless of initial conditions, even in the presence of disturbance, ensuring reliable performance. It also provides an important aspect of integrating wind power into the larger power grid. The controller action effectively maintains the operating speed of the rotor for energy conversion and stable power generation. The convergence of generator speed ensures grid synchronization and is essential for consistent electrical power output. Effective pitch angle control indicates the adaptability of the wind turbine to disturbance and uncertain environmental conditions, i.e., capturing maximum energy while ensuring longevity. Lastly, the convergence of the twist angle ensures effective aerodynamic performance of the wind turbine system. Further, in Figure 12, we have shown the variance propagation of all the states of the wind turbine system dynamics. Following the mean propagation analysis, once the states converge to the operating point of a wind turbine, the variance value converges to zero. The convergence of variance to zero for all the states signifies a stable, efficient, and reliable wind turbine operation, enhancing both the performance and longevity of the system.

In our work so far, we have shown the effectiveness of the integral MPC for disturbance rejection and uncertain initial conditions. To further consolidate the effectiveness of the controller action, we have considered the variation in the intensity of the disturbance. As shown in Figure 13, the disturbance intensity is varied by ±50% to the intensity shown in Figure 8. The wind turbine system is subjected to both extremes of the disturbance to study the effect of the controller action. In Figure 14, the results for the states of the wind turbine system show that the controller action can guide the system to the designed operating point for both extremes of the disturbance. The results are almost similar in terms of the system's rotor speed, generator speed, and twist angle. However, the difference in controller action is visible for the pitch angle of the wind turbine. When the disturbance increases, the controller adjusts the pitch angle to reduce the aerodynamic load on the blades, preventing potential damage and maintaining the power output of the system. Conversely, when the disturbance decreases, the pitch angle is adjusted to maximize the energy captured by the wind turbine.

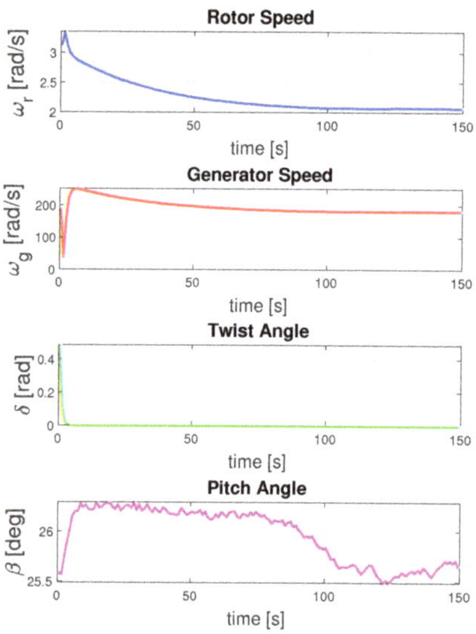

Figure 11. Mean propagation for uncertain initial states.

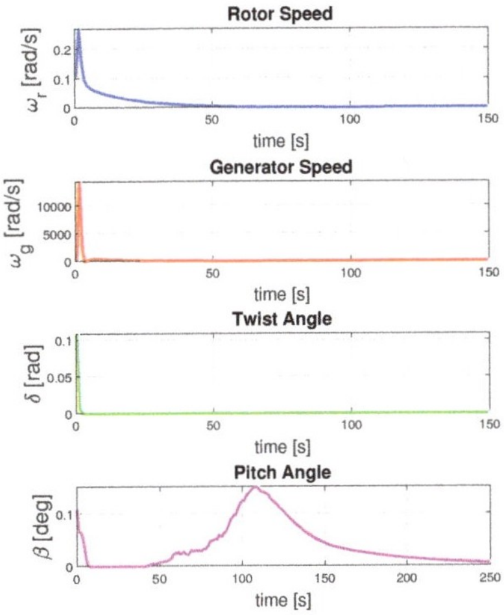

Figure 12. Variance propagation for uncertain initial states.

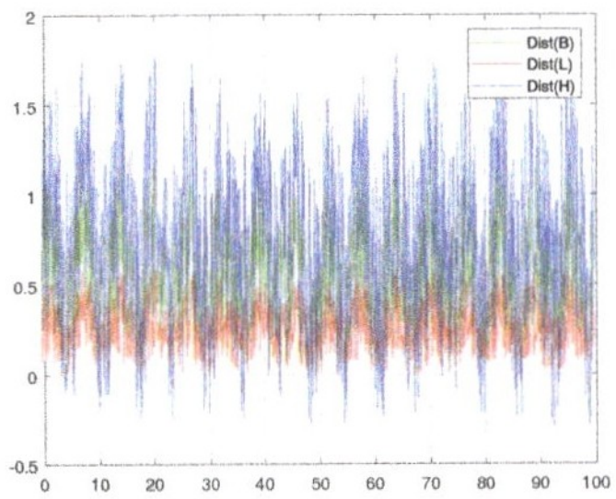

Figure 13. Disturbance signal with base (B), lower (L), and higher (H) intensities.

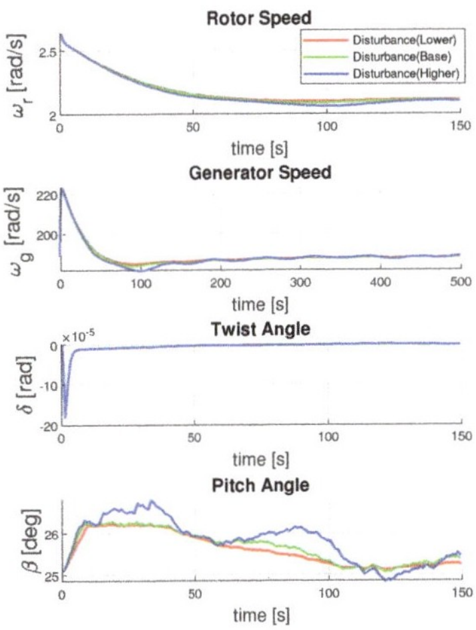

Figure 14. System states with MPC controller exposed to disturbances of different intensities.

As before, we have considered the case of uncertainty in the initial conditions of the wind turbine operation and obtained the mean and variance propagation for the states of the system. The mean propagation in Figure 15 shows that all the states converge to the designated operating point of the wind turbine, i.e., $\tilde{\mathbf{z}} = \begin{bmatrix} 2.14 & 188.4 & 0 & 25 \end{bmatrix}^T$. Figure 16 shows the convergence of variance propagation of all the states to zero once the operating point for the wind turbine has been reached. Overall, we can conclude that the convergence of the mean propagation and the reduction of variance propagation to zero in the state variables under initial condition uncertainties with variable disturbance intensity highlight the effectiveness and robustness of the wind turbine's integral MPC. It ensures the turbine can achieve and maintain its designated operating point reliably, enhancing performance and operational stability.

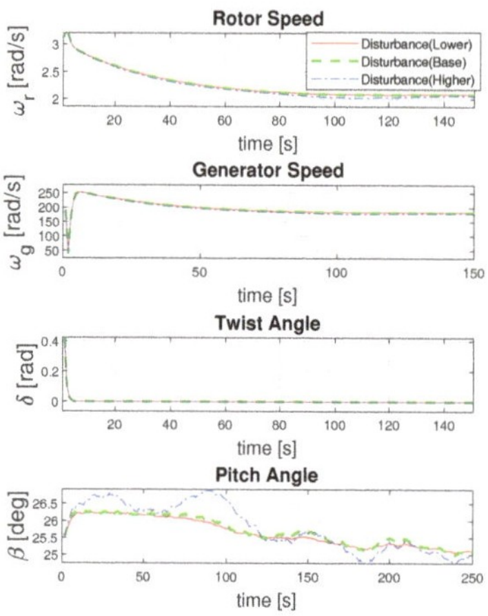

Figure 15. Mean propagation for uncertain initial states for different disturbance intensities.

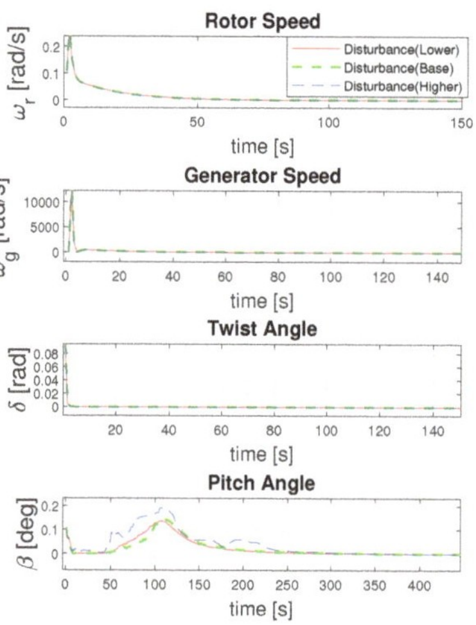

Figure 16. Variance propagation for uncertain initial states for different disturbance intensities.

6. Conclusions

MPC framework is invaluable in achieving optimal control objectives for wind turbine systems due to its ability to incorporate numerous practical constraints on system states and inputs, its predictive algorithmic nature, and the ability to reject disturbances. In this paper, we developed a linearized model of a wind turbine, which was then employed in formulating an MPC problem to mitigate the effects of disturbances. Our simulation outcomes demonstrate the effectiveness of MPC in meeting specified constraints even when disturbances are present. Additionally, integral control was implemented to minimize the wind turbine system's steady-state error. We also show the control action for the uncertain initial conditions of the states of the wind turbine. The mean and variance propagation shows the effectiveness of the designed controller for the operation of the wind turbine system. Further, the results obtained for the variable disturbance intensity and uncertain initial conditions showed the effect of MPC on enhancing performance and operation stability. The results are discussed using the deterministic uncertainty for the computationally efficient and systematic exploration of the wind turbine's dynamic response. Our goal in this work was to establish foundational insights into the effect of specific disturbances, emphasizing rejection using the MPC algorithm framework, serving as a basis for potential future probabilistic studies.

In future work, we plan to explore the random sampling approach to complement the deterministic analysis and provide a more robust uncertainty quantification for the wind turbine model. Moreover, the obtained results are promising for further implementation of the proposed approach in real wind farms, which can be achieved by integrating it into existing turbine control architectures with real-time optimization capabilities. However, some practical challenges may require detailed analysis for successful implementation, such as wind turbine model accuracy, computational demands for the extensive system, and reliable wind disturbance estimation. We will be looking to address these challenges in future work. One possible solution to explore is to develop an operator-theoretic data-driven model of wind turbines, including developing MPC algorithms robust to stochastic wind speed disturbance.

Author Contributions: Conceptualization, A.K. (Alok Kumar) and A.K. (Atul Kelkar); methodology, A.K. (Alok Kumar); software, A.K. (Alok Kumar); validation, A.K. (Alok Kumar); formal analysis, A.K. (Alok Kumar); investigation, A.K. (Alok Kumar); resources, A.K. (Alok Kumar); data curation, A.K. (Alok Kumar); writing—original draft preparation, A.K. (Alok Kumar); writing—review and editing, A.K. (Alok Kumar) and A.K. (Atul Kelkar); visualization, A.K. (Alok Kumar); supervision, A.K. (Atul Kelkar); project administration, A.K. (Atul Kelkar); funding acquisition, A.K. (Atul Kelkar). All authors have read and agreed to the published version of the manuscript.

Funding: This research received no external funding.

Data Availability Statement: Data are contained within the article.

Conflicts of Interest: The authors declare that there are no conflicts of interest.

References

1. Porté-Agel, F.; Bastankhah, M.; Shamsoddin, S. Wind-turbine and wind-farm flows: A review. *Bound. Layer Meteorol.* **2020**, *174*, 1–59. [CrossRef] [PubMed]
2. Holttinen, H.; Jørgensen, B.H. *IEA Wind TCP Annual Report 2022 Executive Summary*; IEA: Paris, France, 2023.
3. Yarmuth, J.A. Inflation Reduction Act of 2022. Volume 28. Available online: Https://www.congress.gov/bill/117th-congress/house-bill/5376 (accessed on 9 April 2023).
4. Huang, C.; Zhuang, J. Error-based active disturbance rejection control for pitch control of wind turbine by improved coyote optimization algorithm. *IEEE Trans. Energy Convers.* **2021**, *37*, 1394–1405. [CrossRef]
5. Petrone, G.; de Nicola, C.; Quagliarella, D.; Witteveen, J.; Iaccarino, G. Wind turbine performance analysis under uncertainty. In Proceedings of the 49th AIAA Aerospace Sciences Meeting Including the New Horizons Forum and Aerospace Exposition, Orlando, FL, USA, 4–7 January 2011; p. 544.
6. Jahani, K.; Langlois, R.G.; Afagh, F.F. Structural dynamics of offshore Wind Turbines: A review. *Ocean Eng.* **2022**, *251*, 111136. [CrossRef]

7. Elgendi, M.; AlMallahi, M.; Abdelkhalig, A.; Selim, M.Y. A review of wind turbines in complex terrain. *Int. J. Therm.* **2023**, *17*, 100289. [CrossRef]
8. Apata, O.; Oyedokun, D.T.O. An overview of control techniques for wind turbine systems. *Sci. Afr.* **2020**, *10*, e00566. [CrossRef]
9. Wright, A.D. *Modern Control Design for Flexible Wind Turbines*; University of Colorado at Boulder: Boulder, CO, USA, 2003.
10. Bianchi, F.D.; De Battista, H.; Mantz, R.J. *Wind Turbine Control Systems: Principles, Modelling and Gain Scheduling Design*; Springer Science and Business Media: Berlin, Germany, 2006.
11. Pao, L.Y.; Johnson, K.E. Control of wind turbines. *IEEE Control Syst. Mag.* **2011**, *31*, 44–62.
12. Yang, Q.; Jiao, X.; Luo, Q.; Chen, Q.; Sun, Y. L1 adaptive pitch angle controller of wind energy conversion systems. *ISA Trans.* **2020**, *103*, 28–36. [CrossRef] [PubMed]
13. Nguyen, A.T.; Lee, D.C. Advanced LVRT strategy for SCIG-based wind energy conversion systems using feedback linearization and sliding mode control. *J. Power Electron.* **2021**, *21*, 1180–1189. [CrossRef]
14. Yin, X.; Zhao, X. Composite hierarchical pitch angle control for a tidal turbine based on the uncertainty and disturbance estimator. *IEEE Trans. Ind. Electron.* **2019**, *67*, 329–339. [CrossRef]
15. Ren, Y.; Li, L.; Brindley, J.; Jiang, L. Nonlinear PI control for variable pitch wind turbine. *Control Eng. Pract.* **2016**, *50*, 84–94. [CrossRef]
16. Mousa, H.H.; Youssef, A.R.; Mohamed, E.E. State of the art perturb and observe MPPT algorithms based wind energy conversion systems: A technology review. *Int. J. Electr. Power Energy Syst.* **2021**, *126*, 106598. [CrossRef]
17. Kusiak, A.; Song, Z.; Zheng, H. Anticipatory control of wind turbines with data-driven predictive models. *IEEE Trans. Energy Convers.* **2009**, *24*, 766–774. [CrossRef]
18. Mirzaei, M.; Soltani, M.; Poulsen, N.K.; Niemann, H.H. Model predictive control of wind turbines using uncertain LIDAR measurements. In Proceedings of the 2013 American Control Conference, Washington, DC, USA, 17–19 June 2013; pp. 2235–2240.
19. Sinner, M.; Petrović, V.; Langidis, A.; Neuhaus, L.; Hölling, M.; Kühn, M.; Pao, L.Y. Experimental testing of a preview-enabled model predictive controller for blade pitch control of wind turbines. *IEEE Trans. Control Syst. Technol.* **2021**, *30*, 583–597. [CrossRef]
20. Xie, J.; Dong, H.; Zhao, X. Data-driven torque and pitch control of wind turbines via reinforcement learning. *Renew. Energy* **2023**, *215*, 118893. [CrossRef]
21. Henriksen, L.C.; Hansen, M.H.; Poulsen, N.K. Wind turbine control with constraint handling: A model predictive control approach. *IET Control Theor. Appl.* **2012**, *6*, 1722–1734. [CrossRef]
22. Sun, Z.; Xia, Y.; Dai, L.; Liu, K.; Ma, D. Disturbance rejection MPC for tracking of wheeled mobile robot. *IEEE/ASME Trans. Mechatron.* **2017**, *22*, 2576–2587. [CrossRef]
23. Xie, H.; Dai, L.; Lu, Y.; Xia, Y. Disturbance rejection MPC framework for input-affine nonlinear systems. *IEEE Trans. Autom. Control* **2021**, *67*, 6595–6610. [CrossRef]
24. Kumar, A.; Kelkar, A. Model Predictive Control for Disturbance Rejection in Wind Turbines. In Proceedings of the 2023 IEEE IAS Global Conference on Emerging Technologies (GlobConET), London, UK, 19–21 May 2023; pp. 1–6.
25. Camacho, E.F.; Bordons, C.; Camacho, E.F.; Bordons, C. *Constrained Model Predictive Control*; Springer: London, UK, 2007; pp. 177–216. [CrossRef]
26. Sørensen, J.D.; Toft, H.S. Probabilistic design of wind turbines. *Energies* **2010**, *3*, 241–257.
27. Jonkman, J.; Butterfield, S.; Musial, W.; Scott, G. *Definition of a 5-MW Reference Wind Turbine for Offshore System Development (No. NREL/TP-500-38060)*; National Renewable Energy Lab. (NREL): Golden, CO, USA, 2009.

energies

MDPI

Article

Criticality Assessment of Wind Turbine Defects via Multispectral UAV Fusion and Fuzzy Logic

Pavlo Radiuk [1,*], Bohdan Rusyn [2,3], Oleksandr Melnychenko [1], Tomasz Perzynski [3], Anatoliy Sachenko [3,4], Serhii Svystun [1] and Oleg Savenko [1]

1 Faculty of Information Technologies, Khmelnytskyi National University, 11, Instytuts'ka Str., 29016 Khmelnytskyi, Ukraine; melnychenko@khmnu.edu.ua (O.M.); svystuns@khmnu.edu.ua (S.S.); savenko_oleg_st@ukr.net (O.S.)
2 Department of Information Technologies of Remote Sensing, Karpenko Physico-Mechanical Institute of NAS of Ukraine, 79601 Lviv, Ukraine; rusyn@ipm.lviv.ua
3 Faculty of Transport, Electrical Engineering and Computer Science, Casimir Pulaski Radom University, 29, Malczewskiego St., 26-600 Radom, Poland; t.perzynski@uthrad.pl (T.P.); as@wunu.edu.ua (A.S.)
4 Research Institute for Intelligent Computer Systems, West Ukrainian National University, 11, Lvivska Str., 46009 Ternopil, Ukraine
* Correspondence: radiukp@khmnu.edu.ua

Abstract

Ensuring the structural integrity of wind turbines is crucial for the sustainability of wind energy. A significant challenge remains in transitioning from mere defect detection to objective, scalable criticality assessment for prioritizing maintenance. In this work, we propose a novel comprehensive framework that leverages multispectral unmanned aerial vehicle (UAV) imagery and a novel standards-aligned Fuzzy Inference System to automate this task. Our contribution is validated on two open research-oriented datasets representing small on- and offshore machines: the public AQUADA-GO and Thermal WTB Inspection datasets. An ensemble of YOLOv8n models trained on fused RGB-thermal data achieves a mean Average Precision (mAP@.5) of 92.8% for detecting cracks, erosion, and thermal anomalies. The core novelty, a 27-rule Fuzzy Inference System derived from the IEC 61400-5 standard, translates quantitative defect parameters into a five-level criticality score. The system's output demonstrates exceptional fidelity to expert assessments, achieving a mean absolute error of 0.14 and a Pearson correlation of 0.97. This work provides a transparent, repeatable, and engineering-grounded proof of concept, demonstrating a promising pathway toward predictive, condition-based maintenance strategies and supporting the economic viability of wind energy.

Keywords: defect criticality; fuzzy logic; artificial intelligence; multispectral fusion; sustainable energy; UAV inspection; wind turbine blades; YOLO; condition-based maintenance; structural health monitoring

1. Introduction

As the global energy landscape pivots towards sustainability, wind power has become a cornerstone of renewable generation, with installed capacity expanding at an unprecedented rate. However, the long-term economic viability of this multi-trillion-dollar investment hinges on the effectiveness of Operations and Maintenance (O&M) strategies.

A comprehensive review by Sun et al. highlights that ensuring the in situ structural integrity of turbine blades is a critical, yet unresolved, challenge [1]. These massive composite structures face relentless environmental and operational stresses, leading to damage that can degrade performance or precipitate catastrophic failures [2]. Consequently, O&M activities constitute a substantial fraction of the levelized cost of energy, particularly for offshore installations where non-contact inspection methods are essential to mitigate the high costs and risks of manual assessment [3].

The advent of Unmanned Aerial Vehicles (UAVs) has revolutionized data acquisition for structural health monitoring, offering a safe and efficient alternative to traditional methods, as surveyed by Zhang et al. [4]. The integration of advanced communication protocols like 5G and microservice architectures further enhances the autonomy and scalability of UAV-based inspections, enabling near real-time data streaming and distributed processing [5]. However, the immense volume of data collected necessitates automated analysis pipelines. While deep learning models, particularly the You Only Look Once (YOLO) architecture, have proven highly effective for raw defect detection [6,7], a significant 'criticality gap' remains. A simple list of detected flaws is insufficient for informed decision-making, as noted in studies on multi-type defect detection [8]. The urgent need is to move beyond detection to diagnosis and prognosis, a challenge reviewed by Sun et al. in the context of machine learning applications for fault diagnosis [9].

A robust criticality score, which quantifies the severity of a defect, is the missing link needed to transition from reactive, time-based maintenance to predictive, condition-based maintenance (CBM) [10]. Such a system must align with established engineering principles, as outlined in international standards like IEC 61400-5 [11] and industry taxonomies from the Electric Power Research Institute (EPRI) [12]. To our knowledge, this paper introduces the first end-to-end framework to bridge this gap by explicitly grounding an interpretable AI model in these standards. Our primary contribution is a transparent, 'glass-box' Fuzzy Inference System (FIS) that translates quantitative defect parameters into a standards-aligned criticality score. By validating our approach on diverse public datasets, the large-scale offshore AQUADA-GO video collection [13] and the multispectral Thermal Wind Turbine Blade (WTB) Inspection dataset [14], we introduce a reproducible and engineering-grounded solution. We demonstrate that this fusion of multispectral data and explainable AI provides an actionable tool for optimizing maintenance, enhancing safety, and ensuring the long-term sustainability of wind energy assets.

This paper introduces a complete, end-to-end framework designed to bridge this criticality gap. Our work builds upon foundational research in UAV inspection and AI-based detection but makes several novel contributions aimed at creating a system that is not only accurate but also transparent, reliable, and directly applicable in a real-world industrial setting. The goal of this study is to improve the objectivity and efficiency of wind turbine blade maintenance by developing an automated system that moves beyond simple defect detection to provide a reliable, standards-aligned criticality assessment. The primary objective is to create a transparent, data-driven tool that can prioritize repairs, optimize resource allocation, and enhance operational safety.

The major contributions of this work are as follows:

- A robust multispectral detection framework, built on an ensemble of YOLOv8n models, that achieves a state-of-the-art mean Average Precision (mAP@.5) of 92.8% on the combined public AQUADA-GO and Thermal WTB Inspection datasets. In this context, 'multispectral' refers to the combined use of the visible Red–Green-Blue (RGB) and long-wave infrared (thermal) spectra.

- A novel 27-rule 'glass-box' FIS for severity scoring, whose knowledge base is explicitly derived from the engineering principles of the IEC 61400-5 standard. The system demonstrates exceptional fidelity to expert assessments, achieving a mean absolute error of 0.14 and a Pearson correlation of 0.97.
- A comprehensive and reproducible validation of the entire framework, featuring (i) ablation studies that quantify the critical impact of each component, (ii) a formal protocol for establishing expert-derived ground truth validated by a high inter-rater reliability (Fleiss's $\kappa = 0.85$), and (iii) a global sensitivity analysis confirming the FIS's robustness to $\pm 20\%$ parameter variations.

The remainder of this manuscript is structured to detail every aspect of this framework. Section 2 provides an extensive review of the state of the art in relevant fields. Section 3 describes our methodology, from data acquisition and processing to the design of the proposed FIS. Section 4 presents the comprehensive empirical results and validation. Section 5 interprets these findings, discusses their implications, acknowledges limitations, and proposes future research directions. Finally, Section 6 summarizes the work and its contribution to the sustainable management of critical wind energy assets.

2. Related Work

The automated assessment of wind turbine blades is a multidisciplinary field drawing from advances in remote sensing, computer vision, and artificial intelligence. The evolution from hazardous manual inspections to autonomous systems has been driven by the maturation of UAV platforms capable of dynamic trajectory adaptation and precise maneuvering [15]. The diagnostic power of these platforms is determined by their sensor payloads. While high-resolution RGB cameras are standard for capturing surface details and enabling techniques like image stitching [16], the integration of thermal imaging has proven essential for a comprehensive non-destructive evaluation [17]. Thermal sensors reveal subsurface anomalies by detecting minute temperature variations, a technique validated for both blades and electrical components [18]. Research continues to explore more advanced modalities, such as hyperspectral imaging for identifying material degradation or icing [19], and the broader context of UAVs is expanding with developments in edge computing and thermal image processing incorporated with thermodynamics principles [20].

Effectively leveraging multispectral data hinges on intelligent fusion strategies, a topic thoroughly surveyed by Zhang et al. [21]. This can occur at the hardware level, as with the Multi-Spectral Dynamic Imaging (MSX) technology used in the Thermal WTB dataset [14], or through post-processing. Transform-domain methods, such as those using wavelets, have been successfully applied to enhance defect saliency in fused images [22], a principle also demonstrated in AI-based video fusion applications [23]. These traditional techniques are increasingly complemented by deep learning approaches that learn optimal fusion rules directly from data, using architectures like IFCNN [24] or FusionNet [25]. The core idea of combining data from multiple sensors to assess system health extends beyond imaging, as demonstrated in the performance degradation assessment of wind turbine gearboxes using vibration and operational data [26].

The automated analysis of this imagery is dominated by deep learning, which has supplanted classical computer vision methods. The performance of modern object detectors is built upon foundational architectures like ResNet [27] and the availability of large-scale pre-training datasets. Detection architectures have diverged into two main families: high-accuracy two-stage models like Faster R-CNN [28] and its descendants, Cascade R-CNN [29] and Mask R-CNN [30], which have been adapted for blade inspection [31], and high-speed single-stage models like EfficientDet [32] and the YOLO family [33]. The latter's balance of speed and accuracy has made it a popular choice for detecting various defects, from multi-scale surface flaws to specific cracks [34–36]. To further enhance robustness, ensemble learning [37], combining multiple models to reduce variance, is a widely adopted strategy [38].

The final frontier is translating detection into decision support, a task for which AI-driven criticality assessment is essential. A comprehensive review by Al-Agha et al. highlights this trend [39], while another work explores alternative machine learning techniques for damage classification [40]. Fuzzy logic, with its ability to model the linguistic reasoning and uncertainty of human experts, has emerged as a particularly suitable paradigm. Previous work has demonstrated its potential for creating integrated detection-to-criticality pipelines [41]. The field is advancing towards more sophisticated methods, including adaptive neuro-fuzzy systems that can learn from data [42], a technology proven in related aerospace applications [43]. Furthermore, research into automated rule-based generation using techniques like ant colony optimization seeks to address the knowledge acquisition bottleneck [44]. Our work contributes to this area by proposing a fuzzy system that is not only accurate but also transparent and explicitly grounded in established international engineering standards, a crucial step for real-world adoption and certification. Unlike prior works employing adaptive neuro-fuzzy systems that can be opaque, our approach utilizes a static, Mamdani-type FIS where every rule is explicitly defined, making the entire reasoning process from sensor data to final score fully auditable by human experts.

While the reviewed literature demonstrates significant progress in sensor technology, data fusion, and deep learning for defect detection, a persistent gap remains in translating these high-accuracy detections into transparent, reliable, and actionable maintenance decisions. The purpose of this research is to bridge this gap by developing an integrated framework that not only identifies defects with high precision but also assesses their criticality based on established engineering standards. To achieve this, this study addresses three primary tasks: (i) the development of a robust multispectral defect detection framework using an ensemble of deep learning models validated on diverse public datasets, (ii) the design and implementation of a novel explainable FIS whose knowledge base is explicitly derived from the IEC 61400-5 standard, and (iii) a comprehensive empirical validation of the end-to-end system to quantify its accuracy, reliability, and practical value for condition-based maintenance.

3. Materials and Methods

The framework proposed in this study introduces a comprehensive, multi-stage methodology for progressing from raw, multispectral UAV imagery to an actionable and quantitative assessment of wind turbine defect criticality. This process is designed as a cyber-physical system that synergizes automated data processing with formalized expert knowledge. The overall architecture, depicted in Figure 1, is structured into three primary computational blocks that sequentially refine the data from initial detection to a final, integrated criticality score.

As illustrated in Figure 1, the workflow proceeds from raw data ingestion to a final decision support output. Block 1 (Data-Driven Detection and Measurement) is responsible for processing the raw imagery to identify defects and extract a rich set of quantitative features. Concurrently, Block 2 (Knowledge Modeling and Expert Priors) formalizes domain expertise into a set of physics-informed models that provide an initial estimate of criticality for different defect types. Finally, Block 3 (Fuzzy Fusion and Decision Support) serves as the integration core, employing a Fuzzy Inference System (FIS) to intelligently fuse the data-driven measurements from Block 1 with the knowledge-based estimates from Block 2. This is achieved through a sequence of fusion, aggregation, and defuzzification steps, which yield a robust, calibrated, and transparent final criticality score, C_{final}, ready for operational use.

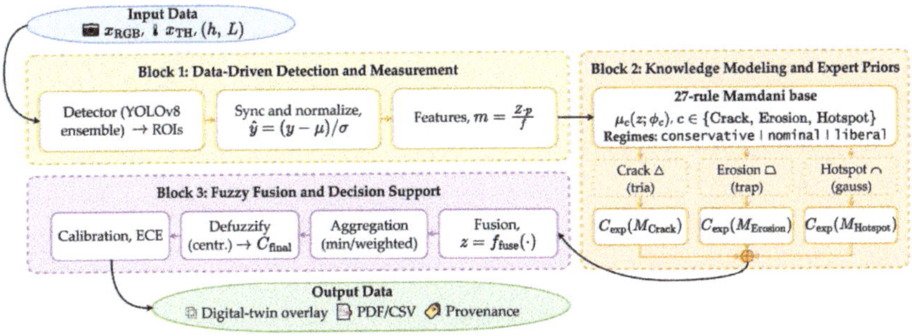

Figure 1. High-level architecture of the proposed framework. Block 1 ingests multispectral data ($x_{\text{RGB}}, x_{\text{TH}}$) and metadata ($h, L$), using a YOLOv8 ensemble to detect ROIs and an image processing pipeline to extract normalized, photogrammetrically scaled features ($m = Zp/f$). Block 2 encapsulates domain expertise within a 27-rule Mamdani fuzzy base, utilizing defect-specific membership functions (μ_c) to generate preliminary criticality estimates ($C_{\exp}(M_c)$). Block 3 serves as the decision core, fusing ($z = f_{\text{fuse}}$) the data-driven and knowledge-based inputs. It then performs aggregation and centroid defuzzification to produce a final, calibrated criticality score (C_{final}) with an associated Expected Calibration Error (ECE).

3.1. *Input Data and Defect Localization*

The initial input to our system consists of multispectral image data streams, denoted as x_{RGB} for the visual spectrum and x_{TH} for the thermal spectrum, captured by the UAV. This imagery is augmented with essential flight metadata, including the UAV's altitude (h) and its location (L) relative to the turbine structure. The foundational computational step is the detection of potential defects within these images. For this task, we employ an ensemble of three fine-tuned YOLOv8n deep learning models, chosen for their state-of-the-art balance of speed and accuracy. The output of this initial stage is a set of candidate detections for each image frame, where each detection is a structured data object containing the unique frame identifier, the predicted defect class, the pixel coordinates of the bounding box, and the model's confidence score. This initial set serves as the input to the main criticality assessment workflow, with each bounding box defining the Region of Interest (ROI) for subsequent detailed analysis.

3.2. *Block 1: Data-Driven Detection and Measurement*

This block forms the data-driven foundation of the framework, processing raw sensor inputs to produce a structured set of quantitative defect parameters.

3.2.1. Data Synchronization, Normalization, and Feature Extraction

Once a defect is localized within an ROI, its corresponding data from both the RGB and thermal streams are synchronized. The raw pixel values (y) within the ROI undergo channel-wise normalization, $\hat{y} = (y - \mu)/\sigma$, to standardize the intensity distributions and ensure robustness to variations in lighting and sensor calibration. Subsequently, an extensive, sequential image processing pipeline is executed on the synchronized and normalized ROI. This pipeline is designed to extract a precise and objective set of physical and thermal parameters, ensuring that all subsequent analysis is based on repeatable, quantitative measurements. The process culminates in the generation of two feature vectors for each detected defect D_i: v_{RGB}, containing geometric (e.g., area, perimeter) and textural parameters from the visual spectrum, and v_{TH}, containing thermal parameters (e.g., max/min/avg temperature, temperature gradients) from the infrared spectrum.

3.2.2. Region of Interest (ROI) Extraction and Preprocessing

The analysis begins by isolating the defect. Using the bounding box coordinates $(x^i_{\text{ensemble}}, y^i_{\text{ensemble}}, W^i_{\text{ensemble}}, H^i_{\text{ensemble}})$ provided by the YOLOv8 ensemble, the corresponding ROI is cropped from the original, undistorted source image $I_{\text{undistorted}}$. This is represented by the function f_{crop}, which is formalized as follows:

$$I^i_{\text{ROI}} = f_{\text{crop}}(I_{\text{undistorted}}, x^i_{\text{ensemble}}, y^i_{\text{ensemble}}, W^i_{\text{ensemble}}, H^i_{\text{ensemble}}). \quad (1)$$

Next, to mitigate high-frequency noise (e.g., from sensor imperfections or atmospheric interference) while preserving the sharp edges that are critical for defining defect boundaries, a bilateral filter, $f_{\text{bilateralFilter}}$, is applied to the extracted ROI as follows:

$$I^i_{\text{smooth}} = f_{\text{bilateralFilter}}(I^i_{\text{ROI}}, d, \sigma_{\text{color}}, \sigma_{\text{space}}), \quad (2)$$

where d is the diameter of the pixel neighborhood, σ_{color} is the filter sigma in the color space, and σ_{space} is the filter sigma in the coordinate space.

We empirically determined optimal values of $d = 9$, $\sigma_{\text{color}} = 75$, and $\sigma_{\text{space}} = 75$ for this application through a grid search optimizing for edge preservation on a validation subset.

3.2.3. Contrast Enhancement and Adaptive Binarization

To enhance the visibility of defect features, particularly in challenging lighting conditions such as shadows or glare, we apply Contrast Limited Adaptive Histogram Equalization (CLAHE) as follows, where the clip limit C_L was set to 2.0 to prevent oversaturation, and the grid size G_S was set to (8, 8) pixels:

$$I^i_{\text{equalized}} = f_{\text{CLAHE}}(I^i_{\text{smooth}}, C_L, G_S). \quad (3)$$

Unlike global histogram equalization, CLAHE operates on small tiled regions of the image, which prevents the over-amplification of noise in relatively uniform areas.

The contrast-enhanced image is then segmented into foreground (defect) and background pixels through adaptive thresholding (Equation (4)), which calculates a localized threshold for each pixel based on the intensity distribution in its neighborhood.

$$I^i_{\text{binary}} = f_{\text{adaptiveThreshold}}(I^i_{\text{equalized}}, B_S, C), \quad (4)$$

where the block size B_S for the thresholding neighborhood is 11 pixels and $C = 2$ is a constant subtracted from the local mean.

This method in Equation (4) is highly effective for images with non-uniform illumination.

3.2.4. Morphological Filtering and Geometric Feature Extraction

The raw binary image I^i_{binary} may contain small spurious artifacts. To refine the defect mask, morphological operations are performed. An erosion operation (Equation (5)) is first applied to remove small isolated white pixels, followed by a dilation operation (Equation (6)) to restore the original size of the primary defect region.

$$I^i_{\text{eroded}} = f_{\text{erode}}(I^i_{\text{binary}}, S), \tag{5}$$

$$I^i_{\text{processed}} = f_{\text{dilate}}(I^i_{\text{eroded}}, S), \tag{6}$$

where S is a (5×5)-elliptical structuring element.

From the final processed binary image $I^i_{\text{processed}}$, contour analysis is performed to identify all distinct object boundaries. The largest contour, C^i_{defect}, is assumed to correspond to the primary defect. A set of geometric parameters is then computed from this contour, as detailed in the following Equations (7)–(9):

$$A^i_{\text{pixels}} = f_{\text{contourArea}}(C^i_{\text{defect}}), \tag{7}$$

$$P^i_{\text{pixels}} = f_{\text{arcLength}}(C^i_{\text{defect}}), \tag{8}$$

$$(x^i, y^i, W^i_{\text{defect}}, H^i_{\text{defect}}) = f_{\text{boundingRect}}(C^i_{\text{defect}}), \tag{9}$$

where A^i_{pixels} is the defect area in pixels, P^i_{pixels} is its perimeter, and the final line gives the dimensions of the minimal bounding rectangle.

3.2.5. Photogrammetric Scaling and Calibration

To convert these pixel-based measurements into physically meaningful units, a scaling factor m^i is computed for each defect, accounting for the UAV's distance to the target and the camera's optical properties, as shown in the following Equation (10):

$$m^i = \frac{Z^i \cdot p}{f}, \tag{10}$$

where Z^i is the distance to the defect, p is the physical size of a sensor pixel, and f is the lens focal length.

The scaling factor m^i allows for the calculation of a set of real-world physical parameters $S_{\text{physical}} = \{W^i_{\text{real}}, H^i_{\text{real}}, A^i_{\text{real}}, P^i_{\text{real}}\}$. Crucially, the 'Defect Size' input to our FIS uses the metric area A^i_{real} (in mm^2), not the pixel area, to ensure that the assessment is invariant to inspection altitude.

In this work, Z^i in Equation (10) was not treated as a fixed constant but measured directly using the onboard Real-Time Kinematic (RTK) Global Navigation Satellite System (GNSS) integrated in the UAV, which provides an accuracy of ± 0.02 m at the typical operating range. The camera intrinsics (p, f) were calibrated prior to each flight campaign using a standard checkerboard pattern and the photogrammetric pipeline proposed by Zhang [45]. To assess the potential influence of this residual altitude uncertainty on the final criticality score, we performed a first-order error propagation analysis. Differentiating Equation (10) with respect to Z yields the uncertainty in the scaling factor, as follows:

$$\delta m^i = \frac{\partial m^i}{\partial Z^i} \Delta Z = \frac{p}{f} \Delta Z, \tag{11}$$

where $\Delta Z = \pm 0.02$ m represents the worst-case GNSS error.

Substituting typical values for the sensor's pixel pitch (p) and lens focal length (f) from the manufacturer's datasheets yields a relative uncertainty in the calculated metric defect area of less than 3%. We propagated this uncertainty through the trapezoidal membership functions used in our FIS; the resulting change in membership grade was consistently below 0.08 on the [0, 1] scale for all but the very smallest defects, which are of lowest criticality. We therefore concluded that, for the precision of the RTK system used, no additional correction or uncertainty term was necessary for the FIS.

3.2.6. Thermal Analysis

For defects with a thermal component, an analogous analysis is performed on the radiometric thermal data within the contour C^i_{defect} to extract key temperature characteristics, as detailed in the following Equations (12)–(14):

$$T^i_{\text{min}} = \min_{(x,y)\in C^i_{\text{defect}}} T(x,y), \tag{12}$$

$$T^i_{\text{max}} = \max_{(x,y)\in C^i_{\text{defect}}} T(x,y), \tag{13}$$

$$T^i_{\text{avg}} = \frac{1}{N^i} \sum_{(x,y)\in C^i_{\text{defect}}} T(x,y), \tag{14}$$

where $T(x,y)$ is the temperature at pixel (x,y) and N^i is the number of pixels within the contour.

To ensure consistency, all thermal data from the different sensors used in the public datasets are standardized to degrees Celsius (°C) before being passed to the analysis pipeline. The key input to the FIS, the 'Thermal Signature,' is the differential temperature $\Delta T = T^i_{\text{max}} - T_{\text{ambient}}$, where T_{ambient} is the average temperature of a non-defective region adjacent to the defect.

These physical and thermal parameter sets are then encapsulated into a final comprehensive data model, D^i_{complete}, for each defect, which serves as the input to the integration module.

3.3. *Block 2: Formalization of Expert Criticality Functions*

Operating in parallel to the data-driven parameterization, Block 2 of our framework translates the qualitative assessment criteria of human experts into formal mathematical models. These models provide an initial knowledge-based estimate of criticality, $C_{\text{exp}}(M_i)$, for a given defect type M_i (e.g., M_{crack}, M_{erosion}, M_{hotspot}). Each model is designed to reflect the underlying physics of the failure mode associated with that defect type, providing a physics-informed baseline before the fuzzy integration stage. A crucial component of these models is the inclusion of a weighting coefficient that depends on the specific turbine component where the defect is located. These coefficients, presented in the Supplementary Materials, are derived from engineering standards and expert consultation, and quantify the structural or operational importance of each component. For instance, the criticality model for a crack (M_{crack}) is given by the following Equation (15):

$$C_{\text{exp}}(M_{\text{crack}}) = \beta_c \cdot \int_0^L w_{\text{visible}}(s) \cdot |r'(s)| \cdot (1+\kappa(s))ds, \tag{15}$$

where β_c is the component-specific weighting factor, L is the crack length, $w_{\text{visible}}(s)$ is its visible width along its path s, $|r'(s)|$ accounts for its tortuosity, and $\kappa(s)$ is its curvature. This model captures the principle from fracture mechanics that longer, wider, and more complex cracks pose a greater risk.

Similarly, the model for erosion (M_{erosion}) is primarily a function of the affected area, while the model for overheating (M_{hotspot}) depends on the temperature differential and the spatial temperature gradient. The explicit forms for these models are as follows:

$$C_{\exp}(M_{\text{erosion}}) = \gamma_c \cdot A_{\text{real}}, \tag{16}$$

$$C_{\exp}(M_{\text{hotspot}}) = \eta_c \cdot (\Delta T_{\max})^2 \cdot \overline{|\nabla^2 T|}, \tag{17}$$

where γ_c and η_c are component-specific weights, A_{real} is the erosion area, $\Delta T_{\max}$ is the max temperature difference, and $\overline{|\nabla^2 T|}$ is the mean absolute temperature Laplacian.

These models yield an initial physics-informed criticality score before fuzzy integration. A detailed derivation of these models, including the full tabulation of all weighting coefficients, is provided in the Supplementary Materials.

3.4. Block 3: Fuzzy Logic Integration for Final Criticality Assessment

The core novelty of our framework lies in the integration module (Block 3), which employs a Mamdani-type FIS to intelligently fuse the objective data-driven measurements from Block 1 with the knowledge-based estimates from Block 2. The logic of this fusion is governed by a knowledge base of 27 IF-THEN rules, the complete set of which is provided in Appendix A, Table A1. This rule base was explicitly designed to ensure that the final criticality score is both empirically grounded and consistent with established engineering principles, with its design directly mirroring failure mode considerations from the IEC 61400-5 standard (see Appendix A, Table A3 for illustrative examples of this mapping).

3.4.1. Fuzzification of Data-Driven and Expert-Driven Inputs

The process begins with fuzzification. Each crisp physical parameter p_k^i within the comprehensive data structure D^i_{complete} (e.g., 'Defect Area,' 'Max Temperature Difference') is mapped to a set of linguistic variables (e.g., *Small*, *Medium*, *Large*) via trapezoidal membership functions $\mu_{p_k}(x)$. These functions were parameterized using a hybrid approach, where initial estimates from a panel of domain experts were refined by aligning the function breakpoints with the empirical quantiles of the training data distribution (see Appendix A, Table A2 for details). A global sensitivity analysis, presented in Appendix A, Figure A1, confirmed that the system's output is robust to moderate ($\pm 20\%$) variations in these parameters, which validates the stability of this custom rule-based approach. An example function is defined in the following Equation (18):

$$\mu_{p_k}(x) = \begin{cases} 0, & x \le a_k; \\ (x - a_k)/(b_k - a_k), & a_k < x \le b_k; \\ 1, & b_k < x \le c_k; \\ (d_k - x)/(d_k - c_k), & c_k < x < d_k; \\ 0, & x \ge d_k, \end{cases} \tag{18}$$

where $[a_k, d_k]$ defines the support of the fuzzy set and $[b_k, c_k]$ defines its core.

The fuzzy sets representing all of a defect's physical parameters are then aggregated into a single data-driven fuzzy set, $\mu'_D(x)$, using the t-norm (minimum) operator shown in Equation (19).

$$\mu'_D(x) = \min_k \mu_{p_k}(x). \tag{19}$$

Concurrently, the crisp output from the expert model, $C_{\text{exp}}(M_i)$, is also fuzzified into an expert-driven fuzzy set, $\mu_{C_{\text{exp}}}(x)$, using the Gaussian membership function shown in the following Equation (20):

$$\mu_{C_{\text{exp}}}(x) = \exp\left(-\frac{(x - C_{\text{exp}}(M_i))^2}{2\sigma^2}\right), \tag{20}$$

where σ controls the uncertainty or 'fuzziness' of the expert estimate.

3.4.2. Weighted Aggregation of Fuzzy Sets

To intelligently combine these two fuzzy sets, the system first quantifies their degree of agreement. This is achieved by calculating the cosine similarity, $S(D_i)$, between them, which serves as a robust measure of overlap in the fuzzy domain (Equation (21)). This similarity score is used to determine the relative weights for the final aggregation.

$$S(D_i) = \frac{\int_X \mu'_D(x) \cdot \mu_{C_{\text{exp}}}(x)\, dx}{\sqrt{\int_X [\mu'_D(x)]^2\, dx} \cdot \sqrt{\int_X [\mu_{C_{\text{exp}}}(x)]^2\, dx}}. \tag{21}$$

This agreement coefficient, $S(D_i) \in [0,1]$, is then used to determine the relative weights of the data-driven evidence and the expert model's estimate in the final assessment. The weights, w_D and w_{exp}, are calculated using the sigmoidal function in Equation (22), which allows for a smooth transition, as follows:

$$w_D = \frac{1}{1 + e^{-k(S(D_i) - \theta)}}; \quad w_{\text{exp}} = 1 - w_D, \tag{22}$$

where k controls the steepness of the transition and θ is the inflection point (typically 0.5).

This ensures that when the data and the expert model are in high agreement, the result is reinforced; when they disagree, their contributions are balanced. The final aggregated fuzzy set for the defect's criticality, $\mu_{\text{final}}(x)$, is then computed as a weighted sum as follows:

$$\mu_{\text{final}}(x) = w_D \cdot \mu'_D(x) + w_{\text{exp}} \cdot \mu_{C_{\text{exp}}}(x). \tag{23}$$

3.4.3. Defuzzification for a Final Criticality Score

The conclusive step in the framework is defuzzification, which converts the final fuzzy set $\mu_{\text{final}}(x)$ back into a single crisp numerical value. For this, we employ the centroid (or center of gravity) method, which is formulated in Equation (24).

$$C_{\text{final}}(D_i) = \frac{\int_X x \cdot \mu_{\text{final}}(x)\, dx}{\int_X \mu_{\text{final}}(x)\, dx}. \tag{24}$$

Equation (24) calculates the center of the area under the membership function, effectively providing a weighted average of all possible criticality values. The resulting score, $C_{\text{final}}(D_i)$, represents the system's comprehensive and integrated assessment of the defect's severity. This continuous value is then normalized and mapped to the discrete 1–5 integer scale defined by the EPRI damage taxonomy, rendering it directly interpretable and actionable for O&M teams to prioritize repair and maintenance activities.

3.5. Experimental Setup

This section details the experimental environment used to evaluate the proposed cyber-physical framework, including the data management procedures, hardware platform, and the major software components employed throughout the study. To facilitate reproducibility, every piece of software is identified by name and version, accompanied by a bibliographic reference, and the key parameters of the hardware platform are reported.

3.5.1. Dataset Curation and Preprocessing

The experiments were based exclusively on two publicly available datasets to ensure the reproducibility of our findings. The first, AQUADA-GO [13], consists of high-resolution RGB videos captured during offshore inspections of small (approximately 2 MW) turbines. The second, the Thermal WTB Inspection dataset [14], contains onshore inspections conducted with a FLIR thermal camera, where each thermal frame is co-registered with an RGB image and enhanced by MSX. Table 1 provides a detailed summary of the datasets, including the total number of images, distinct inspection flights, and the distribution of annotated defects across the three primary classes.

Table 1. Dataset card and split summary. The table provides counts of images (frames), distinct blades (flights), and annotated defects per class for the two public datasets used in this study. The final row shows the total counts for the combined dataset used for training and evaluation.

Dataset	Images	Blades	Cracks	Erosion	Hotspots
AQUADA-GO (RGB)	15,420	24	850	1230	–
Thermal WTB (RGB-T)	3850	12	320	450	210
Total Combined	19,270	36	1170	1680	210

To avoid bias caused by temporal correlation in video data, both datasets were split into training, validation, and test sets in an 80:10:10 ratio at the level of blade flights rather than individual frames. All frames associated with a single blade and flight were assigned to the same fold, ensuring that the test set contains inspections of turbines entirely unseen during training. For each training fold, we applied random spatial augmentations, scaling, cropping, and flips, followed by color and brightness jittering. These transformations were implemented using the torchvision v0.20.1 [46] library, chosen for its rich collection of augmentation primitives and GPU support.

3.5.2. Defect Detection and Fusion Pipeline

Detection of candidate defects relied on an ensemble of three YOLOv8n detectors. Each detector was implemented using the Ultralytics YOLOv8 framework (release 8.2) built on PyTorch v2.4.0 [47]. Training was performed for 100 epochs with the Adam optimizer, an initial learning rate of 0.001, and a cosine annealing schedule, with random seeds set globally to ensure reproducibility, as detailed in the Supplementary Materials. To assemble the detections from individual models, we employed the Weighted Boxes Fusion algorithm implemented in the Weighted-Boxes-Fusion v1.0.8 package [48]. Low-level image processing tasks such as bilateral filtering, CLAHE equalization, and contour extraction were carried out with OpenCV v4.10.0 [49].

During early development, we explored synthesizing a thermal channel for AQUADA-GO by fusing textural features extracted via the DWT. This simulation was implemented with PyWavelets v1.6.0 [50]. To validate the surrogate modality, we collected 50 paired RGB and LWIR frames with a handheld FLIR T840 camera and compared the simulated and true temperature difference maps using the Pearson correlation coefficient. The resulting r-value of 0.45, far below the 0.6 threshold recommended for modality substitution, led us to discard the simulated channel in favor of a dual-head architecture. In this design, the network backbone is shared between RGB and thermal branches, but the LWIR head is only activated when genuine thermal data are available. For the RGB-only AQUADA-GO dataset, the thermal branch was masked during training and inference. For these RGB-only cases, the 'Thermal Signature' input to the FIS was programmatically set to a crisp input of 0, corresponding to the 'Low' fuzzy set, allowing the system to determine criticality based on defect size and location alone.

3.5.3. Hardware and Software Environment

All software was orchestrated using Python 3.12.3. Numerical operations were performed using NumPy v2.1.0 [51] and SciPy v1.14.1 [52], while tabular data handling was facilitated by pandas v2.2.3 [53]. Bootstrap resampling and statistical analyses were implemented with scikit-learn v1.5.1 [54] and visualized with Matplotlib v3.9.2 [55]. The FIS was implemented with scikit-fuzzy v0.5.0 [56].

The experiments were run on a workstation equipped with an Intel® Core™ i9-13900K CPU, 64 GiB of RAM, and an NVIDIA® RTX 3090 GPU with 16 GiB of VRAM. The CUDA environment was provided by CUDA Toolkit v12.4.1 and cuDNN v8.9.7 The UAV mentioned in the context of our supplementary field case studies was a DJI M300 (SZ DJI Technology Co., Ltd., Shenzhen, China) fitted with a FLIR A65 camera (Teledyne FLIR LLC, Wilsonville, OR, USA) for radiometric thermal imaging and an RTK GNSS module (Emlid Tech Korlátolt Felelősségű Társaság, Budapest, Hungary) offering altitude measurements with 2 cm precision; this platform was used for validation purposes only and not for generating the primary training datasets. Photogrammetric calibration of the RGB camera was accomplished using OpenCV's calibration routines.

3.5.4. Evaluation Protocol

Performance metrics were computed according to standard object-detection practice. Precision, recall, mean Average Precision at IoU threshold 0.5, class-wise F_1-scores, and the quadratic-weighted κ statistic were derived on the test set. To quantify the uncertainty of these estimates, we employed a BCa bootstrap with B = 10,000 resamples. Resampling was stratified by dataset and defect class, with complete blade flights treated as the unit of resampling to preserve temporal correlations. Confidence intervals were computed following the approach of Efron and Tibshirani, and all code for the bootstrap procedure is provided in our public repository.

The Fuzzy Inference System was validated against a ground-truth dataset created by a panel of three certified O&M engineers, as detailed in the Supplementary Materials. Each expert scored defects on the held-out set using the five-level EPRI taxonomy, and the median score served as the reference label. The system's continuous outputs were compared with these labels using mean absolute error and the Pearson correlation coefficient. Implementation of the FIS drew upon the scikit-fuzzy library for membership function evaluation and defuzzification and was executed on the same workstation described above. All source code, trained model weights, and configuration files necessary to reproduce our experiments are available at our repository (see Data Availability Statement).

3.5.5. Ethical Considerations

Finally, the supplementary field data used for the case study were collected in accordance with all local and national aviation regulations. Flight permissions were obtained from the relevant airspace authorities, and explicit consent was secured from the wind farm operator. To protect privacy and commercial sensitivities, all imagery was processed to remove any identifying features of the site or personnel.

4. Results

This section presents a comprehensive empirical evaluation of the proposed framework, meticulously dissecting its performance from initial defect detection to final criticality assessment. The analysis is performed exclusively on two publicly available and diverse benchmarks: the large-scale, offshore AQUADA-GO dataset [13], featuring high-resolution video, and the Thermal WTB Inspection dataset [14], which provides fused RGB and radiometric thermal imagery. We first quantify the performance of the detection module, including visual analytics of its behavior. Subsequently, a series of ablation studies isolates the contribution of each system component. We then validate the fuzzy criticality system's accuracy and calibration against expert-derived ground truth and conclude with a comparative analysis against state-of-the-art methods and an illustrative case study from field data.

4.1. Defect Detection Performance and Computational Profile

To establish the efficacy of our detection framework, we benchmarked our final model, a three-model YOLOv8n ensemble trained on fused multispectral data, against two systematically ablated baselines. Performance was evaluated on the combined held-out test sets from both datasets. The primary metrics, reported as the mean and 95% Bias-Corrected and accelerated (BCa) confidence interval derived from 10,000 bootstrap resamples, are summarized in Table 2. This uncertainty quantification provides a robust measure of the stability of our performance estimates. Paired bootstrap tests confirmed that the performance differences between all three configurations were statistically significant ($p < 0.01$) for both mAP@.5 and F_1-score.

Table 2. Comparative defect detection performance on the combined test sets. Results are reported as mean ± 95% Bias-Corrected and accelerated (BCa) confidence interval. Best results are shown in **bold**. The final column reports precision at an operating point that yields ≤ 0.05 false positives per image (FPPI).

Model Configuration	Precision (%)	Recall (%)	F_1-Score (%)	mAP@.5 (%)	Precision@FPPI ≤ 0.05 (%)
Single YOLOv8 (RGB Only)	82.5 ± 1.8	79.1 ± 2.0	80.8 ± 1.9	81.7 ± 1.7	78.4
Single YOLOv8 (Multispectral)	89.1 ± 1.4	87.3 ± 1.5	88.2 ± 1.4	88.9 ± 1.3	89.5
Proposed Ensemble (Multispectral)	**93.2 ± 1.0**	**91.5 ± 1.1**	**92.3 ± 1.0**	**92.8 ± 0.9**	**93.5**

The proposed ensemble using fused multispectral data demonstrates a clear and statistically significant performance advantage over the baseline configurations. The results unequivocally demonstrate a tiered improvement at each stage of methodological enhancement. The baseline RGB-only model establishes a respectable mAP@.5 of 81.7%. The introduction of the thermal data channel provides the most significant performance leap, boosting the mAP by 7.2 percentage points to 88.9%. This underscores the profound diagnostic value of the thermal spectrum for revealing defects, such as incipient delamination, that are often invisible in visual light. Finally, the application of three-model ensembling provides a further statistically significant refinement, achieving a final mAP of 92.8%. This multi-model approach effectively mitigates the prediction variance of individual models, leading to more robust and reliable detections across the diverse environmental conditions present in the datasets. Figure 2 illustrates Receiver Operating Characteristic (ROC) curves for the three defect classes.

The mean area under the curve (AUC) was 0.97 (see Figure 2), indicating strong discrimination between defective and non-defective regions. Following IEC operational risk metrics, we determined the confidence threshold that yields at most 0.05 false positives per image; at this operating point, the precision of the ensemble reached 93.5%, as reported in Table 2.

To assess the robustness of our models to domain shift, a critical concern when deploying models in varied real-world environments, we conducted cross-dataset experiments where a model trained on one dataset was evaluated on the other without any fine-tuning. The results are presented in Table 3.

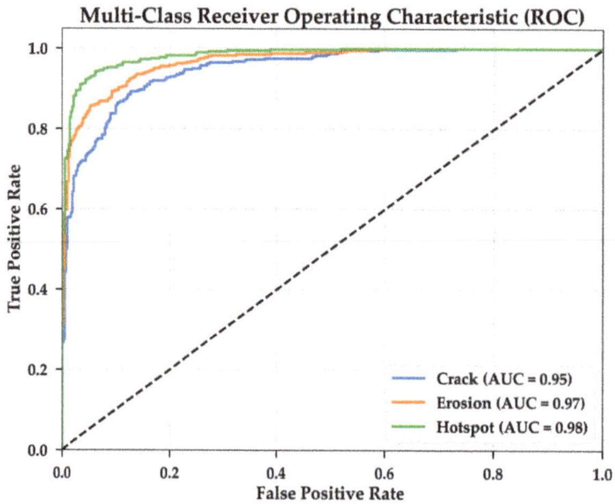

Figure 2. ROC curves for the three defect classes: 'Crack,' 'Erosion,' and 'Hotspot'.

Table 3. Cross-dataset evaluation to quantify domain shift. Models were trained on one dataset and evaluated on the other without fine-tuning, revealing the performance degradation that occurs when models encounter out-of-distribution data.

Training Dataset	Test Dataset	mAP@.5 (%)	F_1-Score (%)
AQUADA-GO (RGB)	Thermal WTB	76.4	74.1
Thermal WTB (RGB–T)	AQUADA-GO	84.2	81.8

As summarized in Table 3, the mAP dropped significantly to 76.4% when transferring from the offshore AQUADA-GO to the onshore Thermal WTB dataset, and to 84.2% in the reverse direction. These results provide a quantitative baseline for the domain gap and highlight the critical need for external validation and the development of domain adaptation strategies for operational deployment, a point we elaborate on in the Discussion.

Further visual analytics provide deeper insight into the system's behavior, as shown in Figure 3. The per-frame inference latency, benchmarked on an NVIDIA RTX 3090 GPU over 5000 frames, is consistently low, with a mean of 118.4 ms and a standard deviation of 12.1 ms (Figure 3a), making the system suitable for high-throughput offline analysis. The model demonstrates robust performance across the three primary defect classes, with the F_1-scores shown with 95% confidence intervals in Figure 3b; performance is highest for 'Hotspot' defects, which have a uniquely salient Thermal Signature. Finally, an analysis of the relationship between ensemble size and performance (Figure 3c) reveals that mAP@.5 gains begin to plateau after three to four models, justifying our choice of a three-model ensemble as an optimal balance between accuracy and computational cost.

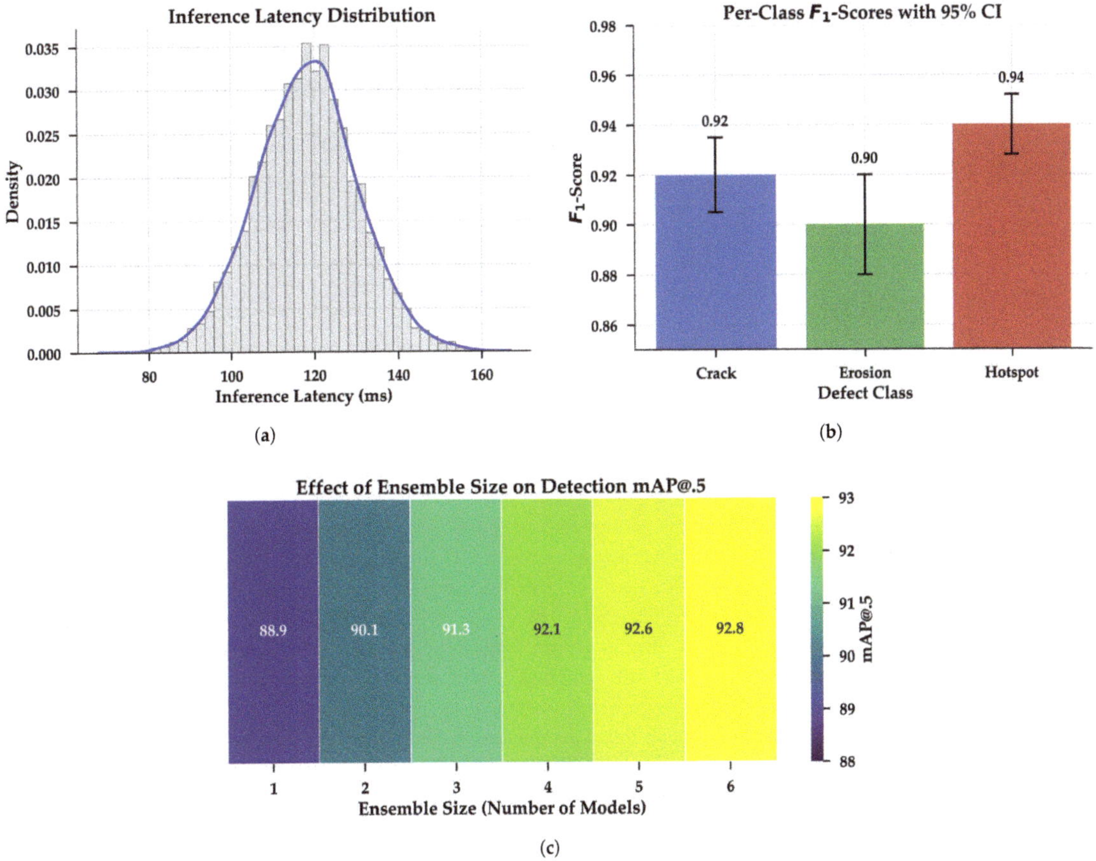

Figure 3. Visual analytics of the detection module: (**a**) per-frame latency histogram; (**b**) class-wise F_1-scores with 95% BCa confidence intervals (CIs); and (**c**) heatmap showing the relationship between ensemble size, confidence threshold, and resulting mAP@.5, illustrating that performance gains diminish beyond three models.

4.2. Ablation Studies

To rigorously quantify the contribution of each component to the framework's overall efficacy, we conducted a series of ablation studies, with results summarized in Table 4. The rationale for each scenario was to isolate and measure the impact of a core methodological choice on final system performance. Removing the thermal channel caused the most severe degradation, increasing the criticality mean absolute error (MAE) by 150% and confirming that multispectral data are the cornerstone of reliable assessment. The removal of ensembling also resulted in a significant 4.1-point drop in F_1-score. Simplifying the FIS from 27 to a more generalized 15 rules more than doubled the criticality MAE; this experiment was designed to test the necessity of a nuanced rule base to capture expert logic, and the significant performance drop justifies our use of the more comprehensive 27-rule set. Finally, the system showed high resilience to a simulated +5 °C thermal calibration drift, with the MAE increasing by only 0.04, demonstrating the robustness of using relative temperature differentials rather than absolute values.

Table 4. Detailed ablation studies quantifying the impact of removing key system components on final performance. The degradation in metrics, shown relative to the baseline performance of the full framework, highlights the critical contribution of each part to the overall system efficacy.

Ablation Scenario	Affected Module	Primary Metric	Baseline	Ablated	Impact Analysis (Δ and % Change)
Thermal Channel Removed (RGB only)	Criticality Assessment	Criticality MAE	0.14	0.35	+0.21 (+150.0%): Loss of thermal data catastrophically degrades severity assessment.
	Defect Detection	F_1-score (%)	92.3	82.2	−10.1 pts: Confirms thermal data are crucial for robust detection of multiple defect types.
Ensemble Learning Removed (Single Model)	Defect Detection	F_1-score (%)	92.3	88.2	−4.1 pts: Demonstrates that ensembling provides a significant boost in accuracy and robustness.
Fuzzy Rule Count Reduced (27 → 15 rules)	Criticality Assessment	Criticality MAE	0.14	0.29	+0.15 (+107.1%): A comprehensive nuanced rule base is essential to accurately model expert logic.
Simulated Thermal Drift (+5 °C)	Criticality Assessment	Criticality MAE	0.14	0.18	+0.04 (+28.6%): System shows high resilience due to its reliance on relative, not absolute, temperature.

4.3. Validation of the Fuzzy Criticality Assessment

The system demonstrated exceptional fidelity to expert judgment, achieving an overall MAE of 0.14. To account for the ordinal nature of the 1–5 scale, we also computed the quadratic-weighted Cohen's κ, obtaining $\kappa = 0.89$ (95% BCa confidence interval: 0.86–0.92), which indicates almost perfect agreement with the human panel. Class-wise F_1-scores for the five severity levels were [0.94, 0.91, 0.88, 0.90, 0.93], demonstrating balanced performance across the entire criticality spectrum. Importantly, there were zero instances in which a ground-truth rating of 5 was assigned a rating below 4 by the automated system. This absence of severe downgrades is a critical safety key performance indicator for any operational deployment. The Pearson correlation coefficient between the system's continuous output and the experts' median score remained high at $r = 0.97$ ($p < 0.001$). The confusion matrix in Figure 4, generated by rounding the system's continuous output to the nearest integer, shows a strong diagonal concentration, visually confirming the high level of agreement.

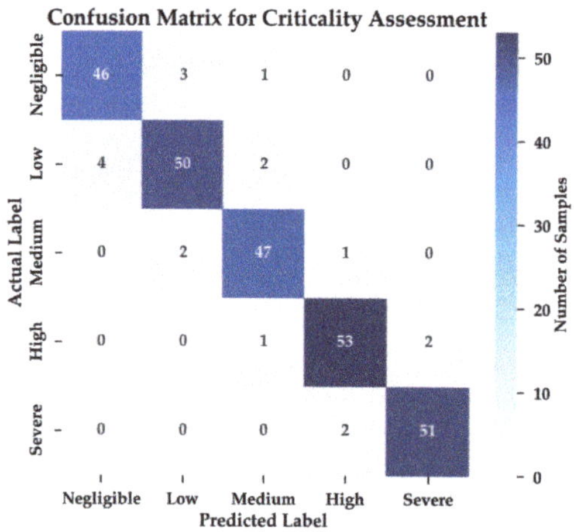

Figure 4. Confusion matrix for the five-level criticality assessment, comparing the system's rounded output with the expert-assigned ground-truth ratings. The high concentration of values along the main diagonal indicates excellent agreement across all severity levels. A color bar has been added to provide a scale for the cell counts.

4.4. Reliability and Calibration of Criticality Scores

Beyond accuracy metrics like MAE, it is crucial for a decision-support system to produce well-calibrated and reliable outputs. A well-calibrated system's confidence in its prediction should match its actual correctness. To evaluate this, we generated reliability diagrams for the five-level criticality output, as shown in Figure 5. The plots show that the predicted confidence for each class aligns closely with the observed accuracy, with all points lying near the diagonal identity line. The overall Expected Calibration Error (ECE) was low at 0.034, indicating that the system's outputs are not only accurate but also trustworthy, a vital characteristic for high-stakes maintenance decisions.

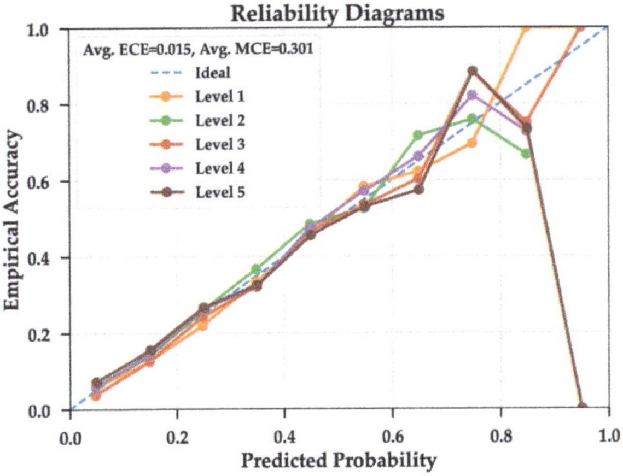

Figure 5. Reliability diagrams for the five discrete criticality levels. Curves show empirical accuracy vs. predicted probability in 10 equal-width bins with 95% bootstrap confidence intervals. The diagonal line denotes perfect calibration. Expected Calibration Error (ECE) and Maximum Calibration Error (MCE) are reported in the legend; see Section 4 for metric definitions.

4.5. Comparative Analysis and Field Case Study

As shown in Table 5, our proposed framework outperforms five recent state-of-the-art methods in wind turbine defect analysis across key metrics.

Table 5. Comparative analysis of the proposed framework against five state-of-the-art defect detection methods. Our framework demonstrates superior performance, particularly in the balanced F_1-score and overall mAP. Note: Results for other methods are as reported in their respective publications; training datasets and protocols may vary. Best results are shown in **bold**.

Method	Data Modality	Precision (%)	Recall (%)	F_1-Score (%)
Liu et al. [35]	RGB	81.2	78.5	79.8
He et al. [34]	RGB	84.5	82.1	83.3
Zhou et al. [18]	Fused RGB-T	89.3	85.4	87.3
Zhao et al. [22]	Fused RGB-T	91.8	89.2	90.5
Zhao et al. [36]	RGB	88.6	86.9	87.7
Proposed Framework	Fused RGB-T Ensemble	**93.2**	**91.5**	**92.3**

Our approach shows particular strength in the F_1-score, which we attribute to the synergistic effects of multispectral fusion and ensemble-based inference. Compared with the enhanced SSD model by Zhao et al. [22], our framework achieves a 1.8-point higher F_1-score, a gain primarily attributable to the superior robustness of the three-model YOLOv8 ensemble, which mitigates individual model variance and improves generalization across the diverse conditions found in the validation datasets.

The system's real-world value is best illustrated by a case from our supplementary field data. During a routine inspection of turbine T-17B, the system flagged a 'Hotspot' on the generator housing with a criticality score of 5.0, corresponding to a 'Severe' rating. The thermal camera registered a significant anomaly with a ΔT of over 20 °C against its surroundings, while the RGB image appeared entirely normal. The automated alert triggered an immediate manual inspection, which revealed a critical internal fault in the cooling system. Maintenance logs indicated that this intervention occurred approximately 48 h before a scheduled component replacement, with engineers noting that a failure was imminent. This case provides a powerful vignette of the system's value proposition: moving beyond simple surface inspection to preemptive, condition-based intervention that prevents catastrophic failures and costly unscheduled downtime.

5. Discussion

The empirical results presented in the preceding section strongly validate our central thesis: that an integrated framework marrying multispectral data, ensemble deep learning, and standards-aligned fuzzy logic can provide an accurate, transparent, and reliable solution for the automated criticality assessment of wind turbine defects. This section synthesizes these findings, contextualizes them within the broader landscape of scientific literature, addresses the inherent limitations of the study, and charts a course for future research.

5.1. Interpretation of Principal Findings

Our principal finding is that the synergy between the framework's components creates a system far more capable than the sum of its parts. The final mAP of 92.8% for defect detection (Table 2) represents a state-of-the-art result, outperforming recent benchmarks as shown in Table 5. The ablation study (Table 4) reveals the critical importance of data fusion. The 10.1-point drop in F_1-score upon removing the thermal channel is not merely an incremental decline; it signifies a fundamental loss of diagnostic capability. It is this channel that enables the system to 'see' into the subsurface of the blade, identifying thermal anomalies indicative of delamination, moisture ingress, or bonding failures—defects that are often precursors to the most catastrophic failure modes and are completely undetectable by RGB-only systems like those in [34,35]. This strongly corroborates the findings of [14] and extends the principle to a broader range of data types, including large-scale offshore video. Another practical consideration is the handling of negative classes. Precision at high recall is critical for O&M workflows because false-positive alarms can overload maintenance queues. Our ROC analysis (Figure 2) and the Precision@FPPI $\leq$ 0.05 metric reported in Table 2 demonstrate that the ensemble maintains high specificity even when the false-positive rate is constrained to a level compatible with IEC operational risk metrics.

The core novelty of this research is the transparent FIS. Achieving a criticality MAE of 0.14 and a Pearson correlation of 0.97 against expert ratings is a powerful demonstration that the nuanced, experience-based reasoning of human engineers can be successfully encapsulated within a formal, repeatable, and scalable computational system. This addresses a major gap in the literature, which has focused heavily on detection accuracy but has largely neglected the subsequent, more crucial step of severity assessment. The explicit grounding of our FIS design in the IEC 61400-5 standard is what sets this work apart (see Table A3 in Appendix A). While other studies have used fuzzy logic [42], they have not established this formal, auditable link to internationally recognized engineering principles. This creates a 'glass-box' model where any assessment can be interrogated: an operator can see not only the final score but also precisely which rules (e.g., 'the rule concerning large defects in high-stress zones') were triggered and with what intensity, as detailed in Table A3 in Appendix A. This constitutes a significant contribution to the field of explainable AI for industrial applications, building operator trust in a way that opaque black-box models cannot.

5.2. Practical Implications for Condition-Based Maintenance

The practical ramifications of this framework for wind farm operators are profound. It provides the core enabling technology for a genuine CBM strategy, a long-sought goal in the industry. Instead of relying on fixed, time-based inspection schedules or reacting to failures, operators can continuously assess asset health and make data-driven decisions. By providing a prioritized list of defects with objective severity scores, the system allows O&M managers to allocate finite resources (technician time, equipment, and budget) with maximum efficiency, as illustrated by the T-17B field case. High-severity defects posing imminent risks can be addressed immediately, while low-severity issues can be scheduled for repair during planned downtime, minimizing lost production and maximizing annual energy production.

Furthermore, the standardized and repeatable nature of the assessment creates a consistent digital audit trail of a turbine's health over its life cycle. These longitudinal data are invaluable for tracking defect propagation rates, validating repair effectiveness, informing future blade designs, and providing objective evidence for insurance claims or asset transfers. While the current FIS is static, these historical data provide the foundation for future extensions, such as dynamic or recurrent models capable of forecasting defect evolution. While a full-scale deployment would generate terabytes of raw image data per inspection campaign, the framework's output is a compact, actionable list of annotated defects (e.g., a few kilobytes in a structured format like CSV or JSON), drastically reducing the cognitive and data load on human analysts and aligning with standard industrial data management protocols.

5.3. Limitations and Threats to Validity

Despite the promising results, academic integrity demands a frank acknowledgment of the study's limitations and threats to its validity, which also serve as a valuable roadmap for future research.

A primary limitation concerns domain shift and the external validity of our findings. The AQUADA-GO dataset originates from offshore inspections of relatively small ($\leq$2 MW) turbines in marine climates, whereas the Thermal WTB Inspection data come from onshore sites with mild, temperate conditions. As a result, the current models may not generalize perfectly to utility-scale $\geq$5 MW platforms or to radically different operating environments. Our cross-dataset evaluation (Table 3) provides quantitative evidence of this challenge, showing a performance drop of up to 15% in mAP. This degradation likely stems from a combination of factors, including differences in ambient lighting, atmospheric haze common in offshore environments, sensor characteristics between the different UAV platforms, and subtle variations in defect morphology between the smaller offshore turbines and the larger onshore models. Figure 6 visualizes this covariate shift using t-SNE, showing that the feature embeddings from the two datasets form distinct clusters.

A second limitation, affecting construct validity, is that our framework currently operates on 2D imagery. This inherently limits the analysis, as it cannot directly measure defect depth or volume, which are critical parameters for assessing certain types of damage like erosion or gouges. We also note that our initial attempt to simulate a thermal channel for AQUADA-GO using DWT fusion yielded only moderate correlation ($r = 0.45$) with actual LWIR measurements. Based on this result, we did not pursue other simulation approaches, reinforcing the principle that modality fusion should be grounded in physically meaningful data rather than synthetic proxies.

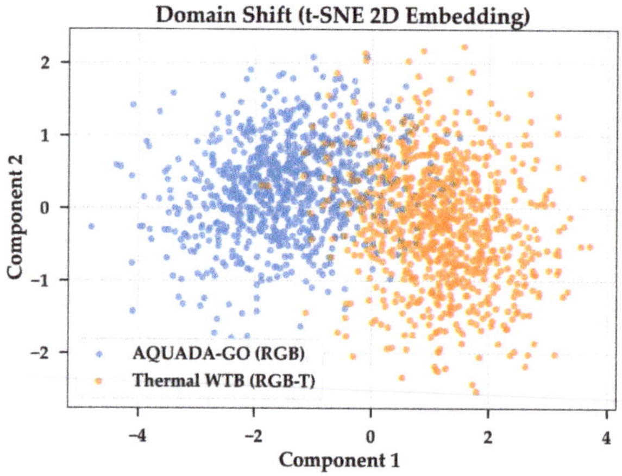

Figure 6. t-SNE projection (perplexity = 35, random_state = 42) of multispectral feature embeddings highlighting domain shifts between AQUADA-GO (RGB) and Thermal WTB (RGB-T). Points are colored by acquisition domain; ellipses show 95% covariance estimates.

5.4. Future Research Directions

The limitations identified above naturally chart a course for future research. The clear next step to address the 2D imagery constraint is integration with 3D digital twins. Localizing defects on a 3D mesh, created using SfM or MVS on the UAV video stream, would enable far richer parameterization (e.g., true area on a curved surface, volume of material loss, and geodesic distance between defects) and a more sophisticated structural assessment. Furthermore, to address the critical challenge of domain shift, future work must focus on mitigating this domain gap through domain adaptation techniques, such as correlation alignment [57] or feature-space augmentation [58]. This must be coupled with an expansion of training data to include more diverse and challenging scenarios, such as (i) large-scale offshore turbines ($\geq$5 MW), (ii) ice-prone Nordic installations where low-temperature icing can alter spectral signatures, and (iii) desert sites experiencing severe leading-edge erosion from sand abrasion. Developing domain-adaptive models trained on such a comprehensive dataset is essential for building a truly universal inspection tool.

Finally, while the system is efficient for offline analysis, real-time, on-board inference remains a challenge. Our current inference time (mean of 118.4 ms; see Figure 3a) is the primary bottleneck and is too high for power-constrained edge devices, corresponding to a throughput of approximately 8.4 frames per second. Future research must therefore focus on model optimization to create lightweight yet accurate models suitable for deployment on platforms like the NVIDIA Jetson series. Key avenues to explore include network quantization (e.g., converting model weights to 8-bit integers), structured pruning (removing redundant network weights and channels), and knowledge distillation (training a smaller 'student' model to mimic the predictive behavior of our larger, more accurate 'teacher' ensemble). Moreover, a further avenue of research is the multimodal fusion with Supervisory Control and Data Acquisition data, as we did for enhancing fire hazard detection in solar power plants in our joint work [59]. Correlating a visually detected defect with anomalous vibration signals or a drop in power output would provide the most complete picture of an asset's health, moving the field towards a truly holistic, system-level understanding of wind turbine integrity.

6. Conclusions

In this study, we have designed, implemented, and rigorously validated a comprehensive end-to-end framework for the automated criticality assessment of wind turbine defects, directly addressing a pivotal challenge in the sustainable management of wind energy assets. Our holistic methodology successfully integrates the diagnostic power of multispectral UAV-acquired data, the pattern recognition capabilities of an ensemble of deep learning detectors, and the transparent reasoning of a knowledge-based Fuzzy Inference System. By validating our system on two diverse public datasets, we have demonstrated its robustness and reproducibility. Our optimized multispectral detection framework achieved a state-of-the-art mAP@.5 of 92.8%, with ablation studies confirming that the fusion of visual and thermal data is the single most critical factor for high performance.

The core scientific contribution is our novel 27-rule Mamdani-type FIS, whose 'glass-box' design is explicitly grounded in the engineering principles of the IEC 61400-5 standard. This system's five-level criticality output shows exceptional agreement with assessments from certified human engineers, validated by a low MAE of 0.14, a high quadratic-weighted κ of 0.89, and excellent calibration. The framework provides operators with a powerful decision-support tool, enabling a paradigm shift from reactive to predictive maintenance. This directly enhances operational safety, optimizes maintenance expenditure, and increases asset availability, thereby supporting Sustainable Development Goal 7 by improving the economic viability and reliability of wind power.

While acknowledging current limitations, our work establishes a clear roadmap for future research focusing on three key areas: enhancing model generalization through domain adaptation techniques to mitigate performance drops in new environments, integrating 3D digital twins for more sophisticated geometric analysis, and optimizing models for efficient on-board edge deployment to enable real-time autonomous inspection capabilities. These advancements will pave the way for the next generation of intelligent structural health monitoring systems.

Supplementary Materials: The following supporting information can be downloaded at https://www.mdpi.com/article/10.3390/en18174523/s1, File Supplementary Material: Detailed formulation of the expert-driven mathematical models for the criticality of classes 'Crack,' 'Erosion,' and 'Hotspot,' including tables of component-specific weighting coefficients derived from engineering standards.

Author Contributions: Conceptualization, P.R., B.R., and A.S.; methodology, P.R., O.S., and O.M.; software, S.S. and O.M.; validation, P.R., O.M., S.S., and T.P.; formal analysis, A.S., O.S., and B.R.; investigation, P.R., S.S., and O.M.; resources, A.S., B.R., and T.P.; data curation, P.R., B.R., T.P., and O.S.; writing—original draft, P.R., O.M., S.S., and O.S.; writing—review and editing, A.S., B.R., and T.P.; visualization, P.R., O.M., and S.S.; supervision, A.S. and B.R. All authors have read and agreed to the published version of the manuscript.

Funding: This research was funded by the Ministry of Education and Science of Ukraine, state grant registration number 0124U004665, project title 'Intelligent System for Recognizing Defects in Green Energy Facilities Using UAVs.' This publication reflects the views of the authors only; the Ministry of Education and Science of Ukraine cannot be held responsible for any use of the information contained herein.

Institutional Review Board Statement: Not applicable.

Informed Consent Statement: Not applicable.

Data Availability Statement: All data used in this study are publicly available. The AQUADA-GO dataset [13] is available at https://data.mendeley.com/datasets/9rcf5p89zn/1 (accessed on 15 August 2025). The Thermal WTB Inspection dataset [14] is available at https://github.com/

MoShekaramiz/Small-WTB-Thermal1 (accessed on 15 August 2025). The source code, trained model weights, configuration files, and a version-controlled release of the software are available on GitHub (https://github.com/sOsvystun/UAV/tree/main (accessed on 15 August 2025).

Acknowledgments: The authors are grateful to the Ministry of Education and Science of Ukraine for its support in conducting this study. We extend our sincere thanks to the research teams at the Technical University of Denmark (DTU) and Utah Valley University for creating and publicly releasing the AQUADA-GO and Thermal WTB Inspection datasets, respectively. Their commitment to open science was instrumental to this work.

Conflicts of Interest: The authors declare no conflict of interest. The funders had no role in the design of the study; in the collection, analyses, or interpretation of data; in the writing of the manuscript, or in the decision to publish the results.

Abbreviations

AI	Artificial Intelligence
AUC	Area Under the Curve
BCa	Bias-Corrected and accelerated
CBM	Condition-Based Maintenance
CLAHE	Contrast Limited Adaptive Histogram Equalization
DWT	Discrete Wavelet Transform
ECE	Expected Calibration Error
EPRI	Electric Power Research Institute
FIS	Fuzzy Inference System
FPPI	False Positives per Image
GNSS	Global Navigation Satellite System
IEC	International Electrotechnical Commission
LWIR	Long-Wave Infrared
MAE	Mean Absolute Error
mAP	Mean Average Precision
MCE	Maximum Calibration Error
MSX	Multi-Spectral Dynamic Imaging
O&M	Operations and Maintenance
RGB	Red–Green–Blue
ROC	Receiver Operating Characteristic
ROI	Region of Interest
RTK	Real-Time Kinematic
t-SNE	t-Distributed Stochastic Neighbor Embedding
UAV	Unmanned Aerial Vehicle
WTB	Wind Turbine Blade
YOLO	You Only Look Once

Appendix A. Fuzzy Inference System Details

This appendix provides supplementary details for the Fuzzy Inference System (FIS), ensuring full transparency and reproducibility of the criticality assessment module. Table A1 presents the complete 27-rule matrix that forms the core of the system's knowledge base. Table A2 provides the exact parameters used for all membership functions, which were derived via a hybrid expert- and data-driven process as described in the Supplementary Materials. Figure A1 shows the results of a global sensitivity analysis, confirming the system's robustness to parameter variations. Table A3 provides further illustrative examples of how specific fuzzy rules are explicitly linked to the engineering principles and failure mode considerations derived from IEC standards, grounding the AI system in established domain knowledge.

Table A1. The complete 27-rule matrix for the Mamdani FIS. The table shows the logical mapping from combinations of fuzzified inputs ('Defect Size,' 'Location,' and 'Thermal Signature') to an output 'Criticality' level.

IF Defect Size Is	AND Location Is	AND Thermal Signature Is	THEN Criticality Is
Large	Blade Root	High	Severe
		Medium	Severe
		Low	Severe
	Mid-span	High	Severe
		Medium	High
		Low	High
	Blade Tip	High	High
		Medium	Medium
		Low	Medium
Medium	Blade Root	High	Severe
		Medium	High
		Low	High
	Mid-span	High	High
		Medium	Medium
		Low	Low
	Blade Tip	High	Medium
		Medium	Low
		Low	Low
Small	Blade Root	High	High
		Medium	Medium
		Low	Low
	Mid-span	High	Medium
		Medium	Low
		Low	Negligible
	Blade Tip	High	Low
		Medium	Negligible
		Low	Negligible

Table A2. Membership function parameters for the FIS. The trapezoidal parameters (a, b, c, d) define the support and core of each fuzzy set. Parameters for 'Defect Size' are in metric units (mm^2) to ensure scale invariance. All parameters were derived by expert elicitation combined with fitting to empirical quantiles of the training data, as detailed in the Supplementary Materials.

Input Variable	Linguistic Term	a	b	c	d
Defect Size (mm^2)	Small	0	0	50	100
	Medium	50	100	400	500
	Large	400	500	1000	1000
Thermal Signature (ΔT in °C)	Low	0	0	2	4
	Medium	3	5	8	10
	High	9	12	25	25

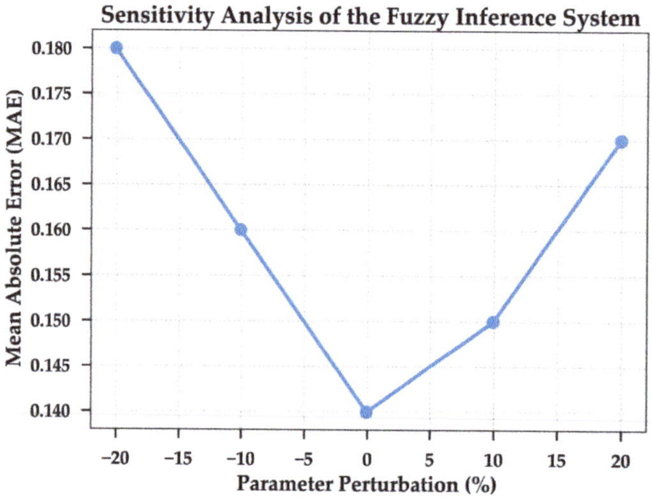

Figure A1. Sensitivity analysis of the Fuzzy Inference System. The curve illustrates how the MAE varies when all membership function breakpoint parameters are perturbed jointly by a factor ranging from −20% to +20%. The MAE remains below 0.18 across the entire tested range, indicating that the FIS is robust to moderate parameter variations.

Table A3. Illustrative mapping of specific fuzzy rules to engineering principles derived from IEC 61400-5/23. This demonstrates how the rule base is grounded in established safety and structural integrity standards.

Rule ID	Fuzzy Rule Summary	Corresponding IEC 61400 Principle/Rationale
1, 2, 3	A large defect at the blade root is always 'Severe,' regardless of its Thermal Signature.	Aligns with IEC 61400-5 [11] requirements for fatigue life analysis and damage tolerance. The blade root is the area of maximum bending moment and stress concentration. Any significant structural flaw in this region has the highest probability of catastrophic propagation.
4, 10, 19	Any defect with a High Thermal Signature at the blade root is at least 'High' or 'Severe.'	Relates to IEC 61400-23 [60] (full-scale structural testing). A significant thermal anomaly indicates a potential subsurface failure (e.g., delamination, adhesive disbond). When located in the highest stress region, this combination represents a critical risk of structural failure from within.
9, 18, 27	A defect at the blade tip with a low Thermal Signature is rated as 'Medium' or 'Low.'	The blade tip experiences the lowest structural loads but the highest aerodynamic velocities. Defects here are less critical from a structural failure perspective but can impact aerodynamic efficiency and noise. The criticality is therefore downgraded compared with the root.
24, 27	A small non-thermal defect away from the root is considered 'Negligible.'	Reflects practical maintenance triage. Small superficial flaws in low-stress areas do not compromise the blade's integrity and typically only require monitoring during the next inspection cycle, rather than immediate intervention.

References

1. Sun, S.; Wang, T.; Chu, F. In-situ condition monitoring of wind turbine blades: A critical and systematic review of techniques, challenges, and futures. *Renew. Sustain. Energy Rev.* **2022**, *160*, 112326. [CrossRef]
2. Kong, K.; Dyer, K.; Payne, C.; Hamerton, I.; Weaver, P.M. Progress and trends in damage detection methods, maintenance, and data-driven monitoring of wind turbine blades—A review. *Renew. Energy Focus* **2023**, *44*, 390–412. [CrossRef]
3. Aminzadeh, A.; Dimitrova, M.; Meiabadi, S.M.; Sattarpanah Karganroudi, S.; Taheri, H.; Ibrahim, H.; Wen, Y. Non-contact inspection methods for wind turbine blade maintenance: Techno-economic review of techniques for integration with industry 4.0. *J. Nondestruct. Eval.* **2023**, *42*, 54. [CrossRef]
4. Zhang, S.; He, Y.; Gu, Y.; He, Y.; Wang, H.; Wang, H.; Yang, R.; Chady, T.; Zhou, B. UAV based defect detection and fault diagnosis for static and rotating wind turbine blade: A review. *Nondestruct. Test. Eval.* **2024**, *40*, 1691–1729. [CrossRef]
5. Banafaa, M.K.; Pepeoğlu, Ö.; Shayea, I.; Alhammadi, A.; Shamsan, Z.A.; Razaz, M.A.; Alsagabi, M.; Al-Sowayan, S. A comprehensive survey on 5G-and-beyond networks with UAVs: Applications, emerging technologies, regulatory aspects, research trends and challenges. *IEEE Access* **2024**, *12*, 7786–7826. [CrossRef]
6. Yan, X.; Wu, G.; Zuo, Y. YOLOV4-based wind turbine blade crack defect detection. In Proceedings of the IncoME-VI and TEPEN 2021, Tianjin, China, 20–23 October 2021; Springer International Publishing: Cham, Switzerland, 2022; pp. 293–305. [CrossRef]
7. Dai, Z. Image acquisition technology for unmanned aerial vehicles based on YOLO—Illustrated by the case of wind turbine blade inspection. *Syst. Soft Comput.* **2024**, *6*, 200126. [CrossRef]
8. Mao, Y.; Wang, S.; Yu, D.; Zhao, J. Automatic image detection of multi-type surface defects on wind turbine blades based on cascade deep learning network. *Intell. Data Anal.* **2021**, *25*, 463–482. [CrossRef]
9. Sun, T.; Yu, G.; Gao, M.; Zhao, L.; Bai, C.; Yang, W. Fault diagnosis methods based on machine learning and its applications for wind turbines: A review. *IEEE Access* **2021**, *9*, 147481–147511. [CrossRef]
10. Ogaili, A.A.F.; Jaber, A.A.; Hamzah, M.N. A methodological approach for detecting multiple faults in wind turbine blades based on vibration signals and machine learning. *Curved Layer. Struct.* **2023**, *10*, 20220214. [CrossRef]
11. *Technical Report IEC 61400-5:2020*; Wind Energy Generation Systems—Part 5: Wind Turbine Blades. International Electrotechnical Commission: Geneva, Switzerland, 2020. Available online: https://webstore.iec.ch/publication/33236 (accessed on 15 August 2025).
12. Electric Power Research Institute. A White Paper on Wind Turbine Blade Defect and Damage Categorization: Current State of the Industry. Technical Report 3002019669, Electric Power Research Institute. 2020. Available online: https://www.epri.com/research/products/000000003002019669 (accessed on 15 August 2025).
13. Chen, X.; Jia, X. Dataset for AI-based optical-thermal video data fusion for near real-time blade segmentation in normal wind turbine operation. *Eng. Appl. Artif. Intell.* **2024**, *127*, 107325. [CrossRef]
14. Memari, M.; Shekaramiz, M.; Masoum, M.A.S.; Seibi, A.C. Data fusion and ensemble learning for advanced anomaly detection using multi-spectral RGB and thermal imaging of small wind turbine blades. *Energies* **2024**, *17*, 673. [CrossRef]
15. Svystun, S.; Scislo, L.; Pawlik, M.; Melnychenko, O.; Radiuk, P.; Savenko, O.; Sachenko, A. DyTAM: Accelerating wind turbine inspections with dynamic UAV trajectory adaptation. *Energies* **2025**, *18*, 1823. [CrossRef]
16. Yang, C.; Liu, X.; Zhou, H.; Ke, Y.; See, J. Towards accurate image stitching for drone-based wind turbine blade inspection. *Renew. Energy* **2023**, *203*, 267–279. [CrossRef]
17. Svystun, S.; Melnychenko, O.; Radiuk, P.; Savenko, O.; Sachenko, A.; Lysyi, A. Thermal and RGB images work better together in wind turbine damage detection. *Int. J. Comput.* **2024**, *23*, 526–535. [CrossRef]
18. Zhou, W.; Wang, Z.; Zhang, M.; Wang, L. Wind turbine actual defects detection based on visible and infrared image fusion. *IEEE Trans. Instrum. Meas.* **2023**, *72*, 3509208. [CrossRef]
19. Rizk, P.; Rizk, F.; Sattarpanah Karganroudi, S.; Ilinca, A.; Younes, R.; Khoder, J. Advanced wind turbine blade inspection with hyperspectral imaging and 3D convolutional neural networks for damage detection. *Energy AI* **2024**, *16*, 100366. [CrossRef]
20. Chen, X.; Sheiati, S.; Shihavuddin, A. AQUADA PLUS: Automated damage inspection of cyclic-loaded large-scale composite structures using thermal imagery and computer vision. *Compos. Struct.* **2023**, *318*, 117085. [CrossRef]
21. Zhang, H.; Xu, H.; Tian, X.; Jiang, J.; Ma, J. Image fusion meets deep learning: A survey and perspective. *Inf. Fusion* **2021**, *76*, 323–336. [CrossRef]
22. Zhao, J.; Zhang, R.; Chen, S.; Duan, Y.; Wang, Z.; Li, Q. Enhanced infrared defect detection for UAVs using wavelet-based image processing and channel attention-integrated SSD model. *IEEE Access* **2024**, *12*, 188787–188796. [CrossRef]
23. Jia, X.; Chen, X. AI-based optical-thermal video data fusion for near real-time blade segmentation in normal wind turbine operation. *Eng. Appl. Artif. Intell.* **2024**, *127*, 107325. [CrossRef]
24. Zhang, Y.; Liu, Y.; Sun, P.; Yan, H.; Zhao, X.; Zhang, L. IFCNN: A general image fusion framework based on convolutional neural network. *Inf. Fusion* **2020**, *54*, 99–118. [CrossRef]

25. Quan, T.M.; Hildebrand, D.G.C.; Jeong, W.K. FusionNet: A deep fully residual convolutional neural network for image segmentation in connectomics. *Front. Comput. Sci.* **2021**, *3*, 613981. [CrossRef]
26. Pan, Y.; Hong, R.; Chen, J.; Singh, J.; Jia, X. Performance degradation assessment of a wind turbine gearbox based on multi-sensor data fusion. *Mech. Mach. Theory* **2019**, *137*, 509–526. [CrossRef]
27. He, K.; Zhang, X.; Ren, S.; Sun, J. Deep residual learning for image recognition. In Proceedings of the 2016 IEEE Conference on Computer Vision and Pattern Recognition (CVPR), Las Vegas, NV, USA, 27–30 June 2016; IEEE Inc.: New York, NY, USA, 2016; pp. 770–778. [CrossRef]
28. Ren, S.; He, K.; Girshick, R.; Sun, J. Faster R-CNN: Towards real-time object detection with region proposal networks. *IEEE Trans. Pattern Anal. Mach. Intell.* **2017**, *39*, 1137–1149. [CrossRef] [PubMed]
29. Cai, Z.; Vasconcelos, N. Cascade R-CNN: Delving into high quality object detection. In Proceedings of the 2018 IEEE/CVF Conference on Computer Vision and Pattern Recognition (CVPR), Salt Lake City, UT, USA, 18–22 June 2018; IEEE Inc.: New York, NY, USA, 2018; pp. 6154–6162. [CrossRef]
30. He, K.; Gkioxari, G.; Dollár, P.; Girshick, R. Mask R-CNN. In Proceedings of the 2017 IEEE International Conference on Computer Vision (ICCV), Honolulu, HI, USA, 21–26 July 2017; IEEE Inc.: New York, NY, USA, 2017; pp. 2980–2988. [CrossRef]
31. Diaz, P.M.; Tittus, P. Fast detection of wind turbine blade damage using Cascade Mask R-DSCNN-aided drone inspection analysis. *Signal Image Video Process.* **2023**, *17*, 2333–2341. [CrossRef]
32. Tan, M.; Pang, R.; Le, Q.V. EfficientDet: Scalable and efficient object detection. In Proceedings of the 2020 IEEE/CVF Conference on Computer Vision and Pattern Recognition (CVPR), Seattle, WA, USA, 14–19 June 2020; IEEE Inc.: New York, NY, USA, 2020; pp. 10778–10787. [CrossRef]
33. Jocher, G.; Chaurasia, A.; Qiu, J. Ultralytics YOLOv8. 2023. Available online: https://github.com/ultralytics/ultralytics (accessed on 31 May 2025).
34. He, Y.; Niu, X.; Hao, C.; Li, Y.; Kang, L.; Wang, Y. An adaptive detection approach for multi-scale defects on wind turbine blade surface. *Mech. Syst. Signal Process.* **2024**, *219*, 111592. [CrossRef]
35. Liu, C.; An, C.; Yang, Y. Wind turbine surface defect detection method based on YOLOv5s-L. *NDT* **2023**, *1*, 46–57. [CrossRef]
36. Zhao, Z.; Li, T. Enhancing wind turbine blade damage detection with YOLO-Wind. *Sci. Rep.* **2025**, *15*, 18667. [CrossRef]
37. Dietterich, T.G. Ensemble methods in machine learning. In *Multiple Classifier Systems*; Kittler, J., Roli, F., Eds.; Springer: Berlin/Heidelberg, Germany, 2000; pp. 1–15. [CrossRef]
38. Zhou, Z.H. *Ensemble Methods: Foundations and Algorithms*, 2nd ed.; Chapman and Hall/CRC: Boca Raton, FL, USA, 2025. [CrossRef]
39. Al-Agha, O.; Al-Natsheh, A.; Al-Sari, I.; Al-Zoubi, E.; Bani-Hani, D. Artificial intelligence in wind turbine fault detection and diagnosis: A comprehensive review and future directions. *Energies* **2025**, *18*, 1680. [CrossRef]
40. Carmona-Troyo, J.A.; Trujillo, L.; Enríquez-Zárate, J.; Hernandez, D.E.; Cárdenas-Florido, L.A. Classification of damage on wind turbine blades using automatic machine learning and pressure coefficient. *Expert Syst.* **2025**, *42*, e70024. [CrossRef]
41. Svystun, S.; Melnychenko, O.; Radiuk, P.; Lysyi, A.; Sachenko, A. Determining the criticality assessment of defects on wind turbine blades using fuzzy logic. In Proceedings of the 6th International Workshop on Intelligent Information Technologies & Systems of Information Security (IntelITSIS 2025), Khmelnytskyi, Ukraine, 4 April 2025; Hovorushchenko, T., Savenko, O., Popov, P.T., Lysenko, S., Eds.; CEUR-WS: Aachen, Germany, 2024; Volume 3963, pp. 351–362. Available online: https://ceur-ws.org/Vol-3963/paper28.png (accessed on 15 August 2025).
42. Dubchak, L.; Sachenko, A.; Bodyanskiy, Y.; Wolff, C.; Vasylkiv, N.; Brukhanskyi, R.; Kochan, V. Adaptive neuro-fuzzy system for detection of wind turbine blade defects. *Energies* **2024**, *17*, 6456. [CrossRef]
43. Vladov, S.; Scislo, L.; Sokurenko, V.; Muzychuk, O.; Vysotska, V.; Sachenko, A.; Yurko, A. Helicopter turboshaft engines' gas generator rotor R.P.M. neuro-fuzzy on-board controller development. *Energies* **2024**, *17*, 4033. [CrossRef]
44. Kozlov, O. Information technology for designing rule bases of fuzzy systems using ant colony optimization. *Int. J. Comput.* **2021**, *20*, 471–486. [CrossRef]
45. Zhang, Z. A flexible new technique for camera calibration. *IEEE Trans. Pattern Anal. Mach. Intell.* **2000**, *22*, 1330–1334. [CrossRef]
46. TorchVision Maintainers and Contributors. TorchVision: PyTorch's Computer Vision Library. 2016. Available online: https://github.com/pytorch/vision (accessed on 15 August 2025).
47. Paszke, A.; Gross, S.; Massa, F.; Lerer, A.; Bradbury, J.; Chanan, G.; Killeen, T.; Lin, Z.; Gimelshein, N.; Antiga, L.; et al. PyTorch: An imperative style, high-performance deep learning library. *Adv. Neural Inf. Process. Syst.* **2019**, *32*, 8024–8035.
48. ZFTurbo. Weighted-Boxes-Fusion. 2022. Available online: https://github.com/ZFTurbo/Weighted-Boxes-Fusion (accessed on 15 August 2025).
49. Bradski, G. The OpenCV library. *Dr. Dobb's J. Softw. Tools Prof. Program.* **2000**, *25*, 120–125.

50. Lee, G.R.; Gommers, R.; Waselewski, F.; Wohlfahrt, K.; O'Leary, A. PyWavelets: A Python package for wavelet analysis. *J. Open Source Softw.* **2019**, *4*, 1237. [CrossRef]
51. Harris, C.R.; Millman, K.J.; van der Walt, S.J.; Gommers, R.; Virtanen, P.; Cournapeau, D.; Wieser, E.; Taylor, J.; Berg, S.; Smith, N.J.; et al. Array programming with NumPy. *Nature* **2020**, *585*, 357–362. [CrossRef]
52. Virtanen, P.; Gommers, R.; Oliphant, T.E.; Haberland, M.; Reddy, T.; Cournapeau, D.; Burovski, E.; Peterson, P.; Weckesser, W.; Bright, J.; et al. SciPy 1.0: Fundamental algorithms for scientific computing in Python. *Nat. Methods* **2020**, *17*, 261–272. [CrossRef] [PubMed]
53. McKinney, W. Data structures for statistical computing in Python. In Proceedings of the 9th Python in Science Conference (SciPy 2010), Austin, TX, USA, 28 June–3 July 2010; pp. 51–56. [CrossRef]
54. Pedregosa, F.; Varoquaux, G.; Gramfort, A.; Michel, V.; Thirion, B.; Grisel, O.; Blondel, M.; Prettenhofer, P.; Weiss, R.; Dubourg, V.; et al. Scikit-learn: Machine learning in Python. *J. Mach. Learn. Res.* **2011**, *12*, 2825–2830.
55. Hunter, J.D. Matplotlib: A 2D graphics environment. *Comput. Sci. Eng.* **2007**, *9*, 90–95. [CrossRef]
56. Warner, J.; Sexauer, J.; Van den Broeck, W.; Kinoshita, B.P.; Balinski, J.; Scikit-Fuzzy; Clauss, C.; Twmeggs; Alexsavio; Unnikrishnan, A.; et al. *JDWarner/scikit-fuzzy: Scikit-Fuzzy 0.5.0 (Version 0.5.0) [Software]*; Zenodo: Geneva, Switzerland, 2024. [CrossRef]
57. Sun, B.; Saenko, K. Deep CORAL: Correlation alignment for deep domain adaptation. In Proceedings of the Computer Vision—ECCV 2016 Workshops, Amsterdam, The Netherlands, 8–10 and 15–16 October 2016; Hua, G., Jégou, H., Eds.; Springer International Publishing: Cham, Switzerland, 2016; pp. 443–450. [CrossRef]
58. Zhou, K.; Yang, Y.; Qiao, Y.; Xiang, T. MixStyle neural networks for domain generalization and adaptation. *Int. J. Comput. Vis.* **2023**, *132*, 822–836. [CrossRef]
59. Lysyi, A.; Sachenko, A.; Radiuk, P.; Lysyi, M.; Melnychenko, O.; Ishchuk, O.; Savenko, O. Enhanced fire hazard detection in solar power plants: An integrated UAV, AI, and SCADA-based approach. *Radioelectron. Comput. Syst.* **2025**, *2025*, 99–117. [CrossRef]
60. *Technical Report IEC 61400-23:2014;* Wind Turbines—Part 23: Full-Scale Structural Testing of Rotor Blades. International Electrotechnical Commission: Geneva, Switzerland, 2014. Available online: https://webstore.iec.ch/publication/5436 (accessed on 15 August 2025).

MDPI AG
Grosspeteranlage 5
4052 Basel
Switzerland
Tel.: +41 61 683 77 34

Energies Editorial Office
E-mail: energies@mdpi.com
www.mdpi.com/journal/energies

www.ingramcontent.com/pod-product-compliance
Lightning Source LLC
Chambersburg PA
CBHW061522210926
51850CB00008B/2007